FLUIDS, WAVES AND OPTICS

29th August, 2018

by Roger Moore
Department of Physics
University of Alberta

DEPARTMENT OF PHYSICS, UNIVERSITY OF ALBERTA

Published by the Department of Physics, University of Alberta.

ISBN: 978-1-54674-971-4

Contents

Preface

This book contains the material taught in the first year, calculus-based introductory physics courses at the University of Alberta. It arose out of a severe dissatisfaction with the coverage of oscillations and waves in standard first year course textbooks where much of the mathematical detail was omitted and important details were simplified away. To address this I wrote my own notes for oscillations and waves and, since this covered a large fraction of the course I was teaching, it then seemed sensible to include the remaining parts of the courses on fluids and light in order to save the students the cost of a bulky, expensive textbook. When I switched to teaching a different first-year course aimed at engineers which covered much of the same material I again expanded the material to include optics and optical instruments in far more detail.

To get the most from this book the reader should be already familiar with the basics of Newton's laws of motion both in the linear and rotational forms and also familiar with the integration and differentiation of polynomials, exponentials and trigonometric functions. While some of the mathematical derivations included here push the envelope of what is possible in a first-year physics course in Canada these have been consigned to appendices to act as resources for the more mathematically inclined reader.

CHAPTER 1

Mathematics

Mathematics is the language we use to describe physics. This chapter will cover the maths that will be primarily used to describe oscillations and waves. It is assumed that the reader is already familiar with the calculus of trigonometric and exponential functions.

While many introductory university texts use sine and cosine functions to describe oscillations and waves imaginary numbers can be used in place of these. Although this is a new concept the resulting maths is simpler because it avoids the need to memorize multiple different trig formulae. This approach is also essential to describing quantum mechanical waves and so although we will primarily use trigonometric functions for both oscillators and waves we will introduce the equivalent complex notation throughout the text, especially in the relevant appendices.

Extending mathematical treatment of oscillators to waves we find that these are systems with two independent variables: position and time. To describe these we need to use a calculus which can cope with more than one independent variable and so we will also introduce partial derivatives. While the ultimate application of these will be to describe waves later this is also the same maths which is used to derive the uncertainties in lab experiments and so we will use partial derivatives in this chapter to calculate experimental uncertainties as a way to introduce the topic.

However, to start with we will discuss the small angle approximations for the common trigonometric functions since these approximations occur frequently when dealing with oscillations and waves.

1.1 Small Angle Approximations

Consider the diagram shown in figure 1.1 which shows a sector of a circle with centre O. By definition the value of the angle θ in radians is the ratio of the arc length to the radius of the circle and so the length of the arc AC is:

$$AC = r\theta \tag{1.1}$$

If we now look at the half-chord AC then using simple trigonometry on the right-angled triangle AOB we have:

$$AB = r\sin\theta \tag{1.2}$$

Now consider what happens as the angle θ shrinks. As θ becomes smaller and smaller the length of the half-chord AB will become closer and closer to the length of the arc AC so that in the limit that θ goes to zero the two have the same length and so for small values of θ we can say that:

$$AC \approx AB \tag{1.3}$$

If we now combine this approximation with the expressions for AC in (1.1) and AB in (1.2) we get:

$$\theta \approx \sin\theta \tag{1.4}$$

for small values of θ when θ is measured in radians.

In a similar way we can also consider the length of the lines OB and OC. Again using simple trigonometry on the right-angled triangle AOB we get:

$$OB = r\cos\theta \tag{1.5}$$

and OC is just the radius of the circle, r. In the limit that θ goes to zero we can see that OC and OB become the same length and so for small values of θ we have $OB \approx OC$ which, when we use the value of OB from (1.5) we get:

$$\cos\theta \approx 1 \tag{1.6}$$

when θ is small.

Finally we can consider the small angle approximation for the tangent of θ by using the relationship:

$$\tan\theta = \frac{\sin\theta}{\cos\theta} \tag{1.7}$$

and so in the limit that θ is small this gives:

$$\tan\theta \approx \theta \tag{1.8}$$

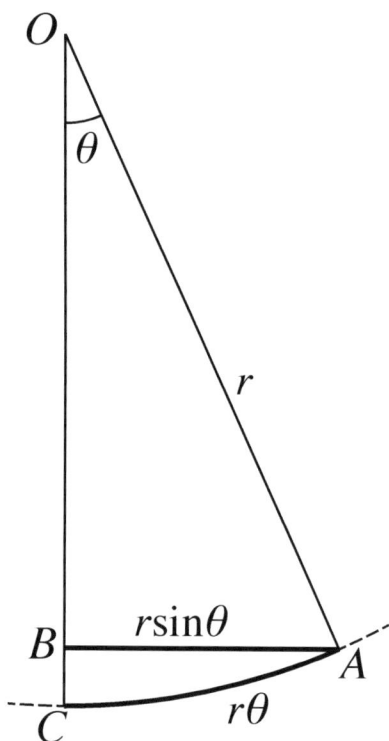

Fig. 1.1: Sector of a circle of radius r which subtends a small angle, θ, at the centre O. The arc AC and the half-chord AC are shown.

1.2 Imaginary Numbers

Consider the square root of a number. For all positive numbers, this is easily defined and calculated and there are always two possibilities e.g. the square roots of 4 are both +2 and -2. However, suppose we ask for the square root of a negative number. Since there is no real number which, when multiplied by itself will give a negative number this appears to have no solution. However, using algebra, we can write down the equation:

$$i^2 = -1 \qquad (1.9)$$

and hence we can write that $i = \sqrt{-1}$. Of course, the problem now comes when we ask what the value of i is: there is no real number that we can write as the solution for i. This is where the true power of algebra lies. It does not matter that we cannot write down a real number as a solution for i, we just keep using i as a quantity which has the property that when it is multiplied by itself the result is -1. Since we cannot write down a real number for i we instead call i an imaginary number.

While this appears to be a simple trick with little apparent value it turns out that imaginary numbers are amazingly useful throughout physics since they have the correct mathematical behaviour to describe many different physical systems. For a start we can now use i to define square roots for all negative numbers using the simple relationship:

$$\sqrt{-n} = \sqrt{-1} \times \sqrt{n} = i\sqrt{n} \quad \text{for } n \in \mathbb{R}_{\geq 0} \qquad (1.10)$$

where n is a positive, real number. For example the roots of -4 are +2i and -2i since:

$$2i \times 2i = 4i^2 = -4 \quad \text{and} \quad -2i \times -2i = 4i^2 = -4$$

Note than engineers will often use the letter j instead of i to denote an imaginary number. This has unfortunately ended up in some programming languages. For example, the Python scripting language, which is commonly used for scientific computation, has j hard coded into the language for imaginary numbers.

1.3 Complex Numbers

We now have two types of numbers: real numbers ($\mathbb{R}$) and imaginary numbers ($\mathbb{I}$). These are simple to multiply and divide:

$$2 \times 4i = 8i , \ 15i \div 5 = 3i \text{ etc.}$$

but what happens if we want to add or subtract them e.g. 3+5i? Since we cannot express i as a real number there is no way to simplify an addition or subtraction operation and so we have to leave the expression as it is i.e. 3+5i. This results in a quantity which has both a real part and an imaginary part and is called a *complex number*.

Since it is sometimes necessary to extract either the real or imaginary parts of the number we also need two new operators: $\Re$ or 'Re' which gives the real part of a complex number and $\Im$ or 'Im' which gives the imaginary part. For example:

$$\Re(8 + 4i) = 8, \ \Im(8 + 4i) = 4, \ \text{Re}(-6 + 3i) = -6, \ \text{Im}(4 - 2i) = -2$$

Note that the result of the imaginary part operator on a complex number is a real number and does not contain the i.

1.3.1 Basic Arithmetic

Basic arithmetic operations on complex numbers are relatively simple with the exception of division. In all cases it is important to remember to treat the real and imaginary parts separately and, at the end, to collect all the real and imaginary parts together to get the final complex number which is the result of the operation.

Addition

$(3 + 2i) + (3 + i)$	$= 6 + 3i$
$(4 + 5i) + (2 + 4i)$	$= 6 + 9i$
$(3 + 2i) + (3 - i)$	$= 6 + i$
$(6 + 3i) + (-2 - 3i)$	$= 4$
$(4 + 2i) + (-4 + 3i)$	$= 5i$

Table 1.1: Examples of complex number addition.

Since the real and imaginary parts have to be treated separately the method is to simply add the real parts together and then add the imaginary parts together with the result generally being another complex number.

$$(a + bi) + (c + di) = (a + c) + (b + d)i \quad \forall \, a, b, c, d \in \mathbb{R} \quad (1.11)$$

Various examples of addition of complex numbers are shown in table 1.1. Note that addition of two complex numbers can result in either a purely real or purely imaginary number as shown in this table.

Subtraction

$(5 + 7i) - (3 + i)$	$= 2 + 6i$
$(4 + 5i) - (2 + 4i)$	$= 2 + i$
$(3 + 2i) - (3 - i)$	$= 3i$
$(6 + 3i) - (8 - 3i)$	$= -2$
$(4 + 2i) - (8 - 2i)$	$= -4 + 4i$

Table 1.2: Examples of complex number subtraction.

Subtraction of complex numbers is performed in exactly the same as addition: the real and imaginary parts are subtracted independently and the result combined which generally gives a complex number.

$$(a + bi) - (c + di) = (a - c) + (b - d)i \quad \forall \, a, b, c, d \in \mathbb{R} \quad (1.12)$$

Examples are shown in table 1.2 and, as before with addition, subtraction of two complex numbers can result in a real or imaginary number.

Multiplication

The same procedure used for addition and subtraction of complex numbers applies to addition. The difference being that multiplication will mix together the real and imaginary parts because the product of the two imaginary parts will give a real number because $i^2 = -1$. To see this we need to expand out the product of the two numbers:

$$
\begin{aligned}
(a + bi) \times (c + di) &= ac + adi + bci + bdi^2 \\
&= (ac - bd) + (ad + bc)i \quad (1.13)
\end{aligned}
$$

where the real part of the answer contains a term bd which is the product of the imaginary parts of the initial numbers. Examples of this type of operation are shown in table 1.3.

Division

Division is more complicated because to simplify the expression we need to get rid of the complex number in the denominator. To do this we have to fall back on our knowledge of multiplication. Looking at equation (1.13) it is clear that if we multiply a complex number by another with the sign of the imaginary part flipped we get a real number.

$$
(a + bi) \times (a - bi) = a^2 + b^2 \quad \forall\, a, b \in \mathbb{R}
$$

This is known as the conjugate of a complex number and is written using a line over the number:

$$
z = a + bi \implies \bar{z} = a - bi \quad (1.14)
$$

Hence to simplify the division of two complex numbers we simply multiply both numerator and denominator by the conjugate of the denominator:

$$
\begin{aligned}
\frac{a + bi}{c + di} &= \frac{a + bi}{c + di} \times \frac{c - di}{c - di} \\
&= \left(\frac{ac + bd}{c^2 + d^2}\right) + \left(\frac{bc - ad}{c^2 + d^2}\right)i \quad (1.15)
\end{aligned}
$$

which will generally give a complex number as a result. Several examples are given in table 1.4.

$$
\begin{aligned}
(5 + 7i) \times (3 + i) &= 8 + 26i \\
(4 + 5i) \times (2 - 4i) &= 28 - 6i \\
(3 + 2i) \times (3 - 2i) &= 13 \\
(3 + 2i) \times (-3 + 2i) &= -13 \\
(5 + 5i) \times (5 + 5i) &= 50i
\end{aligned}
$$

Table 1.3: Examples of complex number multiplication.

$$
\begin{aligned}
\frac{2 + 16i}{2 + 3i} &= 4 + 2i \\
\frac{7 + 6i}{2 + i} &= 4 + i \\
\frac{5 + 5i}{4 + 2i} &= \tfrac{3}{2} + \tfrac{1}{2}i
\end{aligned}
$$

Table 1.4: Examples of complex number division.

$$|4 + 3i| = 5$$

$$\arg(4 + 3i) = 0.644 \,\text{rad}$$

$$|8 - 15i| = 17$$

$$\arg(8 - 15i) = -1.08 \,\text{rad}$$

$$|15 - 8i| = 17$$

$$\arg(8 + 15i) = -0.490 \,\text{rad}$$

Table 1.5: Examples of modulus and arguments for various complex numbers.

1.3.2 Modulus and Argument

The conjugate of a complex number has already been defined by equation (1.14) and we have already seen its use in the division of complex numbers. However, there are two further quantities related to complex numbers which need to be defined.

The modulus of a complex number is denoted using vertical bars on either side of it and is defined as the square root of the sum of the squares of the real and imaginary components:

$$z = x + yi \implies |z| = \sqrt{x^2 + y^2} \tag{1.16}$$

Looking at this definition it is clear to see that we can also relate it to the complex conjugate since:

$$z\bar{z} = (x + yi) \times (x - yi) = x^2 + y^2 = |z|^2 \tag{1.17}$$

Hence the square of the modulus is just the product of the complex number with its conjugate.

The argument of a complex number is denoted by the operator $\arg$ and is defined as:

$$z = x + yi \implies \arg z = \tan^{-1}\left(\frac{y}{x}\right) \tag{1.18}$$

and is always taken using radians, not degrees. It may seem strange to define a quantity related to a complex number as some kind of angle but the reason for this will become clear in the next section!

Examples of various moduli and arguments are shown in table 1.5. Note that two different complex numbers can have the same modulus while having different arguments or vice versa.

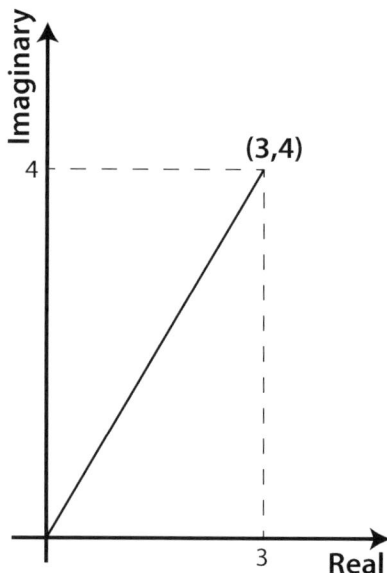

Fig. 1.2: Simple argand diagram showing how a point on the real-imaginary plane can represent a complex number. In this case the number shown is $3 + 4i$.

1.3.3 Argand Diagrams

Having now covered the basic arithmetic operations and definitions associated with complex numbers we need to know how to draw them on a plot. This is achieved using an *argand diagram* which represents all complex numbers as points on a plane. If we take a complex number the corresponding x-coordinate is the real part of the number and the y-coordinate is the imaginary part. For example, the point with coordinates (3,4) represents the complex number $3+4i$ as shown in figure 1.2. Given this mapping of complex numbers to points on the *complex plane* the x axis of an argand diagram is typically labelled as the "real" axis and the y axis as the "imaginary" axis.

Now consider the point on the plane corresponding to the complex number $z = x + yi$ and draw a line from this point to the origin as shown in figure 1.3. Using pythagorus we can easily calculate the length of this line:

$$\text{Length, } A = \sqrt{x^2 + y^2} = \sqrt{z\bar{z}} = |z| \qquad (1.19)$$

Hence the modulus of a complex number is simply its distance from the origin. Next consider the angle which this line makes with the positive real axis. By simple trigonometry:

$$\text{Angle, } \theta = \tan^{-1}\left(\frac{y}{x}\right) = \arg z \qquad (1.20)$$

Hence the angle is simply the argument of the complex number which now makes it clear why we defined it this way!

Looking at the complex plane, as shown in figure 1.3, it is clear that by the modulus and argument of a complex number are essentially polar coordinates. It is possible to uniquely specify a point on the plane by specifying the direction to move from the origin (the argument) and the distance to move along that line (the modulus). Using A and θ as the modulus and argument we can then write down any complex number as:

$$z = x + yi = A(\cos\theta + i\sin\theta) \qquad (1.21)$$

where all we have done is to use basic trigonometry to resolve the modulus into x and y components which, on an argand diagram, correspond to the real and imaginary parts. Whether we use the real and imaginary (cartesian) representation or the modulus and argument (polar) representation is entirely one of convenience.

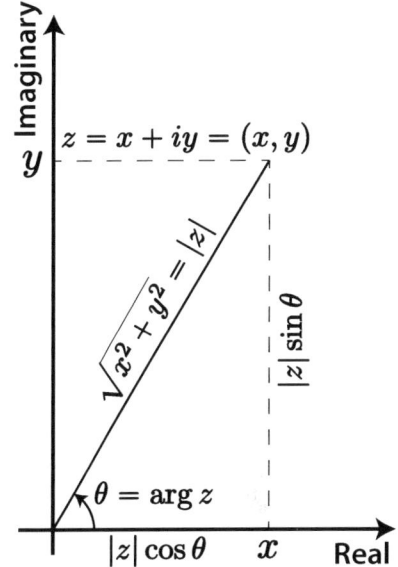

Fig. 1.3: An argand diagram showing how the modulus and argument of a complex number are simply the polar coordinates of the number in the complex plane.

1.4 Complex Exponential Functions

Let's consider what happens when we differentiate an exponential function twice:

$$e^{qx} \underset{d/dx}{\Longrightarrow} qe^{qx} \underset{d/dx}{\Longrightarrow} q^2 e^{qx} \qquad (1.22)$$

Now compare this with sine and cosine functions:

$$\sin(kx) \underset{d/dx}{\Longrightarrow} k\cos(kx) \underset{d/dx}{\Longrightarrow} -k^2\sin(kx) \qquad (1.23)$$

$$\cos(kx) \underset{d/dx}{\Longrightarrow} -k\sin(kx) \underset{d/dx}{\Longrightarrow} -k^2\cos(kx) \qquad (1.24)$$

It is clear from this that after two differentiation operations we recover the same function multiplied by a constant and so it appears that sine and cosine bear some resemblance to an exponent under differentiation. However with a single differentiation there is a clear difference since sine becomes cosine and vice versa whereas the exponent remains the same. So, instead of considering sine and cosine separately lets consider a function consisting of a linear mixture of the two:

$$f(x) = \cos(kx) + a\sin(kx) \tag{1.25}$$

where a is a constant. Note that we do not need a constant for the cosine term since we can always multiply $f(x)$ by any constant and it will behave exactly the same under differentiation i.e. if $f(x)$ behaves like an exponent and remains unchanged by differentiation then $Cf(x)$ will too. If we now differentiate (1.25) we get:

$$\begin{aligned}
\frac{df}{dx} &= -k\sin(kx) + ak\cos(kx) \\
&= k\left[a\cos(kx) - \sin(kx)\right] \\
&= ak\left[\cos(kx) - \frac{1}{a}\sin(kx)\right]
\end{aligned}$$

So for this to be a multiple of the original function given in (1.25) we require that:

$$a = -\frac{1}{a} \implies a^2 = -1 \implies a = i \tag{1.26}$$

where i is an imaginary number. Substituting this back into equation (1.25) and differentiating we get:

$$\begin{aligned}
f(x) &= \cos(kx) + i\sin(kx) \tag{1.27} \\
\frac{df}{dx} &= -k\sin(kx) + ik\cos(kx) = ik\left[\cos(kx) + i\sin(kx)\right] \\
\frac{d^2f}{dx^2} &= -k^2\cos(kx) - ik^2\sin(kx) = -k^2\left[\cos(kx) + i\sin(kx)\right]
\end{aligned}$$

Now compare this to our first differentiation of an exponential (1.22).

$$e^{qx} \underset{d/dx}{\Longrightarrow} qe^{qx} \underset{d/dx}{\Longrightarrow} q^2 e^{qx} \tag{1.28}$$

Now if we make the substitution $q = ik$ this becomes:

$$e^{ikx} \underset{d/dx}{\Longrightarrow} ike^{ikx} \underset{d/dx}{\Longrightarrow} -k^2 e^{ikx} \tag{1.29}$$

This has exactly the same behaviour as (1.27) under differentiation and so the two functions must be equal. The result is called Euler's formula and is one of the most important and fundamental mathematical relationships for physics:

$$e^{ikx} = \cos(kx) + i\sin(kx) \tag{1.30}$$

This was first proven around 1740 by Euler by showing that the Taylor series for each side of the equation are the same. Like many mathematical relationships which are fundamental to physics it can also be used to derive Euler's identity which mathematicians consider remarkable for its beauty. If we make the substitution $kx = \pi$ then:

$$e^{i\pi} = -1 + i0 \implies e^{i\pi} + 1 = 0 \tag{1.31}$$

This identity links some of the most fundamental mathematical constants together: 0, 1, e, i and π.

1.4.1 de Moivre's Theorem

The relationship between complex exponentials and trigonometry functions can also be used to quickly derive another useful relationship. Starting with Euler's formula, (1.30), we make the substitution $k = n$ where n is an integer. This gives:

$$e^{inx} = \cos(nx) + i\sin(nx) \tag{1.32}$$

However e^{inx} can also be written as $(e^{ix})^n$. If we now substitute Euler's formula (1.30) into this we get:

$$e^{inx} = (e^{ix})^n = (\cos x + i\sin x)^n \tag{1.33}$$

Finally we simply combine (1.32) and (1.33) to get a relationship which is known as de Moivre's Theorem:

$$\cos(nx) + i\sin(nx) = (\cos x + i\sin x)^n \tag{1.34}$$

This simple relationship can be used to very quickly derive various trigonometric relationships. For example if we consider the case where $n = 2$ then we have:

$$\cos(2x)+i\sin(2x) = (\cos x+i\sin x)^2 = (\cos^2 x-\sin^2 x)+2i\sin x\cos x \tag{1.35}$$

Since both the real and imaginary parts of the complex number of each side of the equation must be equal we can immediately read out the double angle formulae:

$$\cos(2x) = \cos^2 x - \sin^2 x \tag{1.36}$$
$$\sin(2x) = 2\sin x\cos x \tag{1.37}$$

So we have very quickly and simply proven these well known relationships without having to resort to geometric arguments. In addition we have a far more powerful technique that is easily extendable to large value of n. For example to calculate the formula for $\sin 4x$ all that is required is to expand $(\cos x + i \sin x)^4$ using the binomial theorem and then take the imaginary terms.

1.4.2 Modulus, Argument and Exponential Functions

Returning to our alternative expressions for a general complex number given in (1.21) it is now possible to use Euler's formula to rewrite the polar form of a complex number:

$$z = x + yi = A(\cos \theta + i \sin \theta) = Ae^{i\theta} \tag{1.38}$$

Hence we can write any, general complex number in the form $Ae^{i\theta}$ where A is the modulus of the complex number and θ is the argument. In other words we can write any complex number, z, as:

$$z = |z|e^{i \arg z} \tag{1.39}$$

Example 1.1

Express the complex number $z = 4 - 3i$ in the form $Ae^{i\theta}$ where A and θ are real numbers.

Solution:
To do this we need to calculate the modulus and argument of z:

$$|z| = \sqrt{4^2 + (-3)^2} = 5 \quad and \quad \arg(z) = \tan^{-1}\frac{-3}{4} = -0.644\,\text{rad}$$

We can then write the number in the desired form because $A = |z|$ and $\theta = \arg(z)$ and so we have:

$$z = 4 - 3i = 5e^{-0.644i}$$

1.4.3 Small Angle Approximations

Using imaginary exponentials we can also quickly derive a power series expansion for sine and cosine and then use this to derive the small angle approximations for them without resorting to the geometric proof we used in section 1.1. Starting with the formula for a Taylor series expansion of a function $f(x)$ about a point $x = a$ we have:

$$f(x) = f(a) + f'(a)(x - a) + \frac{f''(a)}{2!}(x - a)^2 + \frac{f'''(a)}{3!}(x - a)^3 + \cdots$$

(1.40)

where each prime denotes a differentiation with respect to x. If we now perform a Taylor series expansion for e^{ix} around $a = 0$ then we get:

$$e^{ix} = 1 + ix - \frac{x^2}{2!} - i\frac{x^3}{3!} + \frac{x^4}{4!} + \cdots$$

(1.41)

because the nth derivative of e^{ix} is just $i^n e^{ix}$. If we now use Euler's identity (1.31) we get:

$$\cos x + i \sin x = 1 + ix - \frac{x^2}{2!} - i\frac{x^3}{3!} + \frac{x^4}{4!} + \cdots$$

(1.42)

For this equation to be true the real part on both sides must be equal which gives:

$$\cos x = 1 - \frac{x^2}{2!} + \frac{x^4}{4!} - \frac{x^6}{6!} + \cdots$$

(1.43)

and so for small values of x where terms of x^2 and higher are negligibly small we end up with the small angle approximation for cosine (1.6). Returning to equation (1.42) we can also see that for this to be true the imaginary parts of both sides must be equal too and so, cancelling out a factor of i we get:

$$\sin x = x - \frac{x^3}{3!} + \frac{x^5}{5!} - \frac{x^7}{7!} + \cdots$$

(1.44)

and again for small values of x where terms of x^2 and higher are negligible we again end up with the small angle approximation for sine (1.4).

Hence we have again derived the small angle approximations using a simple Taylor expansion of an imaginary exponential function. This trick of taking an equation with complex numbers on both sides and splitting it into two separate equations, one equating the real and the other equating the imaginary, is a powerful technique which can often be quite helpful in solving physics problems.

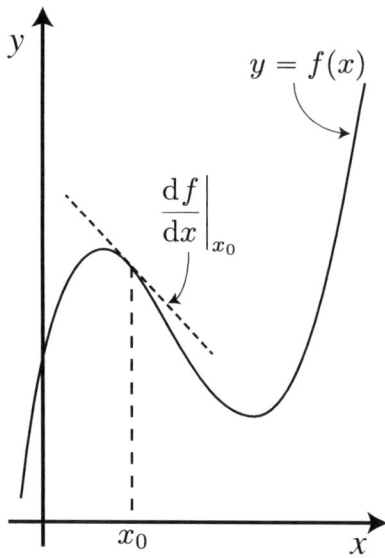

Fig. 1.4: Ordinary differentiation gives the gradient of a curve at a particular point.

1.5 Partial Derivatives

Consider a continuous function of a single variable, x: $f(x)$. If we differentiate this with respect to x the resulting ordinary differential gives the gradient of the function as a function of x. If we plot this graphically then df/dx is the gradient of the line $y = f(x)$ as shown in figure 1.4. This is a simple, unambiguous concept to define because for any point on a continuous, smooth (without kinks) line there is a well-defined concept of what the gradient is at that point.

However what if our function depends on not just one variable but two or more? Consider a function, $h(x, y)$, of two independent variables x and y. To make the concept less abstract we can consider this a height map of a region of land i.e. the value of h for a given value of x and y is simply the height of the land at the point (x, y). In this context, if we want to have a single value for the gradient at a point then we must know the path that we are following. For example, if at one particular point on a hillside, you follow a path that goes parallel to the contours of the hill then the gradient you experience will be zero. However if from the same point, you head off straight down the hill your gradient could be quite large.

This ambiguity means that it is impossible to differentiate the expression with respect to one variable to find a slope: an ordinary derivative has no meaning because there are two unconstrained variables: x and y. However it *is* possible to define a gradient at a point! If we consider the height map then at a given point the land there will have a slope but the slope not only has a magnitude but also a direction. What we need is a vector quantity and not a simple scalar one.

To determine the x and y components for our gradient vector we need to define a new type of derivative. This is called a *partial derivative* and is calculated by differentiating with respect to one variable while keeping all other variables constant. So, for the case of out height map the gradient, denoted by an upside down capital delta, called a 'del', is:

$$\nabla h(x, y) = \left(\left. \frac{\partial h}{\partial x} \right|_y , \left. \frac{\partial h}{\partial y} \right|_x \right) \tag{1.45}$$

where the curly 'd' (∂) denotes a partial derivative and the optional vertical bar lists the variables taken to be constant.

Calculating a partial derivative is very simple: just differentiate with respect to the given variable and assume that everything else is a constant. For example:

$$\frac{\partial}{\partial x}(x^2 + 2xy + y^3) = 2x + 2y$$

$$\frac{\partial}{\partial y}(x^2 + 2xy + y^3) = 2x + 3y^2$$

Partial derivatives are essential to describe waves mathematically because these involve two variables: position and time.

> **Example 1.2**
>
> A wave is travelling along a surface such that the height of the surface a distance x from the origin at a time t is given by:
> $$h(x, t) = A\sin(x - ct)$$
> where A and c are constants. What is the velocity of the surface at a distance x from the origin?
>
> **Solution:**
> *The velocity of the surface is just the rate of change of height with respect to time. We are asked for the velocity at a fixed point x and so we want the rate of change of height with respect to t keeping x constant i.e. we need the partial derivative of $h(x,t)$ with respect to t*
>
> $$\left.\frac{\partial h}{\partial t}\right|_x = -Ac\cos(x - ct)$$
>
> *and so this is the expression for the velocity of the surface at a distance x from the origin.*

1.6 Uncertainties

Partial derivatives can also be used to propagate errors when performing measurements. For example, suppose we were using a GPS device which measured our position in conjunction with the contour map shown in figure 1.5 to determine our height given our GPS position (we'll ignore the fact that GPS can also give a rough measurement of height).

Now GPS devices are only accurate to a few metres so if we are using this, along with the map, to measure our height we need to know how an uncertainty in our x and y coordinates translates into an uncertainty in our height, h. Clearly, this will

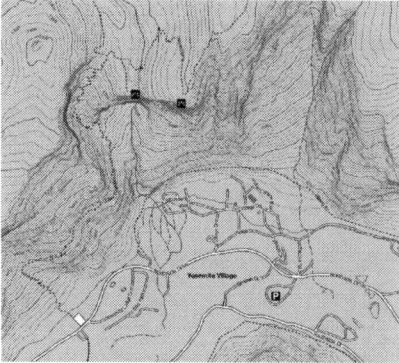

Fig. 1.5: Contour map of the terrain near Yosemite Village in Yosemite National Park in the US.
`http://www.openstreetmap.org/`

depend on where we are located. For example, in the village itself on the valley floor, we know our height very accurately because the land is flat and so a small uncertainty in position makes no difference. However, if we were standing at one of the viewpoints north of the village then a slight inaccuracy in our y coordinate could make a huge difference in our height since it would move us from the top to the bottom of the cliff there. At the same time, a small change in our x coordinate makes very little difference because there is no gradient in the x direction.

To find a quantitative way to calculate the uncertainty due to an error in our x and y coordinates consider the slope of the land as we vary x keeping y constant. If we multiply this by our uncertainty in our x coordinate it will give us an estimate of the possible change in height due to the uncertainty in x. Similarly if we multiply the slope of the land as we vary y, keeping x constant, we get the possible change in height due to the uncertainty in y. This gives us two contributions to the uncertainty in h:

$$\Delta h_x = \left.\frac{\partial h}{\partial x}\right|_y \Delta x \qquad \Delta h_y = \left.\frac{\partial h}{\partial y}\right|_x \Delta y \qquad (1.46)$$

To combine these we make the assumption that the two uncertainties in x and y are uncorrelated. In this situation the contributions are added in quadrature i.e.

$$(\Delta h)^2 = (\Delta h_x)^2 + (\Delta h_y)^2 \qquad (1.47)$$

Now in physics rather than a map which tells us a height at a particular value of x and y coordinate we have equations which tell us the value of a quantity given certain input variables. However we can still think of these equations as acting just like the contour map: we can use them to convert our input variables into a derived quantity. For the map our inputs are just x and y coordinates and the output is the height. For Newton's second law, $F = ma$, the inputs are the mass and acceleration and the output is the force. Hence the same principle applies and we can write the uncertainty in the magnitude of the force as:

$$(\Delta F)^2 = \left(\left.\frac{\partial F}{\partial m}\right|_a \Delta m\right)^2 + \left(\left.\frac{\partial F}{\partial a}\right|_m \Delta a\right)^2 \qquad (1.48)$$

where F and a represent the magnitudes of the force and acceleration respectively. This can be generalized to any equation. If we have a quantity f which is a function of variables $x_0, x_1, \ldots x_n$ then the uncertainty in f is:

$$\Delta f = \sqrt{\sum_{i=0}^{n} \left(\frac{\partial F}{\partial x_i} \Delta x_i\right)^2} \qquad (1.49)$$

Problems

Q1.1: Evaluate the following expressions
(a) $(-6 - 10i) + (18 + 16i)$
(b) $(-20 - i) + (8 + 11i)$
(c) $(-9 + i) + (-19 + 9i)$
(d) $(13 + 15i) + (15 + 18i)$
(e) $(1 + i) + (13 + 14i)$

Q1.2: Evaluate the following expressions
(a) $(-12 + 17i) - (-19 + 17i)$
(b) $(-1 - 19i) - (-6 - 12i)$
(c) $(-17 - 5i) - (-16 + 15i)$
(d) $(6 - 10i) - (4 + 12i)$
(e) $(15 + 7i) - (-16 + 8i)$

Q1.3: Evaluate the following expressions
(a) $(-14 + 10i) \times (9 + 5i)$
(b) $(-3 + 16i) \times (-20 - 7i)$
(c) $(-9 + 8i) \times (-5 - 17i)$
(d) $(5 + 6i) \times (-4 - 9i)$
(e) $(11 + 4i) \times (-20 + 2i)$

Q1.4: Evaluate the following expressions
(a) $(-98 + 14i) \div (-8 - 6i)$
(b) $(40 + 19i) \div (-2 + 7i)$
(c) $(-70 + 38i) \div (-6 + 5i)$
(d) $(48 + 54i) \div (-7 + 3i)$
(e) $(-13 + 19i) \div (-3 - i)$

Q1.5: Evaluate the following expressions
(a) $|9 + 19i|$
(b) $|-2 - 5i|$
(c) $|20 + 18i|$
(d) $|-19 + 4i|$
(e) $|-17 - 16i|$

Q1.6: Evaluate the following expressions in radians limiting your answers to between $-\pi$ to $+\pi$.
(a) $\arg(8 + i)$
(b) $\arg(10 - 20i)$
(c) $\arg(7 - i)$
(d) $\arg(-2 - 3i)$
(e) $\arg(3 - 10i)$

Q1.7: Express the following complex numbers in the form $Ae^{i\theta}$ where A and θ are real numbers and $-\pi < \theta < \pi$.
(a) $14 + 19i$
(b) $-12 + 5i$

(c) $-3 - 3i$
(d) $-8 + 20i$
(e) $12 - 5i$

Q1.8: Evaluate the following partial derivatives where x, y and z are variables.
(a) $\frac{\partial f}{\partial x}$ where $f(x, y) = x^4y^2 + 7x^2 + 14y - 2xy$
(b) $\frac{\partial^2 f}{\partial x^2}$ where $f(x, y) = x^4y^2 + 7x^2 + 14y - 2xy$
(c) $\frac{\partial^2 f}{\partial x \partial y}$ where $f(x, y) = x^4y^2 + 7x + 14y - 2xy$
(d) $\frac{\partial^2 f}{\partial y \partial x}$ where $f(x, y) = x^4y^2 + 7x + 14y - 2xy$
(e) $\frac{\partial f}{\partial x}$ where $f(x, y) = \sin(xy^2) + e^{ixy} - \frac{1}{xy}$

Q1.9: A student doing an experiment measures two quantities A and B and uses these to calculate a result , R, where $R = \sqrt{AB}$. The results of the first two measurements are $A = 12.3 \pm 1.3$ and $B = 3.45 \pm 0.24$. What is the value and uncertainty on the final result, R?

CHAPTER 2

Elasticity

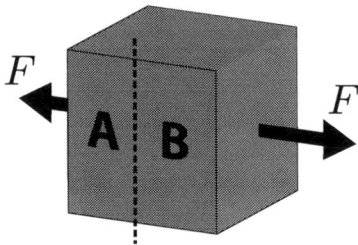

Fig. 2.1: A block in equilibrium with two equal and opposite external forces, F, acting on it. The block is split into two parts A and B by inserting a thin plane.

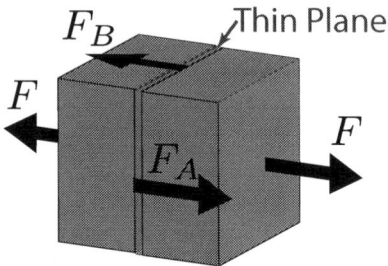

Fig. 2.2: Since the entire block is in equilibrium the contact forces acting on each part of the block must be such that the net force on each part is zero so $F_A = F_B = F$.

In many cases in mechanics, solid objects are treated as ideal rigid bodies that neither deform nor break when forces are applied to them. In real life, this is far from true and our world is full of examples: if you drop a mobile phone the metal case may end up dented and deformed by the fall but the glass front might shatter. Clearly, the way that objects react to forces depends on properties of the materials from which they are made. To be able to quantify how materials behave we need to define some new physical quantities.

2.1 Stress

The internal forces within a material are responsible for any change in the shape of the material and we need to be able to quantify these forces. To start with consider a block sitting in equilibrium with two equal and opposite forces acting on opposite sides along the same line of action as shown in figure 2.1. To determine the internal forces imagine that we divide the cube into two parts and insert a thin, massless plane between the two parts before reattaching everything.

We now need to calculate the forces which are acting on this thin plane since these will be a measure of the internal forces inside the object. Taking each part, A and B, of the block separately we can create a force diagram. Since each part of the block must be in equilibrium for the whole to be in equilibrium the net force on each part must be zero and so the contact force between the inside surface of each part and the thin plane must be equal and opposite to the applied external force, as shown in figure 2.2.

Newton's third law of motion states that for each action there is an equal and opposite reaction and so for each of the contact forces which act on the parts of the block there is an equal and opposite reaction force which will act on the thin plane as shown in figure 2.3. Hence the forces acting on the thin plane will be equal to the applied external forces as shown in figure 2.4. This results in two forces which act on opposite sides of the plane and which are equal and opposite to each other. The forces on the thin plane will *always* cancel in *every* case because the plane is infinitely thin and massless.

This procedure seems rather involved and ends up with the external forces acting on the thin plane so why not just apply the external forces to the plane every time? To see why let's consider a uniform, concrete cylinder which rests on horizontal ground as shown in figure 2.5a. Here the applied external forces are the weight of the cylinder, W, and the contact normal force from the ground which is equal and opposite. If we just used our simple recipe of using the external forces then the force acting on our thin internal plane would be W, the weight of the cylinder, regardless of where the plane is located but in this case that gives the wrong answer.

To understand why let's consider figure 2.5 which shows a plane a distance h above the base of the cylinder. The top cylinder has a weight of W_{top} where $W_{top} < W$ and so to maintain equilibrium the contact force with the plane must exert an upwards force of W_{top} on this cylinder (figure 2.5a). This, in turn, exerts a reaction force downwards on the plane of W_{top}. Now the net force on the plane must be zero so the lower cylinder must exert and upwards contact force of W_{top} on the plane and hence, by Newton's third law, the lower cylinder feels a reaction of W_{top} pushing down on it. The result is that the forces acting on the plane have a magnitude of W_{top} which is less than the full weight, W (figure 2.5b). Indeed the only place where the full weight will be felt is if we put the plane at the base of the cylinder between it and the ground. So it is always best to split the object up and think about the forces acting on a thin plane through the object. This is particularly true for situations where the object is not in equilibrium.

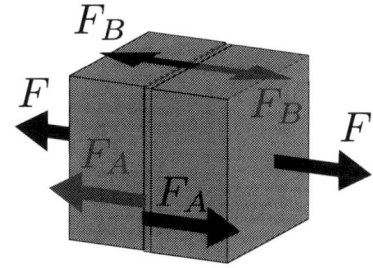

Fig. 2.3: The contact forces acting on parts A and B of the block have equal and opposite reaction forces which act on the thin, massless plane. For the block to be in equilibrium we require that $F_A = F_B = F$.

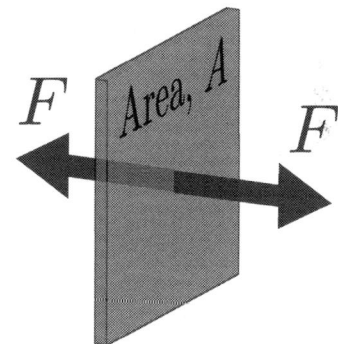

Fig. 2.4: The two contact forces which act on the thin plane. In all cases the net force acting on the thin plane must be zero.

2.1.1 Normal Stress

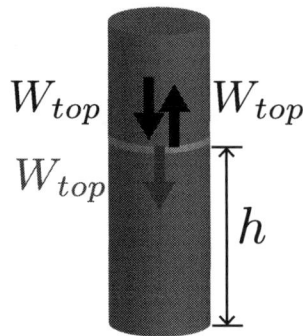

(a) *Top surface contact forces between the plane and the top part of the cylinder.*

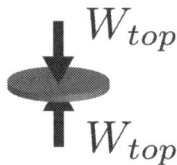

(b) *Both contact forces acting on the plane.*

Fig. 2.5: A vertical cylinder is divided into two parts a distance h above the base by a thin plane. W_{top} is the weight of the top part of the cylinder.

Now that we know how to calculate the forces acting on a thin plane in the middle of some material we now need to convert this into a useful quantity. If we think about rope our intuition tells us that thicker ropes can support more force without deforming or breaking than thin ropes. To account for this we use a new quantity called stress, σ, which is defined as the force per unit area. In both cases, we have discussed so far the forces applied to the plane have been perpendicular to it. This results is what is called normal stress where normal just means perpendicular. There are two types of normal stress: tensile stress where the internal forces are attempting to pull apart the plane (see fig 2.4) and compressive stress where the forces are trying to crush it (see fig 2.5b).

The normal stress is calculated by dividing the perpendicular component of one of the plane's contact forces by the area of the plane:

$$\text{Normal Stress, } \sigma = \frac{F_\perp}{A} \qquad (2.1)$$

where $F_\perp$ is the component of the force perpendicular to the plane and A is the area of the plane. By convention tensile stress is given a positive value and compressive stress a negative one i.e. the positive direction for the perpendicular component of the force is taken to be away from the plane.

Looking at equation 2.1 we have a force divided by an area so it is clear that the SI units for stress are newtons per square metre, $N\,m^{-2}$. Under the SI system can be abbreviated to pascals (Pa) but this is a triviality which hides the underlying units so $N\,m^{-2}$ is generally preferred.

Example 2.1

Consider a circular rod with a uniform density, ρ, which is attached to a support at the top and is hanging vertically at rest in a constant gravitational field, g. If the length of the rod is a what is the tensile stress at a point x below the top of the rod? How will the stress at the support change if (a) the length of the rod is doubled and (b) the radius of the rod is doubled?

Solution:
First, we split the rod at a distance x below the top of the rod. The force acting upwards on the lower part of the rod must be equal to the weight of the lower part of the rod. If we call

the radius r then the weight of the lower part of the rod is:

$$W = \rho \times (a - x) \times \pi r^2 \times g = \pi \rho g r^2 (a - x)$$

This is the force which is acting perpendicular to the imaginary plane which we inserted into the rod. Hence the tensile stress is just the force per unit area which is:

$$\sigma = \frac{W}{\pi r^2} = \rho g (a - x)$$

We can now answer the last two parts which ask about the stress at the support where $x = 0$. Doubling the length of the rod, a, will double the weight of the rod while keeping the cross-sectional area constant hence the stress will also double. However doubling the radius of the rod will increase the weight by a factor of four since the rod's volume increases at πr^2 but it will also increase the cross-sectional area by the same factor of four because the area of the rod scales as πr^2. Hence there will be no change in stress.

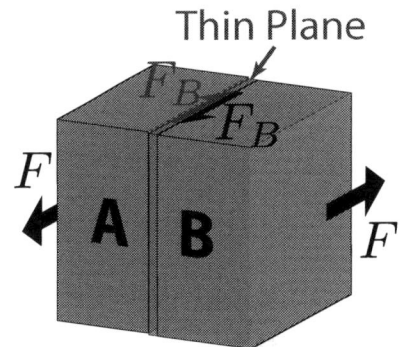

Fig. 2.6: Block being acted upon by two external forces parallel to the thin plane being considered. These generate contact forces between the plane and the block which are parallel to the plane. The contact forces between part B of the block and the plane are shown. For equilibrium $F_B = F$ and a similar condition applies to the contact forces between part A of the block and the plane.

2.1.2 Shear Stress

Having considered the case where the forces on the plane are perpendicular to the plane we also have to consider what happens when the forces are parallel to the plane. Figure 2.6 shows equal and opposite external forces acting on the block such that they are parallel to the thin plane being considered. Going through the same exercise to find the forces acting on the plane we see that, as before, the forces acting on the thin plane precisely cancel, as shown in figure 2.7.

We can now define a new quantity called the shear stress which has a very similar definition to normal stress. The key difference is that the contact force component used is that which acts on one side of the plane and lies in the plane of the cross-section. As a result the definition is:

$$\text{Shear Stress, } \tau = \frac{F_{\parallel}}{A} \qquad (2.2)$$

where $F_{\parallel}$ is the component of the contact force parallel to the plane. Just like normal stress shear stress has units of newtons per square metre (N m^{-2}).

This raises a question though. Looking at figure 2.7 why could we not have just defined our plane as parallel to the face with A-B on it and defined our stress as a normal stress? The key

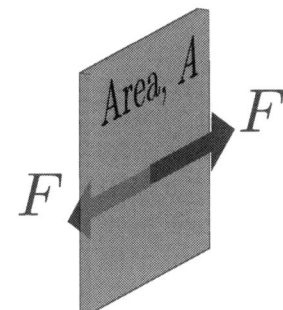

Fig. 2.7: The two contact forces which act on the thin plane in cases of shear stress. As required the net force on the plane is zero.

difference between a normal stress and a shear stress which prevents us from doing this is the line of action of the two forces acting on the object under stress. In figure 2.7 the two forces are clearly opposite to each other but the two lines of action of the forces are separated by the thickness of the block. Compare this to figure 2.3 where both forces are still opposite but clearly act along the same line. This is the key difference between normal and shear stresss: normal stress is caused by opposite forces which act along the same physical line, shear stress is caused by opposite forces which act along parallel, but separated lines.

Example 2.2

A rectangular block of wood with dimensions $5 \times 10 \times 20$ cm has one 10×20 cm face firmly attached to a workbench while the opposite side of the same dimensions is being sanded by a belt sander. The friction force generated by the sander is 30 N. What is the stress on the block of wood?

Solution:
Here the stress on the wood is due to the friction force on the top surface and the contact force between the block and the workbench on the bottom surface. The lines of action of these two forces are not the same and so this must be a shear stress. The stress plane to use must be parallel to the surfaces of the two opposing forces and must lie between them. Hence this plane will have the same area as the face which is being sanded and so the shear stress is just:

$$\sigma = \frac{F}{A} = \frac{30}{0.1 \times 0.2} = 1,500 \, \text{N m}^{-2}$$

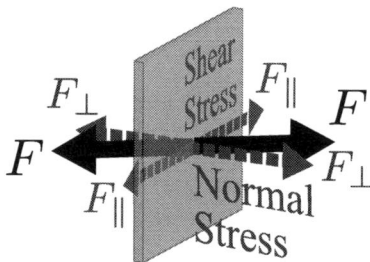

Fig. 2.8: Diagram showing how the contact forces acting on a thin plane can be resolved into parallel and perpendicular components which give rise to shear and normal stress respectively.

2.1.3 Combined and Average Stress

In general, a single plane will have contact forces acting on it that are neither parallel nor perpendicular to the plane. In these cases, the force acting on one side of the plane is resolved into a parallel and perpendicular components. The parallel components giving a shear stress and the perpendicular one giving a normal stress as shown in figure 2.8. These two stresses can then be calculated independently of each other, only the area of the plane will be common to both.

Another point to note is that the definitions we have used for both normal and shear stress calculate the average stress over the plane. In many situations, forces are applied at a point on an object's surface which will concentrate the stress in the region around that point leading to a non-uniform distribution of stress over a plane. Even in cases where the force is applied in a uniform manner over one surface of an object the shape of the object can cause a non-uniform distribution of stress within the object. Understanding the non-uniform distribution of stress is of particular interest to engineers because if too much stress is concentrated at a single point of an object it can lead to mechanical failure of the material at that point.

The Big Picture

Stress is actually a very complex quantity. In this chapter, we only consider simple cases where there is a single cross-section plane whose choice is either obvious or given. However, stress gets a lot more complex for a general case in three dimensions when an object can be divided up using three orthogonal planes. These are typically labelled using the axes which lie in the plane: $x - y$ plane, $y - z$ plane and the $x - z$ plane.

Each of these planes will have a normal and shear stress. The normal stress we label with the axis that is perpendicular to the plane giving: σ_x, σ_y and σ_z. The shear stress we can label with the two axes that define the plane it lies in giving: τ_{xy}, τ_{yz} and τ_{xz}. These six quantities can then be used to create a 3×3 symmetric matrix, called the Cauchy stress tensor which is defined as:

$$\sigma = \begin{pmatrix} \sigma_x & \tau_{xy} & \tau_{xz} \\ \tau_{xy} & \sigma_y & \tau_{yz} \\ \tau_{xz} & \tau_{yz} & \sigma_z \end{pmatrix} \tag{2.3}$$

So stress is neither a scalar nor a vector quantity but a *tensor* quantity. Dealing with tensors is a complex business and so to keep things simple we will only consider problems involving average stresses in a single plane.

2.2 Strain

When a stress is applied to a material it will typically change the shape of the material in a process called deformation. To

(a) *Initial state*

(b) *Final state*

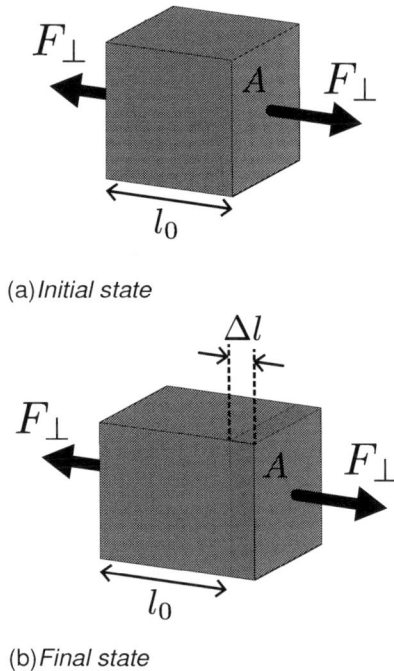

Fig. 2.9: Result of a normal strain deformation of an object caused by a normal stress acting on it. The original length of the object is denoted as l_0 and the change in length due to the strain as Δl and so the strain shown here is $\Delta l / l_0$.

quantify the amount of deformation we need to define a new quantity called strain. Since strain is caused by stress it should come as no surprise that there are two types of strain: normal strain and shear strain each of which is caused by the same type of stress i.e. shear stress causes shear strain.

2.2.1 Normal Strain

The deformation described by normal strain is shown in figure 2.9. The particular example shown is a tensile strain due to a tensile stress and this stress stretches the object to make it longer than it was before. A compressive stress can generate a similar change in length although in such cases the change is negative with the object getting shorter.

If we consider a uniform, tensile stress being applied to a uniform rubber band then each centimetre of the band will stretch by the same amount, e.g. a 1 cm length may become 1.1 cm. This is true because each centimetre of material is under the same stress in identical conditions and so much behave the same. This means that the longer the initial length of the rubber band the longer the final length will be e.g. a 2 cm band will become 2.2 cm long but a 10 cm band will become 11 cm long. Hence the absolute extension of the object depends on its initial length under identical stress conditions. To remove this dependence on the initial length we define the normal strain as:

$$\text{Normal Strain, } \epsilon = \frac{\Delta l}{l} \qquad (2.4)$$

where Δl is the change in the length of the object perpendicular to the plane and l is the original length of the object perpendicular to the plane. Note that since normal strain is just a ratio of two lengths normal strain is a dimensionless quantity and has no units.

2.2.2 Shear Strain

Shear strain is caused by shear stress. Since the forces which generate a shear stress act parallel to the thin plane being considered the extension in shear strain is also in this plane (see figure 2.10) and not perpendicular to it as it was for normal strain. Once again the longer the initial length of the object the larger the total displacement will be and so, just as with normal strain, we need a definition for shear strain which avoids this dependence:

$$\text{Shear Strain, } \gamma = \frac{x}{l} \qquad (2.5)$$

where x is the displacement of the two ends of the object parallel to the cross-section plane and l is the original length of the object perpendicular to the cross-section plane as shown in figure 2.10.

This definition appears very similar to the one for normal strain and, like normal strain, shear strain is a dimensionless quantity without units. The major difference is that the displacement for shear strain is perpendicular to the original length used.

2.3 Moduli of Elasticity

Armed with the two new quantities stress and strain it is easy to see that stresses cause strains so now we need to ask how they are related. Since we have two different types of stress-strain pairings we are clearly going to have at least two different relationships for these two cases and, as we shall see, there is also a third case that we need to consider.

In general, it is easy to see that the more stress (force) we apply to an object the greater the strain (extension): increasing the force of tension applied to rubber band causes it to become longer. Obviously, any material can be stressed enough that it either permanently deforms, e.g. bending a paperclip, or breaks e.g. tearing a sheet of paper. However small stresses create an elastic deformation of the material which is defined as temporary change in the shape of the material so that when the stress is removed the material returns to its original shape.

It is the relationship between stress and strain for these elastic deformations that we will consider here. In such regions we need a new quantity called the modulus of elasticity to relate stress and strain which is defined as the rate of change of stress with respect to strain:

$$\text{Modulus of Elasticity, } \lambda = \frac{\mathrm{d}\sigma}{\mathrm{d}e} \qquad (2.6)$$

which is simply the gradient on a plot of stress vs. strain for the material. However for many materials and types of strain this has a constant value for stresses below some threshold. In this case we can expand the above definition by considering a macroscopic change in the region where λ is constant starting with zero stress and zero strain and ending up at some particular value of stress and strain. This gives:

$$\lambda = \frac{\Delta\sigma}{\Delta e} = \frac{\sigma - 0}{e - 0} = \frac{\sigma}{e} \qquad (2.7)$$

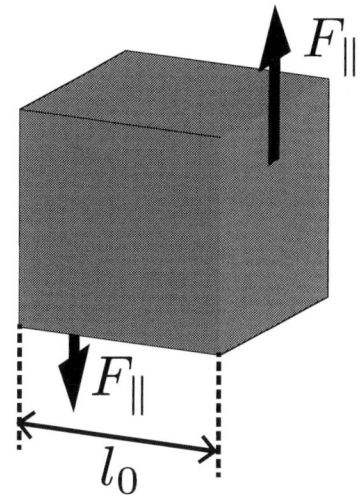

(a) Initial state

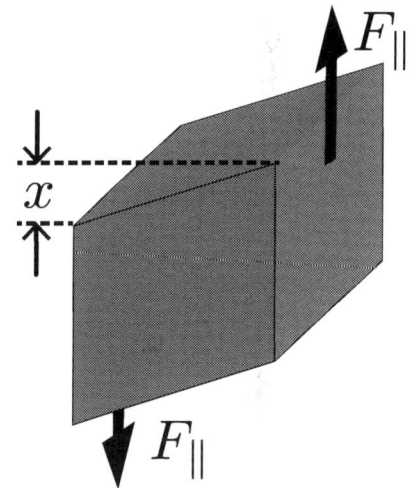

(b) Final state

Fig. 2.10: Result of a shear strain deformation of an object caused by a shear stress acting on it. The original length of the object is denoted as l_0 and the displacement perpendicular to this length is x and so the strain shown here is x/l_0

Hence in this region, starting at zero stress, where the modulus of elasticity is constant it can be defined as:

$$\text{Modulus of Elasticity, } \lambda = \frac{\text{Stress}}{\text{Strain}} \qquad (2.8)$$

However it is important to remember that this is only true in this limited region and it is not the general definition of elastic modulus.

Since there are different types of stress and strain there are also different types of elastic moduli and three such types will be discussed next.

2.3.1 Young's Modulus and Hooke's Law

A contemporary of Sir Isaac Newton, Robert Hooke (1635-1703) discovered that under certain conditions the tensile force and extension are proportional to each other i.e. a doubling of the stress causes a doubling of the strain. This was published in 1678 in a very concise form, "Ut tensio, sic vis", which is latin for "As the extension so the force" and, when applied to springs, the law still bears his name today.

The constant of proportionality between the tension and the extension for a spring in Hooke's law is called the spring constant. However the spring constant is particular to a specific spring and is not a property of the material. The generalization of Hooke's law to stress and strain was described by another British physicist, Thomas Young (1773-1829) (see also chapter 11 for his work on light), who noted that tensile stress and strain were proportional under certain conditions. The constant of proportionality, called Young's modulus , depends on the material in question and is defined as:

$$\text{Young's Modulus, } Y = \frac{\text{Tensile Stress}}{\text{Tensile Strain}} = \frac{\sigma}{\epsilon} \qquad (2.9)$$

Since stress has SI units of newtons per square metre and strain is dimensionless Young's modulus has units of newtons per square metre ($N\,m^{-2}$). Starting with equation 2.9 we can substitute in the values for the stress and strain in terms of the applied force and change of length to get:

$$\text{Young's Modulus, } Y = \frac{\sigma}{\epsilon} = \frac{F_\perp l_0}{A \Delta l} \qquad (2.10)$$

The values of Young's modulus has been measured for many materials, see table 2.1 and provides a simple and convenient way to relate the applied tensile stress to the strain. However it is worth noting that materials only obey Young's modulus over a particular range of loads: above or below this region the ratio of stress to strain is not constant.

Let's now rearrange equation (2.10) to express the force in terms of the extension of the material. Doing this gives:

$$F_\perp = \left(\frac{YA}{l_0}\right)\Delta l \qquad (2.11)$$

Now, since Young's modulus (Y) is a constant as are the cross-sectional area (A) and original length (l_0) we have just derived Hooke's law: the force ($F_\perp$) is proportional to the extension (Δl)! The difference is that now we have the "spring" constant in terms of properties of the material: Young's modulus, the original length and the cross-sectional area:

$$\text{"Spring" Constant, } k = \frac{YA}{l_0} \qquad (2.12)$$

Now this is not really a spring constant - springs involve shear as well as normal stresses so to calculate a spring constant requires more than just Young's modulus but the principle is the same. Springs utilize the constant relationship between stress and strain to generate a device which has a constant ratio of force to extension over a large range of loads.

Material	Young's Modulus (GN/m^2)
Rubber	0.01-0.1
Nylon	2-4
Aluminium	69
Glass	50-90
Copper	117
Steel	200
Beryllium	287
Tungsten	400
Diamond	1,220

Table 2.1: The value of Young's Modulus for various common materials. Source: Wikipedia

2.3.2 Shear Modulus

Shear stresss produce shear strains and so the rate of change of the shear stress with respect to the shear strain is called the shear modulus, G. However for sufficiently small shear stresses the shear modulus is typically constant and so in these cases we can write it as:

$$\text{Shear Modulus, } G = \frac{\text{Shear Stress}}{\text{Shear Strain}} = \frac{F_\parallel l_0}{Ax} \qquad (2.13)$$

which is derived from equations (2.2) and (2.5). Just as with Young's modulus the shear modulus has units of $\text{N}\,\text{m}^{-2}$, the same as those for stress, because strain is a dimensionless quantity.

2.3.3 Bulk Modulus

So far we have only dealt with stress coming from a force applied in a single direction. In these circumstances, materials can deform either by stretching, compressing or shearing with their volume potentially remaining roughly constant. However, if a stress is applied from all directions simultaneously such flexibility is no longer possible and the volume of the material must either increase or decrease depending on the applied stress.

This type of response is different from either normal or shear strain and is called a bulk strain. The stress, which must be applied equally in all directions, is called a bulk stress and for cases where Hooke's law is obeyed and the ratio of the stress to the strain is a constant it should be no surprise to learn the constant of proportionality is called the bulk modulus.

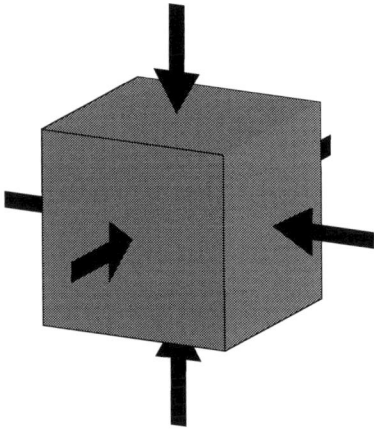

Fig. 2.11: A bulk stress is one applied equally in all directions to an object. This is commonly the result of being immersed in a fluid.

Applying a bulk stress means that the force per unit area must be constant for all directions. This commonly occurs when objects are immersed in a fluid, either liquid or gas. Under these conditions, the fluid exerts a force on every immersed surface in a direction perpendicular to the surface. The larger the surface the larger the force and so we define a quantity called pressure that is the perpendicular force per unit area acting on the surface. Unlike a force pressure has no direction: it acts on all surfaces regardless of their orientation.

Returning now to our object, immersed in a liquid, since the orientation of its surfaces does not matter we can see that the object will feel the same force per unit area on every surface, as shown in figure 2.11, which will result in the same normal stress being applied in every direction simultaneously. This is a bulk, or volume, stress which is the same as the applied pressure.

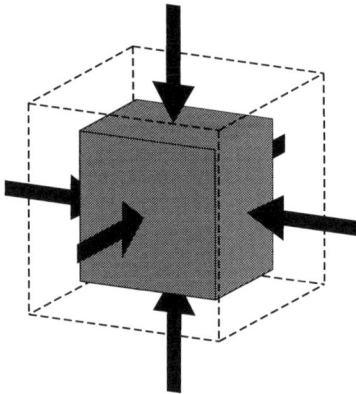

Fig. 2.12: A bulk strain showing that the object shrinks in all directions so that its overall volume is reduced.

The bulk, or volume, strain represents a change in the object's volume. Unlike with normal or shear strain there is no single direction along which we can take a length measurement because the bulk stress is being applied from all directions at once. Hence, rather than a length, we need to consider the volume as shown in figure 2.12. The result is that the bulk strain is defined as the change in volume divided by the original volume:

$$\text{Bulk Strain} = \frac{\Delta V}{V_0} \qquad (2.14)$$

where ΔV is the change in volume and V_0 is the original volume. As with other strains, this is a dimensionless quantity.

The bulk modulus is then defined as the rate of change of the bulk stress with respect to the bulk strain. However, since a positive increase in external pressure will crush an object and result in a negative change in pressure, convention adds a minus sign so that the bulk modulus for materials will be positive as per the other moduli of elasticity. This gives:

$$\text{Bulk Modulus, } B = -V_0 \frac{dp}{dV} \qquad (2.15)$$

where p is the applied external pressure. For macroscopic changes in regions where the bulk modulus is constant as can write this definition as:

$$\text{Bulk Modulus, } B = -\frac{\Delta p}{\Delta V / V_0} \qquad (2.16)$$

where Δp is the change in the applied external pressure. Note that since pressure is a force per unit area and the bulk strain is dimensionless the units for the bulk modulus are newtons per square metre (Nm^{-2}) like all the other moduli of elasticity.

Example 2.3

When a marshmallow is put in a vacuum chamber and all atmospheric pressure removed it expands to a volume twice its original size. If the atmosphere normally exerts a pressure of $101\,kN\,m^{-2}$ what is the bulk modulus of marshmallow?

Solution:
Start with the definition of the bulk modulus given in equation (2.15). Since its new volume is twice the original the change in volume, ΔV, equals the original volume V_0 and so the fractional change in volume is +1 and the change in pressure is -101 $kN\,m^{-2}$ since the pressure was reduced. Putting these numbers into the equation gives:

$$B = -\frac{\Delta p}{\Delta V / V_0} = \frac{101 \times 10^3}{1} = 101\,kN\,m^{-2}$$

2.4 Hysteresis

In general real materials have non-constant moduli of elasticity. A typical stress-strain curve for a rubber band under increasing tension is shown in figure 2.13. This clearly shows that the strain is only proportional to stress for a particular range of loads: below, or above, this limit the ratio is not constant. Suppose now that we plot the same stress-strain curve but instead of starting with zero tension and increasing this time we start with it a high stress and decrease it.

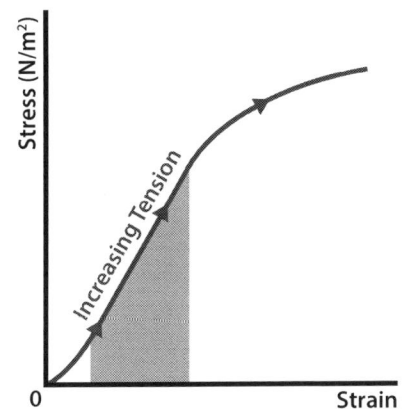

Fig. 2.13: Simulated plot of the deformation (strain) of a rubber band to increasing tensile stress. The shaded area shows the region where the strain is proportional to the stress.

Figure 2.14 shows the stress-strain curve from reducing the stress on a rubber band along with the curve obtained when increasing it. This shows a clear difference between the two curves with the increasing stress (loading) curve generally showing a small strain for the same stress compared to the decreasing stress (unloading) curve. This effect is known as hysteresis.

Now consider the area under the curve. For a rubber band with constant cross-section the tensile stress is proportional to the external tension applied. Similarly, the tensile strain is proportional to the extension in the direction of the applied force. As a result, the area under each curve is proportional to the work done by the externally applied tension. Since the unloading curve has a smaller area under it than the loading curve this means that the energy returned by un-stretching the rubber band is less than the energy it takes to stretch the rubber band in the first place. The difference in energy ends up as heat energy in the rubber. This is why, for example, Formula 1 race cars have a warm-up lap before the race begins: the repeating elastic stretching and relaxing of the tyre's rubber while driving the lap heats the tyre up and increases the friction between it and the road surface.

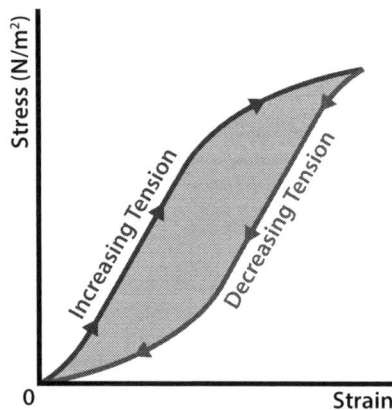

Fig. 2.14: Simulated plot of the deformation (strain) of a rubber band to both increasing and decreasing tensile stress. The exaggerated, shaded area between the lines is proportional to the energy loss of stretching and then relaxing the rubber band. This energy is converted into heat.

2.5 Plastic Deformation and Breaking

For very large stresses the elastic limit of the material can be exceeded. Beyond this limit, the material undergoes plastic deformation and will not return to its original shape, although when the stress is removed it may return some, or even most, of the way. Figure 2.15 shows an idealized stress-strain response curve for a material. An easy example of such behaviour is a paperclip. For small deformations, such as encountered in typical use clipping sheets of paper together, the deformation is elastic: when the paper is removed the paperclip returns to its original shape. However if you try to clip a thick stack of paper, or simply pull apart the two halves of the clip by hand, the clip is permanently deformed and does not return to its original shape.

If a material is stretched well beyond its elastic limit it may reach its breaking strength. At this point the gradient of the stress-strain curve becomes negative i.e. larger strains require less stress than the currently applied amount (see fig 2.15). The result is that unless the applied stress is quickly reduced or removed, the material will rapidly increase its strain until it fractures and breaks.

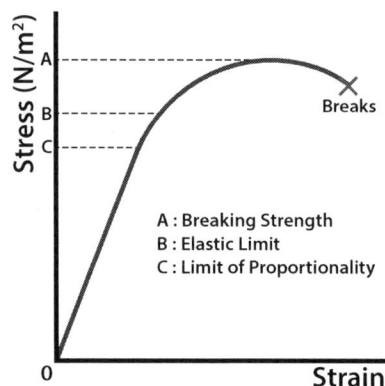

Fig. 2.15: An idealized stress-strain curve for a material showing the limit of proportionality, elastic limit and breaking strength.

The breaking strength of a material is, therefore, the stress which if applied will cause the material to fail. Such a stress will cause the material's strain to eventually increase to the point where it fails but this takes time due to inertia. Hence the breaking strength must be exceeded for sufficiently long enough to reach a strain which causes the material to fail.

In general, materials will have different breaking strengths for different types of stress. For example, concrete has a high compressive strength and can support very large compressive stresses but has a low tensile strength and can be relatively easily pulled apart. Steel has the reverse property in that it has a high tensile strength but low compressive strength. Reinforced concrete combines these two materials to produce a material which has a high breaking strengths for both compressive and tensile loads. Another example would be paper which has a high tensile strength but a comparatively low shear strength.

The separation between the elastic limit and breaking strength determines how a material behaves under large stresses. Ductile materials have a large difference between their elastic limit and their breaking strength and so can easily be permanently deformed by the application of sufficient stress. Examples of ductile materials are copper, aluminium and steel all at room temperature. Brittle materials have little or no difference between their elastic limit and breaking strength and so such materials cannot be permanently deformed and will simply break. Examples of brittle materials at room temperature are glass, china and some types of rigid plastic. Note that whether a material is brittle or ductile can change with temperature. For example, if glass is heated sufficiently it becomes extremely ductile which is how the glass blower, shown in figure 2.16, can deform hot glass using air pressure and a mould to create a vase.

This finally allows us to explain the behaviour of the materials in our mobile phone which we dropped at the start of this chapter. The metal in the case is a ductile material and, when the elastic limit is exceeded by being dropped onto a hard surface, the case undergoes a plastic deformation and is permanently dented. The glass front is a brittle material and so if the stress exceeds its elastic limit the glass will break and shatter otherwise it will return to its original shape and appear unaffected.

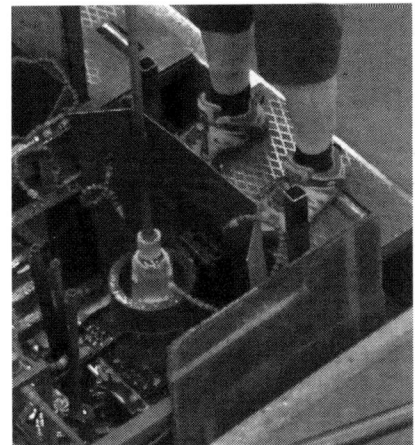

Fig. 2.16: A glass blower at Dartington Crystal in Devon, UK, relies on the ductile nature of glass at high temperature in order to shape it using air pressure and a mould.

Problems

Q2.1: A concrete bridge is supported in one location by a single, circular pillar which has a radius of 0.5 m and a height of 5 m. When a lorry drives over the bridge it applies an additional load of 200 kN to the pillar which causes it to compress by 0.1 mm. What is the modulus of elasticity of the concrete in the pillar?

Fig. P2.1: *Crane hook attached to a pulley which has a steel cable wrapped around it as shown.*

Q2.2: A crane has a hook which is suspended by a steel cable wrapped around an ideal pulley as shown in figure P2.1. The cable has a radius of 1 cm.
(a) If the tensile strength of steel is 500 MN m^{-2} what is the maximum weight that the crane can lift without the cable breaking?
(b) Is the tensile strength of the steel in the cable used (with a suitably generous margin) to determine the maximum load which a crane can safely lift? Explain your answer.

Q2.3: A nylon rope with a diameter of 15 mm is wound around a winch and is used to lift loads vertically up a 50 m high cliff. A load of mass 75 kg is attached to the rope at the base of the cliff and the winch started. The Young's Modulus for nylon is 3 GN m^{-2} and the gravitational field is 9.81 m s^{-2}.
(a) What length of rope does the winch wind in before the load starts to lift?

(b) If the same load was attached to two, synchronized winches each with its own, identical rope how much rope would each winch wind in before the load lifted in this case?

Q2.4: A machine which is operated on a dock contains a spring of length l and constant k. This spring has to be constantly replaced due to rusting so to solve this problem an engineer suggests replacing it with an equal length of rubber. If the modulus of elasticity for rubber is λ what is the cross-sectional area of the length of rubber which will replace the spring assuming that the modulus of elasticity remains constant?

Q2.5: A tow truck is winching a van out of a ditch using a steel cable that is 5 m long and has a radius of 2 mm. Unfortunately, cable has a small defect caused by rust at one point along it and while this defect does not affect the properties of the bulk of the cable it significantly lowers the maximum tensile load which the cable can sustain. The winch applies a tension of 5 kN to the cable which then suddenly snaps. If the Young's Modulus for steel is 25 GN m^{-2} how much energy is released when the cable snaps?

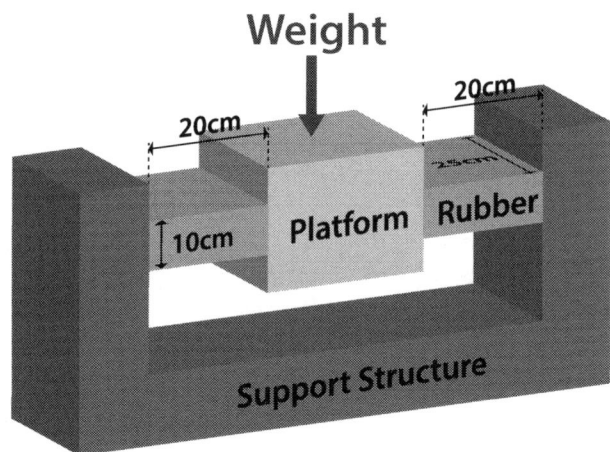

Fig. P2.2: *Design of a platform intended to be isolated from vertical vibrations in the support structure.*

Q2.6: A platform designed to isolate objects from vibrations is attached to two, identical blocks of rubber as shown in figure P2.2.

The rubber blocks are $10 \times 20 \times 25$ cm with the 20 cm dimension separating the platform from its support as shown in the figure. A weight of 2 kN is placed on the platform. If the shear modulus of rubber is 600 kN m^{-2} how far will the platform sink?

Q2.7: A bicycle pump consists of a cylinder with a radius of 1 cm and length 50 cm filled with air. A piston can slide along the length of the cylinder to compress the air and force it into a bicycle tyre. One such pump has its nozzle blocked such that no air can escape. In order to free this blockage, the cyclist pushes down on the piston with a force of 10 N. Assuming that the pump remains blocked how far does the piston move given that the adiabatic (which means assuming no energy gain or loss) bulk modulus of air is 142 kN m^{-2}?

Q2.8: It is often stated that water, which has a bulk modulus of 2.2 GN m^{-2} is incompressible.
(a) If this were a true statement what would be the value of the bulk modulus of water?
(b) Considering a cube of water with a volume of 1 cm^3 what force would need to be applied to each face of the cube in order to reduce the volume of the cube by 1%?

CHAPTER 3

Fluid Statics

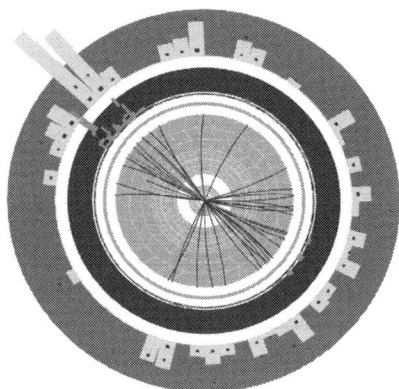

Fig. 3.1: A lead ion collision event from the ATLAS detector on the LHC at CERN showing an asymmetric jet event. The other jet loses its energy to the quark-gluon plasma created by the collision. Quark-gluon plasma is the state of matter that filled the very early universe and has the properties of a fluid.

A fluid is an amorphous substance whose molecules can move past one another. The result is that, unlike solids, fluids do not maintain a rigid structure and can flow. Under gravity, this generally means that they assume the shape of the container that they are placed in although for some fluids this can take a *very* long time. The most common examples of fluids are either liquids, such as water or gases, such as air. These differ in that in gases are easily compressible whereas liquids are far harder, though not impossible, to compress.

Going beyond liquids and gases plasma, which is a gas heated until the collisions between atoms have enough energy to knock off electrons and ionize the atoms, also behaves as a fluid. The same is true for the hottest state of matter known, quark-gluon plasma, where the collisions are so energetic that the nuclei themselves "melt" into a soup of subatomic particles called quarks and gluons. This state has only recently been discovered by the Large Hadron Collider at CERN in collisions between extremely high energy lead nuclei (see figure 3.1). The exact properties of matter in this state are still being determined but current indications are that the strong bindings between the quarks mean that it behaves as an incredibly dense, ideal liquid.

Flow is perhaps the most important property of fluids but in this chapter, we will study the properties of static fluids where there is no flow. To start we need to define some new quantities which are properties of the fluid being considered rather than a particular sample of it.

3.1 Density

Fluids are generally uniform in nature and so the total mass of a particular sample simply depends on the volume of the fluid present. Since the volume depends entirely on the particular circumstances instead of mass a more useful quantity for a fluid is the density which is defined as the mass per unit volume and is usually denoted by the greek letter rho, ρ. From the definition we have the relationship:

$$\text{Density, } \rho = \frac{m}{V} \tag{3.1}$$

Material	Density (kg/m³)
Air	≈ 1.2
Water	1,000
Helium	0.166
Aluminium	2,700
Iron	7,874
Copper	8,960
Lead	11,340
Gold	19,300
Osmium	22,590
Nucleus	2.3×10^{17}

Table 3.1: Densities of some common materials.

where m is the mass of the fluid and V is its volume. This is a property of a fluid and, for liquids, can generally taken to be constant unless there are unusual or extreme circumstances. The SI units for mass are kilograms and for volume cubic metres and so the SI units for density are kilograms per cubic metre ($\text{kg}\,\text{m}^{-3}$). A common, alternative metric, but not SI, unit for density is grams per cubic centimetre ($\text{g}\,\text{cm}^{-3}$). This has a conversion factor to $\text{kg}\,\text{m}^{-3}$ of:

$$1\text{kg/m}^3 = \frac{1,000\,\text{g}}{(100\,\text{cm})^3} = \frac{1,000\,\text{g}}{1,000,000\,\text{cm}^3} = 0.001\,\text{g/cm}^3 \tag{3.2}$$

Another non-metric density unit is density which is a ratio of the density of a fluid to that of water at $4\,^{\circ\circ}\text{C}$ where it is densest. This is sometimes used to measure the concentration of solutions, especially in the brewing and distilling industry, so while it is not used by physicists it is definitely of interest to many of them!

3.2 Pressure

In chapter 2 we saw how by inserting a thin, massless plane into a solid we could measure the internal forces inside it and use this to calculate the stress on the material. We can repeat this exercise with fluids to discover the internal forces inside them. The first thing to note is that fluids are amorphous and can flow. This means that there is no preferred direction: unlike with a solid, there is no need to consider three orthogonal planes and calculate the forces acting on each one because all directions are the same. This greatly simplifies things because it means we only need to consider the forces on any one plane at any given point and, due to the very nature of the fluid, all planes will have identical forces acting on them.

The second effect of a fluid's nature is that there can be no shear stress. The molecules in a fluid are free to move past one another so, unlike in a solid where the molecules and atoms are bound in some form of a lattice, in an ideal fluid, any shear stress will meet with zero resistance and just cause the fluid to flow. By the same argument, fluids cannot support a tensile stress because this will just pull the molecules apart. Therefore there can only be a compressive stress and it must apply uniformly in all directions just like a bulk stress.

Consider a thin plane inserted in a fluid, as shown in figure 3.2. If the area of the plane is small, δA, then we can assume that the small force acting on the plane, $\delta F_\perp$, is approximately constant over the entire area. We can use this to calculate the force per unit area on the plane. However, because fluids are uniform and amorphous they have no preferred direction and so instead of calling this stress as we do for solids we call it pressure. In this case the pressure is given by:

$$\text{Pressure, } P = \lim_{\delta A \to 0} \frac{\delta F_\perp}{\delta A} = \frac{dF_\perp}{dA} \qquad (3.3)$$

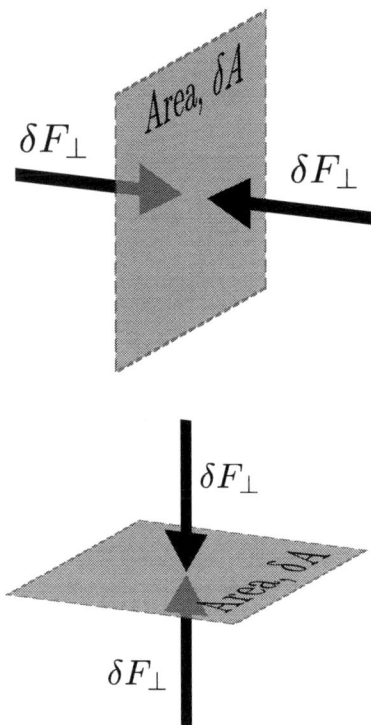

Fig. 3.2: Forces acting on a small area, δA, of a plane in a fluid. The same magnitude of force acts regardless of the orientation of the plane.

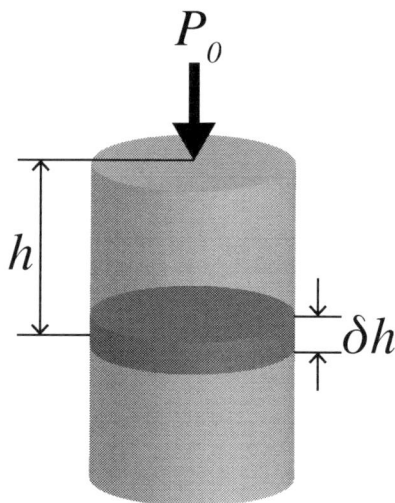

where we have taken the limit of the area going to zero in order to calculate the pressure at a point. Now since the force at this point has the same magnitude regardless of the orientation of the plane due to the properties of the fluid pressure is a scalar quantity, not a vector one. Since pressure is a force per unit area its SI units are newtons per square metre (N/m^2) or, if preferred, pascals can be used (Pa).

3.3 Static Liquid Pressure

Consider a column of an ideal fluid with a pressure P_0 pushing down on the top, as shown in figure 3.3. Since an ideal fluid cannot support any shear stress any force on the fluid in the column from the fluid surrounding the column must be horizontal. Hence if we resolve forces vertically we can ignore the surrounding fluid. Consider a small element of the column a distance h below the surface of the liquid as shown in figure 3.3. This element is in equilibrium and so it must have a larger pressure pushing on its lower surface than on its top surface such that the pressure difference exerts a force that counters the elements weight. The forces on the element are shown in figure 3.4 and if we resolve these forces vertically then equilibrium requires that:

$$A(P + \delta P) = AP + \delta m\, g \implies A\delta P = \delta m\, g \qquad (3.4)$$

Fig. 3.3: A column of fluid which has a pressure P_0 exerted on its surface. An small element of the column is shown a distance h below the surface.

where δm is the mass of the element and g is the gravitational field strength. Now the mass of the volume can be rewritten in terms of the element's volume and the density of the fluid:

$$\delta m = \rho A \delta h \qquad (3.5)$$

Combining equations (3.4) and (3.5) by eliminating δm we get:

$$\frac{A \delta P}{g} = \rho A \delta h \implies \delta P = \rho g \delta h \qquad (3.6)$$

Next we take the limit as δh goes to zero to get an expression we can integrate:

$$\delta P = \rho g \delta h \underset{\delta h \to 0}{\implies} dP = \rho g \, dh \qquad (3.7)$$

This gives the infinitesimal increase in pressure of a fluid for a small drop in height. To calculate the pressure at a distance h below the surface we simply integrate starting and the surface and descending to a distance h below the surface:

$$\int_{P_0}^{P} dP = \int_{0}^{h} \rho g \, dh$$
$$P - P_0 = \rho g h \qquad (3.8)$$

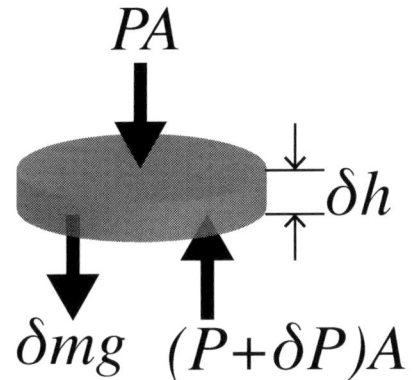

PA

δh

$\delta mg \quad (P+\delta P)A$

Fig. 3.4: A small element of the fluid column shown in fig. 3.3 with the forces acting on it.

where P is the pressure at a depth h below the surface and P_0 is the pressure at zero depth i.e. at the surface. Note that we have made a very important assumption when doing this integral: we have taken the density to be a constant. Hence this calculation will give the static pressure for incompressible fluids which is typically a good approximation for liquids but not gases. Rearranging we get the formula for the static pressure at a depth h below the surface of a liquid:

$$P = P_0 + \rho g h \qquad (3.9)$$

This shows that the pressure of a fluid in a gravitational field increases linearly with depth.

Example 3.1

At what depth below the surface of the ocean is the water pressure twice the pressure at the surface given that atmospheric pressure is $101 \, \text{kN} \, \text{m}^{-2}$, that the density of water is $1{,}000 \, \text{kg} \, \text{m}^{-3}$ and that $g = 9.81 \, \text{m} \, \text{s}^{-2}$?

Solution:
The pressure at a depth h below the surface of the ocean is:

$$P = P_0 + \rho g h$$

where P_0 is the pressure at the surface. Since we want the *pressure to be twice this we have the equation:*

$$2P_0 = P_0 + \rho g h \implies P_0 = \rho g h \implies h = \frac{P_0}{\rho g}$$

Now all we need to do is put in the numbers and evaluate:

$$h = \frac{101 \times 10^3}{1000 \times 9.81} = 10.3\,\text{m}$$

3.4 Pascal's Law and Hydraulics

Looking at equation (3.9) it is clear that the static liquid pressure at any point in a liquid does not just depend upon the depth at that point but also on the applied external pressure on its top surface. Any increase in the externally applied pressure will increase the pressure at every point in the liquid. This observation is called Pascal's law which states that:

"A change in pressure at any point in an enclosed fluid at rest is transmitted undiminished to all points in the fluid."

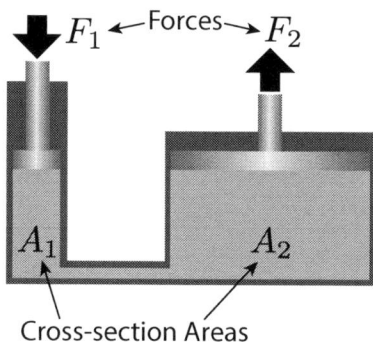

Fig. 3.5: A simple hydraulic system showing an incompressible fluid filling two connected pistons of areas A_1 and A_2. A force of F_1 acts downwards on one piston resulting in a force of F_2 acting upwards on the other.

This simple principle is what allows hydraulic systems to work. Consider a simple hydraulic system consisting of a small piston, with a cross-sectional area A_1 connected to a large piston with cross-sectional area A_2, as shown in figure 3.5. The pistons are at the same height and the volume between the pistons is filled with an incompressible, ideal liquid. If a force, F_1, is exerted downwards on the small piston it will apply a pressure to the liquid in the system where:

$$\text{Pressure, } P = \frac{F_1}{A_1} \tag{3.10}$$

Pascal's law states that this pressure will now be applied throughout the liquid and so this pressure will also be exerted against the larger, second cylinder. Hence we can also write:

$$\text{Pressure, } P = \frac{F_2}{A_2} \tag{3.11}$$

where F_2 is the force exerted on the second cylinder. Since the pressure is equal we can combine (3.10) and (3.11) to eliminate it which gives:

$$\frac{F_1}{A_1} = \frac{F_2}{A_2} \implies F_2 = \left(\frac{A_2}{A_1}\right) F_1 \tag{3.12}$$

However A_2 is a lot larger than A_1 and so this means that F_2 must be larger than F_1 by the same ratio. This "magnification" of the applied force, combined with the flexible positioning of the cylinders which can be connected by a hose, is why hydraulic systems are so very useful.

While this increase in force from a hydraulic system may seem a little too good to be true there is a price to be paid. To show this let's consider the distance moved by the pistons. If the first piston is depressed by a distance x_1 then the volume of liquid displaced is simply the cross-sectional area multiplied by the area of the cylinder. Since the liquid is incompressible this is the same volume by which the larger cylinder expands and so we have:

$$\text{Volume, } V = A_1 x_1 = A_2 x_2 \implies x_2 = \left(\frac{A_1}{A_2}\right) x_1 \qquad (3.13)$$

where x_2 is the displacement of the second cylinder. This shows that while the force has been increased by the ratio of the areas the displacement has been decreased by the same factor. For example, if a particular hydraulic system has a ratio of cylinder areas of 1:10 then the force which can be generated is 10 times the applied force but to move that force through a distance x would require the applied force to be moved through a distance of $10x$.

The result in equation (3.13) can also be shown to result in conservation of energy. If we consider the work done by the applied force then, substituting in the values for F_1 and x_1 from (3.13) and (3.12) respectively we have:

$$\text{Work Done} = F_1 x_1 = \left(\frac{A_1}{A_2}\right) F_2 \times \left(\frac{A_2}{A_1}\right) x_2 = F_2 x_2 \qquad (3.14)$$

Hence the work done by the applied force is equal to the work done by the resulting force and, since no energy is stored in the system, this shows that energy is conserved.

3.5 Absolute and Relative Pressure

All devices which measure pressure actually measure pressure differences. However, nature provides an absolute pressure scale because vacuum has an absolute zero pressure so devices which measure a pressure difference compared to vacuum are said to measure absolute pressure. The simplest example of such a device is the barometer, shown in figure 3.6.

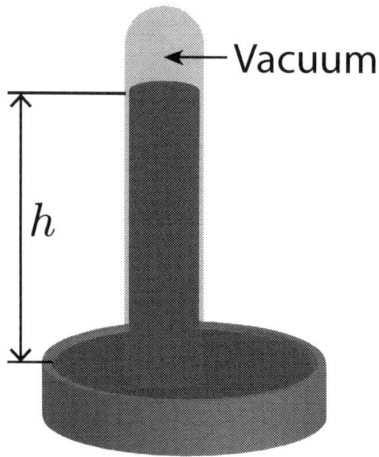

Fig. 3.6: A simple barometer showing the vacuum at the top of the tube. The height of mercury in the tube is a measure of the external, atmospheric pressure.

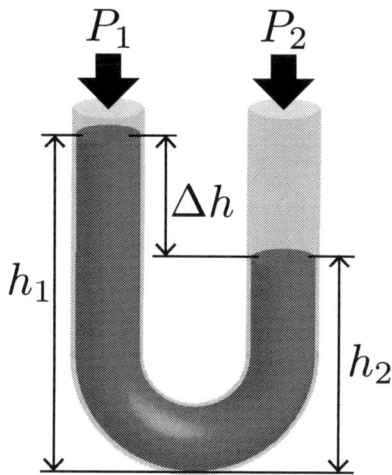

Fig. 3.7: A simple manometer showing the two arms under pressures P_1 and P_2. The height difference in the liquid between the arms is proportional to the pressure difference.

This consists of a liquid (often mercury) filled tube which has the top end sealed and the open end placed just under the surface of a liquid reservoir. When the system is released the level of liquid in the tube will fall leaving a vacuum at the top of the tube.

However, the liquid will not run out of the tube entirely because atmospheric pressure acting on the surface of the liquid reservoir generates an upwards force on the liquid in the tube. When this upwards force is equal to the static pressure of the liquid the level in the tube will cease to fall. To determine when this will be, consider the static pressure of the liquid in the tube at the same level as the surface of the reservoir:

$$\text{Pressure, } P_t = P_0 + \rho g h_t = \rho g h_t \qquad (3.15)$$

where ρ is the liquid density and h_t is the height of the liquid in the tube above the surface of the reservoir. Now consider the pressure at the surface of the reservoir. Since there is no height of fluid above it the pressure here is equal to atmospheric pressure. Now for equilibrium, these two pressures must match otherwise fluid will either flow into or out of the tube until the pressure equalizes. Hence we end up with the relationship:

$$\text{Pressure, } P_{atm} = \rho g h_t \qquad (3.16)$$

Thus by measuring the height of fluid in the tube, we can determine atmospheric pressure. This is an absolute pressure measurement because the pressure difference being measured is that between the external atmosphere and the vacuum at the top of the tube.

A barometer is ideal for measuring an absolute pressure but in many cases what we need is a relative pressure measurement. For example, the pressure in car and bicycle tyres is measured relative to atmospheric pressure because this is what determines how the tyre will perform and a tyre with zero pressure does not mean that it contains a vacuum just that it is at atmospheric pressure. To measure a pressure difference we use a manometer, as shown in figure 3.7. When both arms of the manometer are open to the atmosphere the level of fluid in each arm is equal. However, when one arm is connected to a source of pressure the fluid levels will change. If we consider the pressure at the bottom of the device using the pressure in each of the two arms we have:

$$\text{For arm 1: } P = P_1 + \rho g h_1 \qquad (3.17)$$
$$\text{For arm 2: } P = P_2 + \rho g h_2 \qquad (3.18)$$

where h_1 and h_2 are the heights of the fluid surface in arms 1 and 2 respectively and P_1 and P_2 are the external pressures applied to the surface of the liquid in arms 1 and 2 respectively. Since there can only be one pressure at a point these values must be equal and so we have:

$$P_1 + \rho g h_1 = P_2 + \rho g h_2 \implies P_1 - P_2 = \rho g(h_2 - h_1) \qquad (3.19)$$

This can be simplified to:

$$\Delta P = \rho g \Delta h \qquad (3.20)$$

where ΔP is the difference in pressure between the arms and Δh is the difference in height between the surface of the liquid in each arm with the lower arm pressure having the highest liquid surface.

3.6 Buoyancy

When an object is submerged in a fluid the pressure of the fluid will act on every surface of the object. However, we know that this pressure is not constant and increases with increasing depth of fluid. This means that the pressure pushing upwards on the lower surfaces of the object will be larger than the pressure pushing down on the upper surfaces of the object. The result is that there is a net, upwards force acting on the object due to the pressure of the fluid this is known as buoyancy.

To calculate this force consider figure 3.8 which shows a small volume element of a solid immersed in an incompressible liquid. By symmetry, the horizontal pressure will precisely cancel and so there is no net horizontal force on the element. Vertically this is not the case because the pressure at the top, P_t, is less than the pressure on the bottom, P_b and is given by:

$$\text{Pressure Difference, } \delta P = P_b - P_t = \rho_l g \delta y \qquad (3.21)$$

where ρ_l is the density of the liquid. To convert this into a force we multiply by the area on which the pressure acts which gives the net, upwards force on the small volume element as:

$$\text{Force, } \delta F = \delta P \, \delta x \, \delta z = \rho_l g \, \delta x \, \delta y \, \delta z = \rho_l g \, \delta V \qquad (3.22)$$

where δV is the volume of our element. This is just the force on one volume element. If we consider adding a second volume element we can see that it does not matter where we place it the net force will double. If we stack them vertically then the pressure difference will double and if we stack them horizontally

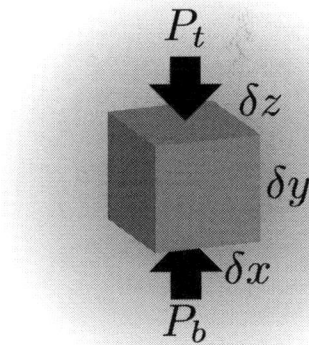

Fig. 3.8: A small volume element immersed in a fluid. The pressures acting on the top and bottom of the element are shown as P_t and P_b respectively.

then the area the pressure difference acts on will double. The next step is to add all the volume elements in a solid object together so we take the limit as δV goes to zero while integrating over the volume of whatever solid object we have:

$$\int dF = \int \rho_l g\, dV \implies F = \rho_l g V \qquad (3.23)$$

where V is the total volume of the object. Now looking at this equation we see that the density of the liquid multiplied by the object's volume is just the mass of liquid which is displaced by the object. This is further multiplied by the gravitational field to give the weight of liquid displaced by the object. Eureka! We now have a simple rule for the buoyancy force:

> "The upwards buoyancy force that is exerted on a body immersed in a liquid is equal to the weight of the fluid that the body displaces."

This is Archimedes' principle. Unfortunately, the apocryphal story of Archimedes being so excited that he exclaimed "Eureka!", leapt from his bath and ran naked down the street occurred when he realized that the volume of fluid dispersed by an object wholly immersed in a fluid was equal to the volume of the object. So we can only imagine what he did in his excitement when he discovered this more important principle!

Buoyancy is used to determine whether objects will float. If the weight of liquid displaced is greater than the weight of the object then the object will float. Another way to say this is that if the *average* density of the object is less that the density of the liquid then the object will float. This is why steel hulled ships float because their average density, which includes the air inside the hull, is less than that of water even though the density of steel used in their construction is well above it.

Example 3.2

A coal barge is a rectangular box with a length of 30 m and a width of 5 m. If the height of the barge is 5 m, allowing some margin for safety, and its empty weight is 200 kN what is the maximum weight of coal which it can transport when in freshwater with a density of 1,000 kg m^{-3}?

Solution:

According to Archimedes' principle the weight of the barge plus the coal will displace a weight of water equal to the combined weight. Hence if the combined weight is W, the density of water ρ then the height by which the barge will lower in the water will be:

$$W = \rho l w d$$

where l is the length of the barge and w is its width. To get the largest weight we need the largest permissible drop in height and so putting in the numbers we get:

$$W = 1000 \times 30 \times 5 \times 5 = 750\,\text{kN}$$

This is the limit on the combined weight of the barge plus the coal so to get the maximum weight of coal we simply subtract the weight of the barge. Hence the maximum weight of coal is 550 kN or about 56 t of mass assuming in the Earth's gravitational field.

Problems

Q3.1: The kilogram is the only SI base unit still defined by an object instead of a fundamental measurement. One approach being studied to replace it is an almost perfectly smooth sphere of absolutely pure silicon which would enable the kilogram to be defined in terms of a length, the radius of the sphere. If the density of silicon is 2,329.0 kg m^{-3} what is the radius of the sphere which has a mass of 1 kg?

Q3.2: A bicycle tyre is inflated to a pressure of 375 kN m^{-2}. If the mass of the bicycle and rider combined is 90 kg and the gravitational field is 9.81 m s^{-2} what is the total area of both tyres which is in contact with the ground?

Q3.3: A large, cubic box of mean density ρ and mass m is placed on an air cushion which is gradually inflated in order to lift the box. At what pressure above atmospheric does the box start to lift if the gravitational field strength is g?

Q3.4: A large, sealed wooden water barrel contains 100 l of water and is rated to withstand a pressure of 200 kN m^{-2} relative to atmospheric pressure. A small hole is drilled in the top of the and a thin, plastic pipe is firmly attached to it. This pipe is then filled with water. At what depth of water in the narrow pipe will the barrel fail?

Q3.5: A simple barometer is made by taking a tube of mercury and up-ending ii in a small mercury bath. The result is a 74 cm column of mercury standing above the surface level of the bath. The density of mercury is 13,594 kg m^{-3} and the gravitational field is 9.81 m s^{-1}.
(a) What is the atmospheric pressure?
(b) Mercury is toxic and so it is suggested to replace it with alcohol as was done with thermometers. How high would the tube need to be for an alcohol barometer given that the density of alcohol is 789 kg m^{-3}?

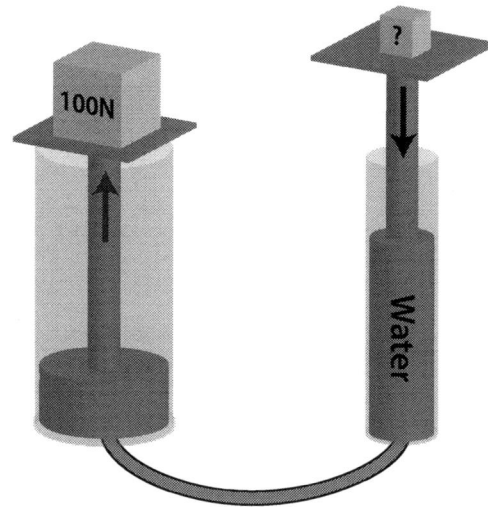

Fig. P3.1: *Two syringes of different radii are connected together and filled with water. A large weight is placed on the platform attached to the large syringe and when a smaller weight is placed on the other platform it depresses causing the large weight to rise.*

Q3.6: A professor devises a demonstration which consists of a light, horizontal platform attached to the top of the plunger of a syringe of radius 0.5 cm. This syringe is attached to a second syringe of radius 2.5 cm by means of a tube and a second light, horizontal platform is attached to the top of the plunger of the second syringe. The who system is filled with water and starts with the plunger of the smaller syringe fully raised and the plunger of the larger syringe fully depressed. The professor then places a weight of 100 N on the platform attached to the large radius syringe as shown in figure P3.1.
(a) What is the smallest weight which the professor can place on the first platform to cause the heavy weight to rise?
(b) If the light weight on the smaller platform falls a distance of 10 cm how much will the heavy weight rise by?
(c) If the system was filled with air instead of water explain qualitatively and with clear reasoning any difference in the behaviour of the system.

Q3.7: A spherical, helium balloon is made of thin, mylar sheets which have a thickness of

12 μm and a density of 1,390 kg m^{-3}.

(a) What is the smallest balloon radius which would float on the surface of Mars where the atmosphere has a density of 0.02 kg m^{-3} if, at the same pressure, the density of helium could be considered negligible?

(b) If an identical balloon were released on Earth where the atmosphere has a density of 1.2 kg m^{-3} what would be the balloon's initial vertical acceleration given that at the pressure of the Earth's atmosphere helium has a density of 0.166 kg m^{-3} and that the gravitational field is 9.81 m s^{-2}?

Q3.8: A cube of copper with sides of 5 cm is attached to a spring scale by a light, inextensible string in a gravitational field of 9.81 m s^{-2}. A light, cylindrical beaker with a radius of 5 cm is filled with a 15 cm depth of water and is placed on a set of scales. The cube, still attached to the spring scale, is then lowered into the beaker such that it is completely submerged without touching. The density of water is 1,000 kg m^{-3} and the density of copper is 8,960 kg m^{-3}.

(a) What is the force measured by the spring scale when the cube is in the water?

(b) What is the force measured by the scales underneath the beaker of water when the cube is in the water?

Q3.9: Two tubes, with radii of 1 cm and 2 cm, are connected by a narrow pipe which has a closed tap in it. The thinner tube is filled to a depth of 10 cm with water which has a density of 1,000 kg m^{-3} and the wider tube is filled to the same depth but with oil with a density of 800 kg m^{-3} as shown in figure P3.2. The gravitational field is 9.81 m s^{-2}.

(a) When the tap is opened and the system has reached equilibrium what is the height difference between the fluid levels in each tube and which tube has the higher level of fluid?

(b) How would things be different if the experiment were repeated on the moon where gravity is one-sixth that on Earth?

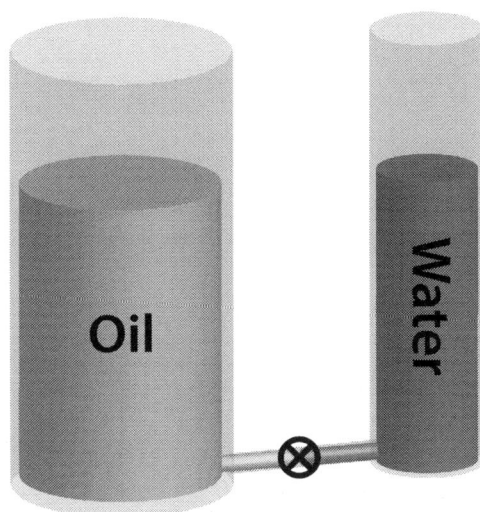

Fig. P3.2: *Two, open tubes of differing diameters are filled with water and oil respectively and are connected by a narrow pipe with a tap in the centre.*

CHAPTER 4

Fluid Dynamics

This chapter will discuss the dynamic behaviour of fluids and so will concentrate on one property which all fluids exhibit: flow. Fluid dynamics is a huge, complex field which is encountered frequently in everyday life from predicting the weather to determining how a mould will fill with liquid to designing faster, more efficient aircraft. The physics needed to describe such detailed behaviour can be extremely complex and this chapter will only scratch the surface of a very broad and deep field of study.

4.1 Surface Tension

Surface tension is a property of liquids which arises from the unbonded molecules on the surface of the liquid as shown in figure 4.1. These are in a higher energy state than molecules in the bulk of the liquid because they have fewer bonds to neighbours. This results in the surface of a liquid having an associated energy per unit area and the liquid acting so as to try and reduce its surface area which is why liquids have smooth, rather than rough, surfaces. The SI units of surface tension are typically newtons per metre ($N\,m^{-1}$) which is dimensionally equivalent to joules per square metre ($J\,m^{-2}$) which are also sometimes used.

Surface tension can have a significant impact on a fluid's dynamical properties. For example, objects which are denser than a liquid may rest on the surface if they are light enough that the surface tension can support them. In addition, the surface tension causes liquids to resist passing through fine meshes which

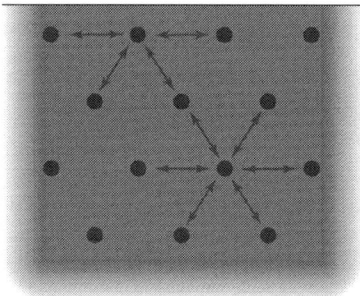

Fig. 4.1: Diagram showing how molecules near the surface of a liquid have fewer neighbour bonds. This increases the energy of their state and so there is an energy cost per unit area of a liquids surface that gives rise to a surface tension.

is why, when washing clothes we use warm, soapy water which has a far lower surface tension than cold, pure water.

Another example where surface tension becomes important is in small scale droplets of a liquid. The effect of surface tension scales with area whereas the effect of pressure forces comes from the bulk and so scales with volume. For a spherical drop, the ratio of area to volume is $\frac{3}{r}$ where r is the drop radius. Hence for small radii, the effect of surface tension can be significant. This is readily noticeable in the shape of rain drops which are spherical, not teardrop shaped, due to the surface tension dominating the pressure forces and so acting to minimize the surface area of the drop.

4.2 Fluid Flow

One of the most important properties of a fluid is that they flow. This is not only one of the key features which distinguish them from solids but we encounter fluid flow, or the results of its application, commonly in everyday life. Aerodynamics, the study of not just air flow but for any gas, is used in the design of many things from bicycle helmets to aeroplanes where the aim is to reduce drag. However, it is also the key physics behind weather forecasting and natural gas pipelines. Hydrodynamics, the study of not just water but any incompressible fluid flow, is similarly commonly encountered in day-to-day life with applications including boat and ship hulls, the design of injection moulds, oil pipelines, car tyre tread patterns and hydro-electric power plants.

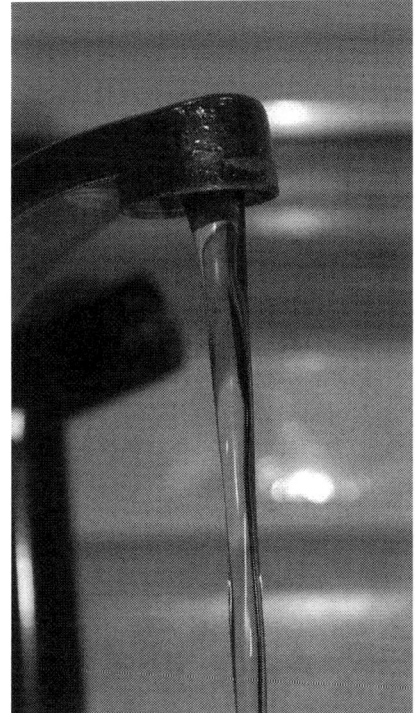

Fig. 4.2: Picture of laminar water flow from a kitchen tap. Note the smooth surface of the water indicating the steady flow of the water in the column.

4.2.1 Types of Flow

Fluid flow can be broadly separated in one of two types: laminar, or steady flow and turbulent, or unsteady, flow.

Laminar Flow

In laminar flow adjacent planes of molecules in the fluid move over each other smoothly. The result is a smooth flow where the fluid velocity at a particular point remains constant over time, hence the alternative name of steady flow. An example of the laminar flow of water from a tap is shown in figure 4.2.

Figure 4.3 shows the velocity profile in a pipe for the water in

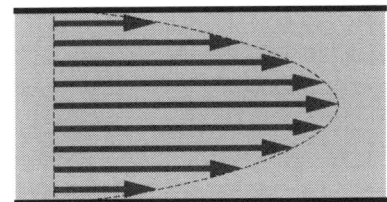

Fig. 4.3: Velocity profile for laminar flow of a real liquid in a pipe. The fluid at the edges is slowed down by friction with the walls and at the point of contact is actually zero.

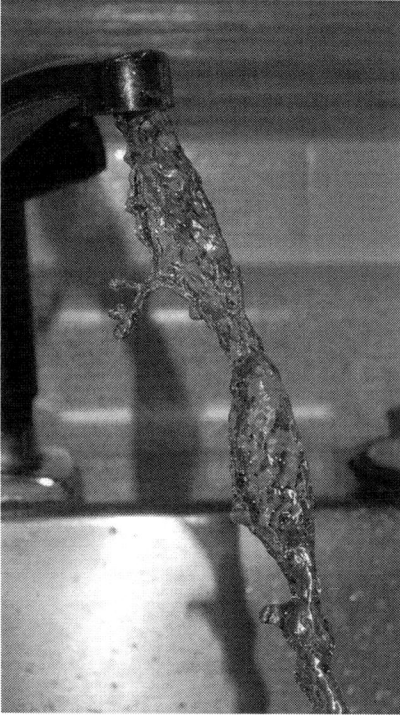

Fig. 4.4: Picture of turbulent water flow from a kitchen tap. The irregular, and changing, surface of the water shows that the flow pattern is constantly shifting and changing.

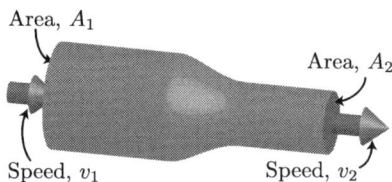

Fig. 4.5: Fluid flowing through a pipe which narrows.

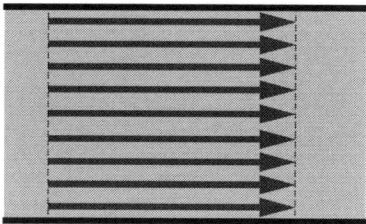

Fig. 4.6: Velocity profile for laminar flow of an ideal fluid in a pipe. Ideal fluids have zero viscosity and so have no friction forces to slow them. The result is that the velocity is the same everywhere.

such a flow. The water is all flowing in the same direction and the velocity remains constant over time at any given point. The velocity difference over the profile of a pipe is due to the viscosity of the water which is a property covered in section 4.7. Laminar flow is the easiest to understand and is the only type of flow which we will discuss in a quantitative manner here. Laminar flow is the type typically found at low fluid velocities.

Turbulent Flow

As the flow velocity of a fluid increases at some point, the flow will transition from laminar flow to turbulent flow. In turbulent flow, the fluid velocity at a given point in the flow is highly variable and may even temporarily reverse at points! Figure 4.4 shows water from a tap undergoing turbulent flow. Although only an instant in time is captured it is clear that the water is chaotic and that the flow at any one, fixed point in space will not be constant. This type of flow is extremely hard to describe mathematically because it is so chaotic and so this chapter will only mention it in a qualitative, not quantitative, manner.

4.3 Continuity Equation

Newtonian mechanics, unlike relativistic mechanics, requires the conservation of mass i.e. the initial mass of a Newtonian system is always equal to the final mass. This has implications for steady state fluid flow in that, since there is no build up of fluid in the system, the rate of mass flowing into the system must equal the rate of mass leaving the system.

Consider figure 4.5 which shows an ideal fluid flowing through a pipe that narrows. For an ideal fluid there is no resistance to motion (zero viscosity) and so the fluid velocity of the fluid at every point in the pipe is the same as shown by the velocity profile in figure 4.6. If the fluid entering the system has a density ρ_1 then the mass per unit time that is flowing into the pipe is:

$$\text{Mass Entering per unit Time, } f_{in} = \rho_1 A_1 v_1 \qquad (4.1)$$

Similarly we can derive an expression for the mass which leaves the system per unit time which is simply:

$$\text{Mass Leaving per unit Time, } f_{out} = \rho_2 A_2 v_2 \qquad (4.2)$$

where ρ_2 is the density of the fluid as it leaves the system shown. Now the requirement that the mass of the system is conserved means that the mass entering per second must be balanced by the mass leaving per second because there is no buildup of mass in the system (we are assuming the flow is in a steady state). Hence combining (4.1) and (4.2) we have:

$$f_{in} = f_{out} \implies \rho_1 A_1 v_1 = \rho_2 A_2 v_2 \qquad (4.3)$$

This is known as the *continuity equation*. This can be simplified further for liquids, such as water, which are incompressible to a good approximation. For these fluids the density is always approximately constant i.e. $\rho_1 = \rho_2 = \rho$. This means that it can be cancelled from either side of the equation and we are left with the continuity equation for incompressible liquids:

$$A_1 v_1 = A_2 v_2 \qquad (4.4)$$

This explains why, when you put your finger over a tap or the end of a hosepipe to partially block it, the water greatly speeds up. As you reduce the area out of which the water flows it has to flow faster in order to satisfy the continuity equation.

Example 4.1

A syringe is connected to a hollow needle with a radius twenty times smaller than the radius of the syringe itself. The syringe is filled with water and the plunger depressed so that the water squirts out of the end of the needle at 10 m/s. What is the speed at which the plunger is being depressed?

Solution:
To a large extent water is incompressible and so the continuity equation will apply. The velocity of the syringe's plunger, which we will call v, is the same as the velocity of the fluid in the body of the syringe. Hence, from the continuity equation, we have:

$$A_s v = A_n \times 10$$

where A_s and A_n are the areas of the cross-sections of the syringe and the needle respectively. However, we can write the areas in terms of the radii to get:

$$\pi r_s^2 v = 10 \pi r_n^2 \implies r_s^2 v = 10 r_n^2$$

In the question we are told that the needle has a radius 20 times smaller than the syringe which gives:

$$(20 r_n)^2 v = 10 r_n^2 \implies 400 r_n^2 v = 10 r_n^2 \implies v = 0.025 \, \text{m/s}$$

Hence the syringe's plunger is depressed at 2.5 cm/s.

4.4 Force on a Hosepipe

Consider the situation shown in figure 4.7 where a gardener is holding a hosepipe such that it rises vertically from the ground and is bent through 90°. Every second there is a mass of water flowing through the pipe that will change its momentum as it moves through the 90° bend. Newton's second law tells us that the rate of change of momentum equals the force applied and so there must be a force acting on this water to cause its change in momentum. Newton's third law says that there is also an equal and opposite reaction to the force and this is the force that the gardener holding the pipe will feel.

Since in the vertical direction there is the additional complication of weights due to gravity we will consider only the horizontal component of the force. Initially, for the water entering the system, there is no horizontal component of momentum. Every second the volume of water leaving the system is:

$$\text{Volume per second} = Av \tag{4.5}$$

where A is the area of the pipe's cross-section and v is the velocity of the ideal fluid. To convert this into a mass per second we simply multiply by the density of the fluid:

$$\text{Mass per second} = \rho Av \tag{4.6}$$

where ρ is the fluid density. The last step is to calculate the momentum per second which, in Newtonian mechanics, is simply the mass multiplied by the velocity. Hence the horizontal momentum leaving the pipe per second is just:

$$\text{Momentum per second} = \rho Av^2 \tag{4.7}$$

Now Newton's second law states that the rate of change of momentum equals the force. Since the initial momentum of the water in a horizontal direction is zero and the final horizontal momentum of the water *each second* is given by equation (4.7) this is equal to the force acting on the water. Newton's third law tells us that there is an equal and opposite reaction on the pipe and so the force on the hosepipe, which the gardener will feel, is simply:

$$F_{\text{pipe}} = \rho Av^2 \tag{4.8}$$

This equation is applicable to many situations where a fluid is expelled from a nozzle, for example in jet and rocket engines although it is often a gross simplification!

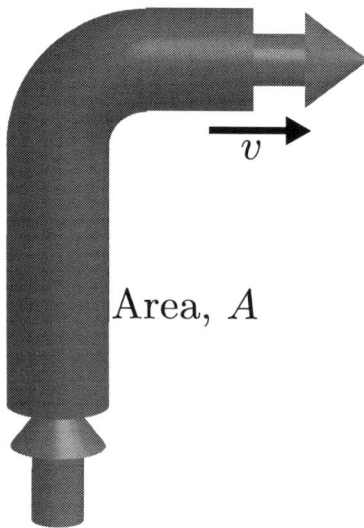

Fig. 4.7: Simple hosepipe of cross-section A showing a 90° bend. Water enters with a vertical velocity and exits with the horizontal velocity of v.

Area, A

v

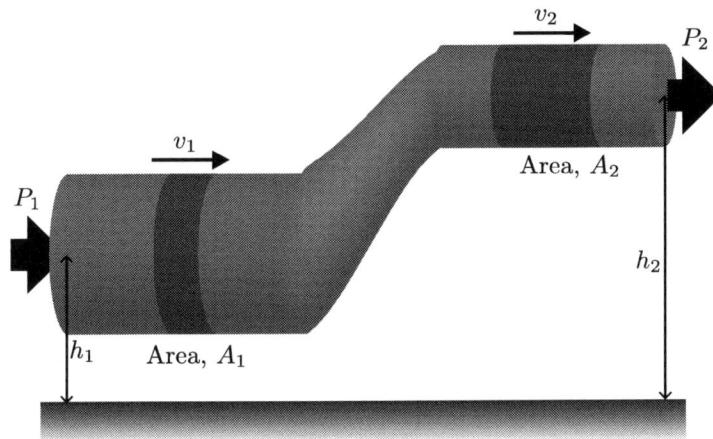

Fig. 4.8: Fluid flowing with a velocity v_1 from a pipe of area A_1 at a height of h_1 into a second pipe of area A_2 at a height of h_2 where the fluid moves with a velocity v_2. The pressure in the first pipe is P_1 and in the second pipe is P_2.

Example 4.2

A jet engine on an aeroplane creates a thrust of 310 kN and has a radius of 3 m. If the density of air is 1.2 kg/m^3 at what velocity is air expelled from the engine?

Solution:
The jet engine creates thrust by expelling air at a high velocity. Hence the force that is generated is given by:

$$F = \rho A v^2$$

We can rearrange this to get an expression for the velocity of the air:

$$v = \sqrt{\frac{F}{\rho A}} = \sqrt{\frac{F}{\rho \pi r^2}}$$

Now all that remains is to stick in the numbers and we get:

$$v = \sqrt{\frac{310 \times 10^3}{1.2\pi \times 3^2}} = 95.6 \text{ m/s}$$

4.5 Bernoulli's Equation

Having applied the simple, Newtonian concept of mass conservation to fluids the other obvious conservation law to apply is that of energy. This fundamental principle is the physics behind an observation made by the Swiss physicist Daniel Bernoulli who, in 1738, noted that an increase in the speed of a fluid is associated with a simultaneous drop in its pressure.

Consider the diagram shown in figure 4.8 which shows a fluid flowing from a wide, low pipe into a higher, narrow pipe. For an ideal fluid undergoing laminar flow the system is in a steady state and so the net work done on the system by forcing fluid into it must equal the change in energy of the fluid plus any work done by the fluid leaving the system. This gives us three energy terms that we need to consider: the net work done on the system by the pressure; the change in the kinetic energy of the fluid and the change in the gravitational potential energy of the fluid

To do this consider the energy input and output which occurs in a single second. Since there is no energy build-up in the system these two values must balance. First the work done by a pressure is the total force acting multiplied by the distance it moves. The force is simply the area multiplied by the pressure and the distance it moves in one second is just the velocity of the fluid. Hence the work done is just:

$$\text{W}_{\text{in}} = P_1 A_1 v_1 \qquad \text{W}_{\text{out}} = P_2 A_2 v_2 \qquad (4.9)$$

for both the work done on the system by P_1 and the work the system does with P_2. Next we need to consider the kinetic energy of the fluid which enters, or leaves, the system in our one second window. This is the standard $\frac{1}{2}mv^2$ formula of classical mechanics where we get the fluid mass from its volume and the fluid density, ρ:

$$\text{KE}_{\text{in}} = \frac{1}{2}\underbrace{\rho A_1 v_1}_{\text{mass}} v_1^2 = \frac{1}{2}\rho A_1 v_1^3 \qquad \text{KE}_{\text{out}} = \frac{1}{2}\rho A_2 v_2^3 \qquad (4.10)$$

Note that we have made an implicit assumption here that the density of the fluid is constant. Hence from this point on we are only considering the flow of fluids under situations where the density is constant. Finally we need to consider the incoming and outgoing gravitational potential energies. Again we get the mass of fluid entering and leaving from the fluid volume and density. Using the standard mgh formula for a constant gravitational field we have:

$$\text{GPE}_{\text{in}} = \underbrace{\rho A_1 v_1}_{\text{mass}} gh_1 \qquad \text{GPE}_{\text{out}} = \rho A_2 v_2 gh_2 \qquad (4.11)$$

So to conserve energy we require that:

$$\text{W}_{\text{in}} + \text{KE}_{\text{in}} + \text{GPE}_{\text{in}} = \text{W}_{\text{out}} + \text{KE}_{\text{out}} + \text{GPE}_{\text{out}}$$

$$P_1 A_1 v_1 + \frac{1}{2}\rho A_1 v_1^3 + \rho A_1 v_1 gh_1 = P_2 A_2 v_2 + \frac{1}{2}\rho A_2 v_2^3 + \rho A_2 v_2 gh_2$$

$$(4.12)$$

Faster flow on top

Aeroplane Wing

Slower flow underneath

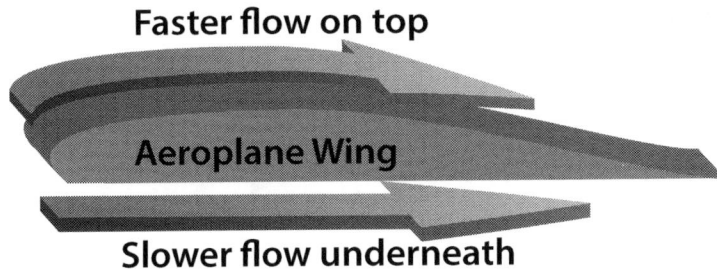

Fig. 4.9: Flow of air over the wing of an aeroplane. The shape of the wing forces the air flowing over the top of the wing to move faster than that underneath. Due to the Bernoulli effect, this causes a lower pressure to form on the top of the wing which generates lift.

Now we need to remember the continuity equation from (4.4) which requires that $A_1 v_1 = A_2 v_2$. This only applies to incompressible fluids but we have already made this assumption. Substituting for $A_2 v_2$ in (4.12) gives:

$$P_1 A_1 v_1 + \frac{1}{2}\rho A_1 v_1^3 + \rho A_1 v_1 g h_1 = P_2 A_1 v_1 + \frac{1}{2}\rho A_1 v_1 v_2^2 + \rho A_1 v_1 g h_2$$

$$P_1 + \frac{1}{2}\rho v_1^2 + \rho g h_1 = P_2 + \frac{1}{2}\rho v_2^2 + \rho g h_2 \qquad (4.13)$$

This is called Bernoulli's equation and is a direct consequence of the conservation of energy for incompressible fluids. It can also be stated, in a form called Bernoulli's principle, that at any point in a system filled with an incompressible fluid:

$$P + \frac{1}{2}\rho v^2 + \rho g h = \text{constant} \qquad (4.14)$$

This principle is the physics behind flight. Consider figure 4.9 showing two points one above and one below the wing of an aircraft. The shape of the wing is such that when it is forced to move through the air it accelerates the air on top to a higher velocity. For a thin wing the height is approximately constant and so Bernoulli's principle requires that the pressure drops with an increase in velocity. This means that the fast moving air above the wing must have a lower pressure than the slow moving air underneath the wing. This pressure difference due to the different velocities of the air is what generates the lift the aeroplane needs to fly.

4.6 Torricelli's Law

In 1643, before Bernoulli discovered his equation, the Italian scientist Evangelista Torricelli discovered a law for the fluid velocity from the hole in the bottom of a tank. This was later shown to be a specific case of Bernoulli's equation but it is worth repeating here both as an example application of Bernoulli but also because it is a common situation.

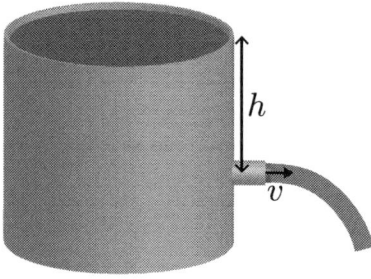

Fig. 4.10: A tank of water with a narrow spout a distance h below the tank's surface. Torricelli's Law gives the relationship between the spout's water velocity and the depth below the surface.

Torricelli's law states that the speed of fluid flowing out of a hole in a tank at depth h below the surface of the fluid in the tank is the same as if the fluid had been dropped at rest from a height h. Using simple constant acceleration equations this gives:

$$v = \sqrt{2gh} \tag{4.15}$$

To prove this consider the tank with a spout a distance h below the surface which is shown in figure 4.10. The figure also shows two points: the top surface of the water in the tank (A) and the water which has just left the tank through the spout (B). Applying Bernoulli's equation (4.13) to these two positions we have:

$$P_0 + \frac{1}{2}\rho v_A^2 + \rho gh = P_0 + \frac{1}{2}\rho v_B^2$$
$$\implies \frac{1}{2}\rho v_A^2 + \rho gh = \frac{1}{2}\rho v_B^2$$
$$\implies \frac{1}{2}v_A^2 + gh = \frac{1}{2}v_B^2$$

where P_0 is atmospheric pressure, ρ is the density of the liquid, v_A is the velocity of the liquid in the tank and v_B is the velocity of the liquid leaving the spout. Now if the tank is very large and the spout is small then the level in the tank will fall extremely slowly and so the velocity of the fluid at the top of the tank is approximately zero. Making this approximation means that we can take v_A to be zero which gives:

$$\rho gh = \frac{1}{2}v_B^2$$
$$\implies v_B = \sqrt{2gh} \tag{4.16}$$

and so since v_B is velocity of the water leaving the tank equation (4.16) agrees with Torricelli's Law given in (4.15). The conclusion is that Torricelli's Law is a consequence of Bernoulli's equation for large tanks with small spouts.

4.7 Viscosity

So far all the fluids we have considered quantitatively have been ideal fluid which have no resistance to shear and tensile stresses. However, almost all real fluids have a non-zero value of a property called viscosity which, in some ways, can be considered as friction for liquids. The larger the viscosity of a fluid the more it will resist shear and tensile stresses and so the harder it will be to make the fluid flow.

In completely general terms viscosity is a resistance to stress applied to the fluid. As a result for completely general cases viscosity, like stress, is a tensor quantity. However, the most common type is shear, or dynamic, viscosity which is often what is referred to when discussing the viscosity of a fluid.

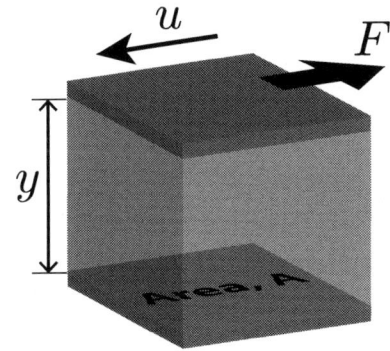

Fig. 4.11: Dynamic, or shear, viscosity is defined using the drag force, F, on a large plate of area, A, which is moving with a speed u relative to another plate when there is a thickness, y, of fluid between the plates.

Dynamic viscosity is defined, and can be measured, by considering a system consisting of two, large plates of area A separated by a thin film of fluid of thickness y, as shown in figure 4.11. When top plate is made to move at a constant velocity, u, relative to the bottom, stationary plate it is found experimentally that the magnitude of the force acting on the plate is proportional to the area of the plates, A, and the velocity and is inversely proportional to the thickness of the fluid. Hence, by introducing a constant of proportionality, η, we can write:

$$F = \eta A \frac{u}{y} \qquad (4.17)$$

where η is a property of the fluid called the dynamic viscosity. Its SI units are the Ns/m^2 or kgm^{-1}s^{-1}. Another unit often encountered for viscosity is the poise (P). However, this is *not* an SI unit and belongs to the cgs unit system which is why the conversion factor between the units is:

$$1\,P = 0.1\,Ns/m^2 \qquad (4.18)$$

The Big Picture

Calculating fluid flow taking into account the viscosity is an extremely challenging task and one that has an entire field devoted to its study. The result of applying Newton's second law to fluid motion are the Navier-Stokes equations, the general form of which is:

$$\rho \left(\frac{\partial \vec{v}}{\partial t} + \vec{v} \cdot \nabla \vec{v} \right) = -\nabla P + \nabla \cdot \mathrm{T} + \vec{f} \qquad (4.19)$$

where ρ is the fluid density, $\vec{v}$ the velocity of the fluid as a function of position and time - called the "velocity field", P the fluid pressure, T a component of the stress tensor of the fluid and $\vec{f}$ represents the "body forces" acting on the fluid.

As you can imagine solving this equation is not at all trivial and far above the scope of this chapter! The equation is non-linear and so complex that mathematicians have not yet even proven that there always is a valid solution in three dimensions for a physical set of initial conditions nor that, if a solution does exist, it will not contain a singularity. Both of these conditions are required if this is a correct description of fluid dynamics.

The solutions to the Navier-Stokes equation encompass all the behaviour of Newtonian fluid flow including turbulence and, when coupled to Maxwell's equations for electromagnetism, it can be used to describe magnetohydrodynamics of plasmas which is increasingly important to understanding the "space weather" which can affect satellites and even terrestrial power grids.

4.7.1 Poiseuille's Law

Although a full description of viscous flow is beyond the scope of this chapter there are certain, common situations where viscous flow can be easily described. One of these is laminar, viscous flow in a straight pipe. The velocity profile for laminar, viscous flow in a pipe has already been shown in figure 4.3. The profile has a parabolic shape with the fluid in contact with the pipe surface remaining at rest. This thin layer of stationary fluid in viscous flow is the reason why fan blades can remain covered in dust despite the flow of air over their surface when the fan is in operation.

The relationship between the pressure drop, ΔP in the pipe (due to the fluid's viscosity) and the physical parameters of the system was discovered experimentally in 1838 before the Navier-Stokes equations were known - by Jean Poiseuille, a French physician. The result is known as Poiseuille's law:

$$\Delta P = \frac{8\eta L Q}{\pi r^4} \tag{4.20}$$

where η is the dynamic viscosity, L is the length of the pipe, Q is the volume flow rate (in m^3/s) and r is the radius of the pipe. The extremely strong dependence on the pipe radius is the reason why even a small narrowing of a pipe can hugely reduce the fluid flow.

Example 4.3

A house is connected to the water mains by a 10 m long, horizontal, straight pipe with a radius of 2 cm. The pressure of the water mains supply is a constant 100 kN m^{-2} relative to the atmosphere and the viscosity of water is 1.0 mN s m^{-2}. (a) What is the pressure of the water where the pipe enters the house when a tap is opened that generates a flow of

10 l/s. (b) What is the maximum flow rate of water into the house? (c) After a few years, the pipe partially collapses so that the radius of the pipe is halved, what is the new maximum flow rate into the house?

Solution:
To answer part (a) we need to be very careful with our units. SI units require that we convert from litres to cubic metres so since 1 l = 1,000 cm^3 =10^{-3} m^3 the volume flow rate in cubic metres per second is 1×10^{-3} m^3/s. We can now use Poiseuille's law to calculate the pressure drop along the pipe:

$$\Delta P = \frac{8\eta L Q}{\pi r^4} = \frac{8 \times 1 \times 10^{-3} \times 10 \times 10 \times 10^{-3}}{\pi \times (0.02)^4} = 1{,}592\,\text{N}\,\text{m}^{-2}$$

This is the pressure drop and so the pressure where the pipe enters the house is:

$$P = 100 \times 10^3 - 1{,}592 = 98{,}408\,\text{N}\,\text{m}^{-2}$$

For part (b) the maximum possible flow rate will occur when the pressure drop is equal to the mains pressure so that the water entering the house is at atmospheric pressure i.e. has a zero pressure relative to the atmosphere. To calculate this we need to rearrange Poiseuille's law to get an expression for the volume flow rate:

$$\Delta P = \frac{8\eta L Q}{\pi r^4} \implies Q = \frac{\pi \Delta P r^4}{8\eta L}$$

Now we have to set the pressure difference to be equal to the mains pressure, which is relative to the atmosphere, and use the values provided for the remaining quantities. This gives:

$$Q = \frac{\pi \times 100 \times 10^3 \times (0.02)^4}{8 \times 1 \times 10^{-3} \times 10} = 0.628\,\text{m}^3/\text{s} = 628\,\text{l/s}$$

For the final part we now have to halve the radius of the pipe to 1 cm and re-run the same calculation. This gives a volume flow rate of:

$$Q = \frac{\pi \times 100 \times 10^3 \times (0.01)^4}{8 \times 1 \times 10^{-3} \times 10} = 0.0393\,\text{m}^3/\text{s} = 39.3\,\text{l/s}$$

This result clearly shows the dramatic dependence of the volume flow rate on the pipe radius: halving the radius reduced the flow rate by a factor of 16.

4.7.2 Terminal Velocity and Stokes' Law

When an object falls through a viscous fluid there is a drag force due to the fluid's viscosity. This drag force increases with velocity until it equals the weight of the object. At this point, there is zero net force on the object and so it will not accelerate further and instead maintains a constant velocity which is referred to as its terminal velocity. This is the reason why going out in a rain storm is generally not fatal! If air had zero viscosity then it is a simple constant acceleration problem to determine the velocity of a raindrop hitting the ground after starting at rest in a cloud 2,000 m above the ground:

$$v^2 = u^2 + 2as \implies v = \sqrt{2 \times 9.81 \times 2,000} = 198\,\text{m/s} \quad (4.21)$$

Hence raindrops would have speeds of around 700 km/h and those in thunderstorms, which have higher clouds, would probably be supersonic! Fortunately for us air is a viscous fluid and so raindrops have a terminal velocity no more than about 9 m/s (or 32 km/h). This is due to the drag force caused by the air flow around the falling object, see figure 4.12, that exerts a force because of the air's viscosity. The size of the drag for large object velocities is given by:

$$\text{Drag Force, } F_d = \frac{1}{2}\rho_f v^2 C_D A \quad (4.22)$$

where ρ_f is the density of the fluid, v is the object's velocity relative to the fluid, A is the cross-sectional area perpendicular to the velocity and C_D is the dimensionless drag coefficient for the object. The drag coefficient depends both on the object's shape and on the viscosity of the fluid and is non-trivial to calculate for the general case.

While this is the case for fast moving objects for very small, slow-moving spheres the drag force is linear with the velocity and is given by Stokes' Law:

$$\text{Drag Force, } F_d = 6\pi\eta r v \quad (4.23)$$

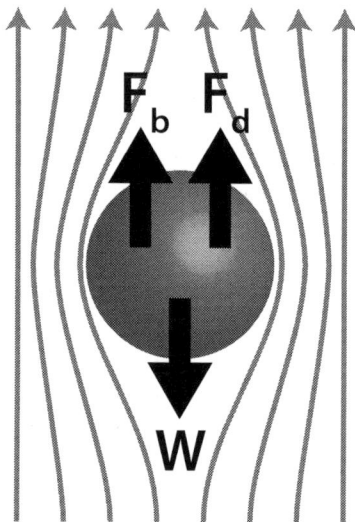

Fig. 4.12: Sphere falling under its own weight, W, through a viscous fluid exhibits a drag force, F_d due to the fluid flow around the sphere. In addition to the drag force the buoyancy force, F_b, also acts on the sphere.

where η is the fluid viscosity, r is the radius of the *small* sphere and v is the sphere's velocity relative to the fluid. This law is the basis for the falling sphere viscometer which measures a fluid's viscosity from the terminal velocity of a sphere of known size and density. However it is important to note that it is only valid for small spheres and low velocities, situations known as *creeping flow*, where the inertia of the fluid flow around the sphere can be neglected. For example, the drag due to a 1 μm sphere moving through water is given by Stokes' Law.

Example 4.4

A sprinkler system for a fire extinguisher produces a fine mist consisting of water droplets with a radius of 30 µm. What is the terminal velocity of such a drop in the air assuming that the buoyancy force of water in air is negligible? [η_{air}=18.3 µNs/m^2, ρ_{water} = 1,000 kg/m^3, g=9.81 m/s^2]

Solution:

At the terminal velocity there is no acceleration and so the net force on the droplet must be zero. If we look at figure 4.12 we can see that there are three forces acting on the drop: its weight, buoyancy and drag. We are told the buoyancy force is negligible (for air and water it is roughly 1,000 times smaller than the weight) so this leaves the weight and the drag and for the net force to be zero these must cancel. Hence we have F_d = W. For a small, slow moving sphere we can use Stokes' law for the drag. This gives:

$$W = 6\pi\eta rv$$

However we can rewrite the weight in terms of the density and radius to get:

$$\frac{4}{3}\pi r^3 \rho g = 6\pi\eta rv \implies \frac{4}{3}r^2 \rho g = 6\eta v$$

which we can rearrange to get and expression for the terminal velocity. We can then evaluate this expression to get the answer:

$$v = \frac{2\rho g r^2}{9\eta} = \frac{2 \times 1,000 \times 9.81 \times (30 \times 10^{-6})^2}{9 \times 18.3 \times 10^{-6}} = 10.7\,\text{cm/s}$$

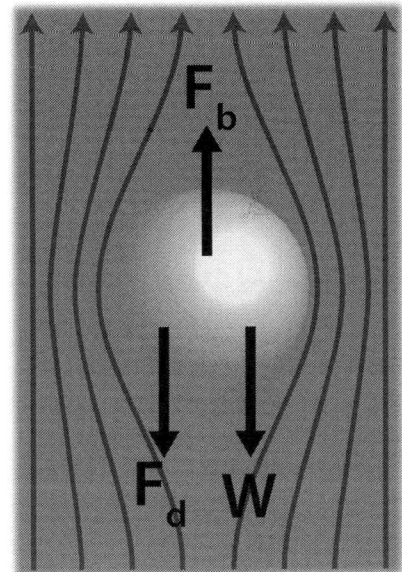

Fig. 4.13: Forces acting on a small, spherical bubble rising through a column of water.

Example 4.5

A glass of water is filled from a tap which has an aerator attachment. The water is initially cloudy due to a large number of very fine, spherical air bubbles which take 103 s to completely clear. If the depth of water in the glass is 15 cm and the bubbles travel at a constant velocity what is the radius of the air bubbles? [η_{air}=18.3 µNs/m^2, ρ_{water} = 1,000 kg/m^3, ρ_{air} = 1.2 kg/m^3, g=9.81 m/s^2]

Solution:

There are three forces acting on each air bubble: weight, drag and buoyancy. If the bubble is travelling at a constant velocity then these forces must all add to give zero. Since the bubble is less dense than the surrounding water the buoyancy force will dominate and so the drag force will act downwards, opposing the motion of the bubble as shown in figure 4.13. If we apply the condition that the forces must cancel then we have:

$$F_b = W + F_d$$

Writing the weight and buoyancy forces in terms of the volume of the bubble and using Stokes' law to calculate the drag force we have:

$$\frac{4}{3}\pi r^3 \rho_{water} g = \frac{4}{3}\pi r^3 \rho_{air} g + 6\pi \eta r v$$

We can now cancel through by a factor of both π and r and rearrange things a little to get:

$$\frac{4}{3} g r^2 (\rho_{water} - \rho_{air}) = 6\eta v$$

Rearranging this further to get an expression for the radius of the bubble we get:

$$r = \sqrt{\frac{9\eta v}{2g(\rho_{water} - \rho_{air})}}$$

Now the only value we are missing from this expression is the velocity of the bubbles. However we know that the glass clears in 103 s and that the furthest a bubble will have to travel is from the bottom of the glass to the surface which is 15 cm and hence we know that the bubble velocity is:

$$v = \frac{\Delta x}{\Delta t} = \frac{0.15}{103} = 1.456 \times 10^{-3} \text{m/s}$$

Finally all we need to do is evaluate our expression for the radius using the velocity we just calculated:

$$r = \sqrt{\frac{9 \times 18.3 \times 10^{-6} \times 1.456 \times 10^{-3}}{2 \times 9.81 \times (1000 - 1.2)}} = 3.50\,\mu m$$

Problems

Q4.1: A large water main has a diameter of 50 cm and can support a maximum water velocity of $10\,\mathrm{m\,s^{-1}}$. If a typical tap has a diameter of 1.25 cm and a flow velocity of $1\,\mathrm{m\,s^{-1}}$ how many open taps can this water main support?

Q4.2: A one-person, streamlined survival tent with a mass of 4 kg is designed to withstand extreme wind conditions. The upper surface area of the tent which is exposed to the wind has an area of $2.5\,\mathrm{m^2}$ and the density of air is $1.2\,\mathrm{kg\,m^{-3}}$. If the gravitational field is $9.81\,\mathrm{m\,s^{-2}}$ and the height of the tent is negligible at what wind speed would a tent with an 80 kg person inside lift off if it were not tied down?

Q4.3: While in deep space an accidental firing of a thruster in leaves an astronaut separated by from his spacecraft and travelling away from it at $10\,\mathrm{m\,s^{-1}}$. Thinking quickly he grabs a sharp tool and makes a small, 4 mm radius, circular hole in his suit on the opposite side of the suit to the spacecraft. The mass of the astronaut in his suit is 100 kg, the absolute pressure of pure oxygen in his suit is $33\,\mathrm{kN\,m^{-2}}$ and the density of the oxygen gas is $0.5\,\mathrm{kg\,m^{-3}}$.
(a) If the suit is large and the pressure remains approximately constant what velocity does the oxygen leave the hole with?
(b) How long should the astronaut leave the hole open to give himself a velocity of $3\,\mathrm{m\,s^{-1}}$ towards his spacecraft?

Q4.4: An experimental power station is designed to use the temperature difference between deep and surface ocean water. To do this water has to be pumped to the surface from 2.5 km down using a pipe with a radius of 0.5 m. The density of the water can be taken to be constant at $1,000\,\mathrm{kg\,m^{-3}}$ and atmospheric pressure is $101\,\mathrm{kN\,m^{-2}}$.
(a) For a pump located at the bottom of the pipe what pressure, relative to the atmosphere, would it need to achieve to produce a flow rate of $100\,\mathrm{m^3\,s^{-1}}$ given that the water starts at rest?
(b) If the pump were located at the surface and was used to suck water up from the depths what would be the maximum volume flow rate that it could achieve?

Q4.5: A geyser can be modelled as a large, underground reservoir of water, of density $1,000\,\mathrm{kg\,m^{-3}}$ under pressure with a narrow opening to the surface. If one such geyser has a reservoir 150 m below the surface and the geyser it produces reaches a height of 35 m what is the pressure in the reservoir relative to the atmosphere?

Q4.6: A water tower consists of a large tank of water whose base is h above the ground. A small leak develops in the side of the tank at the base.
(a) Derive an expression for the distance from the tower where the water hits the ground, x, in terms of h, and the depth of water in the tank, d.
(b) Explain how the result to the first part can be independent of the gravitational field.

Q4.7: Assuming that Stokes' law applies what is the terminal velocity of a raindrop of radius 1 mm given that the density of water is $1,000\,\mathrm{kg\,m^{-3}}$, the viscosity of air is $18.4\,\mathrm{\mu N\,s\,m^{-2}}$ and that the gravitational field is $9.81\,\mathrm{m\,s^{-2}}$ and that the effects of buoyancy can be neglected.

Q4.8: A tiny air bubble is released from a submarine travelling at great depth in the ocean. The densities of air and water at these pressures are $120\,\mathrm{kg\,m^{-3}}$ and $1,000\,\mathrm{kg\,m^{-3}}$ respectively, the viscosity of water is $1.422\,\mathrm{mN\,s\,m^{-2}}$ and the gravitational field is $9.81\,\mathrm{m\,s^{-2}}$.
(a) What is the terminal velocity of an air bubble with a radius of 1 mm?
(b) Explain *qualitatively* what will happen to the terminal velocity of the air bubble as it rises towards the surface and why this occurs.

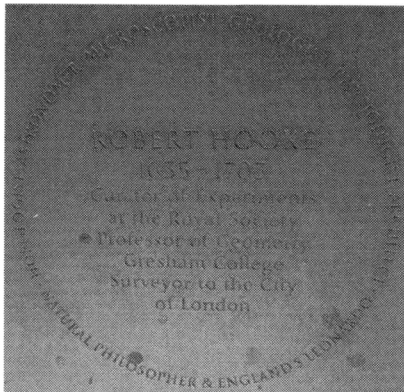

Fig. 5.1: Modern portrait of British physicist Robert Hooke reconstructed from the descriptions of his colleagues. No contemporary portrait of him exists.

Fig. 5.2: Memorial to Robert Hooke in the pavement near the Monument in London, UK commemorating his work as Surveyor of the City of London immediately after the devastation of the Great Fire in 1667.

CHAPTER 5

Simple Oscillations

From the orbit of the earth around the sun to the vibration of atoms in a solid the universe is full of examples of repetitive motion. Understanding how to characterize and predict the behaviour of such systems is fundamental to not just every branch of physics but for most other sciences as well since systems of oscillators are what makes up the phenomenon we call waves which not only play many roles in everyday life but also describe the most fundamental physics we know from atoms all the way down to the fundamental particles of nature such as the Higgs boson.

The first oscillator harnessed by humans was the simple pendulum. Galileo Galilei (1564-1642) was the first to study the pendulum starting around 1602 apparently after having his interest sparked by watching a swinging chandelier in Pisa Cathedral. He got as far as giving a design for a pendulum clock to his son, Vincenzo, who unfortunately died in 1649 before completing it. As a result, the first pendulum clock was constructed in 1656 by the dutch physicist Christiaan Huyghens (1629-1695) whom we will learn more about when we discuss light.

The British physicist Robert Hooke (1635-1703), famous for his law regarding the force of springs discussed briefly in section 2.3.1, studied the conical pendulum in 1666. This is a pendulum where the mass moves in a horizontal circle instead of back and forth. A year later there was the Great Fire of London after which he lead the rebuilding effort and if you visit London today you can find a memorial stone to him in a pavement near the Monument (to the fire) in London, see figure 5.2, where he is remembered for this more than his many contributions to physics! Hooke suggested to Sir Issac Newton (1642-1726) that, like the conical pendulum, the circular motion of the planets was due to a radial force acting towards the centre of a circle which lead to Newton developing his law of gravity.

Oscillations also play a significant role in modern physics. In 2001 the Sudbury Neutrino Observatory (SNO) in Canada (see figure 5.3) solved a long-standing problem in physics, called the solar neutrino problem, by showing that tiny subatomic particles called neutrinos that are emitted by nuclear fusion reactions in the heart of the sun oscillate between three different flavours as they propagate to Earth. Previous experiments only looked for the flavour of neutrino that was produced in the sun and found about a half to a third of the expected number. The Sudbury Neutrino Observatory (SNO) was the first experiment to look for all three flavours and saw a rate consistent with expectation. This discovery proved that neutrinos have a mass and in 2015 the Physics Nobel Prize was awarded to Art McDonald and Takaaki Kajita for the discovery of neutrino oscillations. Unfortunately, this type of oscillation is quite complex because it is quantum mechanical in nature and it takes place between three flavours of neutrinos so we will limit ourselves to the discussion of basic, mechanical oscillations here.

Fig. 5.3: The SNO detector 2.1 km underground in Creighton Mine, Sudbury, Ontario before the cavern was filled with pure water. The black, spherical vessel covered in photomultiplier tubes was filled with heavy water. The deuterium nucleus in heavy water can interact with all three flavours of neutrino in a recognizable way which allowed the total flux of all neutrino flavours from the sun to be measured for the first time.

Photo courtesy of Ernest Orlando, LBNL and SNO.

5.1 Describing Oscillations

Objects undergoing periodic motion in one dimension will oscillate around an equilibrium position, this is the point in their motion where the net force on the object is zero. Away from this equilibrium point, a net force will act on the object so as to restore the object to the equilibrium point.

The maximum displacement from the equilibrium position is called the amplitude of the motion. The units of amplitude vary depending on the type of oscillation being described. For mechanical systems the amplitude is usually a length but, for example, in alternating current electric circuits the amplitude could be a current in amps or a potential difference in volts.

The very definition of an oscillation requires that a motion is being repeated. The time to make one repetition, or cycle, is called the period, T, of the motion. For example, the period of the Earth in orbit around the sun is 365.25 days. Unlike amplitude, period is always a measure of time and so, in SI units, is measured in seconds (s).

Although the amplitude and period are sufficient to describe oscillations for rapid oscillations it is common to give the number of repetitions, or cycles, per second. This is called the frequency of the motion. The frequency, f, of a system is simply the reciprocal of the period:

$$\text{Frequency, } f = \frac{1}{T} \tag{5.1}$$

where f is the frequency and T is the period. Since the period is measured in seconds the units of frequency are inverse seconds (s^{-1}) which have an associated SI unit called the hertz (Hz).

Another type of frequency that is very useful is called the angular frequency which is defined as 2π times the frequency:

$$\text{Angular Frequency, } \omega = 2\pi f = \frac{2\pi}{T} \tag{5.2}$$

Since the factor of 2π comes from the number of radians in one full revolution the units for ω are radians per second (rad s^{-1}) but since radians are dimensionless this is sometimes dropped. While the usefulness of this definition may not be apparent yet keep it in mind as we discuss the mathematical description of an oscillator.

5.2 Simple Harmonic Motion

The most common type of oscillation in physics is known as simple harmonic motion. To study this type of motion let's examine the system in figure 5.4 which shows a mass m resting on a horizontal, frictionless surface. It is attached to a spring that obeys Hooke's law with a spring constant k and which can be either compressed or extended. The mass is initially displaced from its equilibrium position and moves subsequently under the action of the spring such that its displacement at a time t is x.

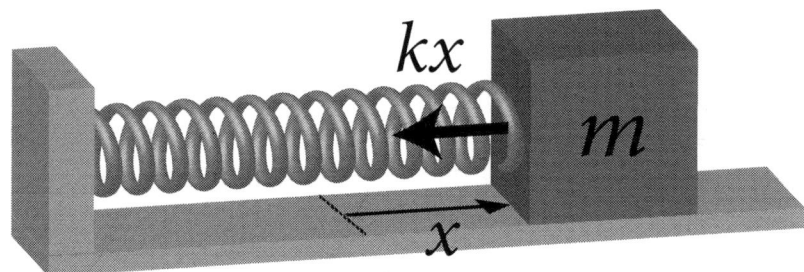

Fig. 5.4: Mass m resting on a frictionless surface and attached to a spring with constant k and displaced a distance x from equilibrium.

Since x will change with time we need to find a function which describes how the displacement x varies with time t and so we need an equation whose solution is a function $x(t)$. This is true for a differential equation, called the equation of motion, which we derive by applying Newton's second law to the system.

5.2.1 Mathematical Solution

The net force acting on the mass in the horizontal direction is given by Hooke's law:

$$\text{Net Force, } F = -kx \qquad (5.3)$$

where the negative sign is because the force is always in the opposite direction to the displacement x. Now Newton's second law applies and so the acceleration of the mass horizontally is simply:

$$F = ma = -kx \quad \implies \quad a = -\frac{k}{m}x \qquad (5.4)$$

where a is the acceleration of the mass. If we write this out using the full, differential form for the acceleration we have:

$$\frac{d^2x}{dt^2} = -\frac{k}{m}x \qquad (5.5)$$

This a second order differential equation whose solution, $x(t)$, is a function which describes the motion of the system and hence this the equation of motion. This form of differential equation is one that is incredibly common in physics and fortunately it is easy to solve!

It is clear from (5.5) that we need a function for $x(t)$ that when differentiated twice gives the same function. The first function with these properties that may come to mind is a simple exponential since this remains is unchanged by differentiation so let's try that first. We start with $x(t) = Ae^{qt}$ where A and q are constants.

$$x(t) = Ae^{qt} \quad \implies \quad \frac{d^2x}{dt^2} = q^2 Ae^{qt} = q^2 x \qquad (5.6)$$

This seems to work well but we have a problem. Comparing this equation to the equation of motion for the spring (5.5) we require that:

$$q^2 = -\frac{k}{m} \qquad (5.7)$$

Both the mass and the spring constant must be positive numbers so our solution for q must be something that when squared gives a negative number! Obviously there is no real number that can satisfy this requirement but, as we saw in section 1.2, an imaginary number does. Hence we simply use the imaginary number, i, which is defined such that:

$$i^2 = -1 \tag{5.8}$$

The mathematics of this imaginary number is described in more detail in section 1.2 but for the moment it is sufficient that we simply use the relationship above. If we return to our problem with the oscillator we can now define a new quantity using the greek letter omega, ω, where $q = i\omega$. Putting this into our attempt at a solution we get:

$$x(t) = Ae^{i\omega t} \implies \frac{d^2x}{dt^2} = -\omega^2 Ae^{i\omega t} = -\omega^2 x \tag{5.9}$$

which works perfectly provided we define:

$$\omega = \sqrt{\frac{k}{m}} \tag{5.10}$$

Now if we look at (5.9) we have a problem. This is a complex number but the solution we require must be a completely real number because it represents the displacement of the mass from the equilibrium position. If we do this experiment in the lab you can measure this distance with a ruler and will always get a real number so what has gone wrong?

The problem we have here is that the solution we have found is not the only one and, if we stop to consider what we have done when finding a solution to the equation of motion, this is actually very easy to see. Equation (5.9) is a second order differential equation which means that in finding a solution we have effectively taken an indefinite integral twice and so there have to be two constants of integration in the solution. However, in our case, we only have one constant, A, which means that we are missing a whole class of solutions! Note that ω is defined by the physical parameters of the system, in this case, the spring constant k and the mass m as shown in (5.10), and so is not a constant of integration.

While this lack of a second constant of integration tells us that we are missing a class of solution, unfortunately, it provides no information at all about what it is. However, if we return to our existing solution we can see that if we flip the sign of the exponent i.e. $i\omega t \to -i\omega t$ we will have a different function which will also satisfy the equation of motion because $(-i)^2 = (-1)^2 \times i^2 = i^2 = -1$. Hence we have a new solution, $x(t) = Be^{-i\omega t}$ with its own constant of integration, B, and so the general solution is the sum of these:

$$x(t) = Ae^{i\omega t} + Be^{-i\omega t} \qquad (5.11)$$

where A and B are constants. Now we have a complete solution but we still have the problem that it is the sum of two complex numbers. Fortunately, the sum of two complex numbers can be a real number and if we choose the values for A and B, which are themselves complex numbers, we can come up with a completely real solution. The precise method to do this is shown in section 5.3 but for now, we will take a less mathematically rigorous approach by returning to the original equation of motion and trying a different function by inspection.

Returning to (5.5) we still need a function that, when differentiated twice, will give the same function back and fortunately we have another candidate to try: the cosine function. If we start with $x(t) = A\cos(\omega t)$ and differentiate twice then we get:

$$x(t) = A\cos(\omega t) \implies \frac{dx}{dt} = -A\omega \sin(\omega t)$$

$$\implies \frac{d^2x}{dt^2} = -A\omega^2 \cos(\omega t) = -\omega^2 x \qquad (5.12)$$

This works! We have the minus sign which we required and again ω is related to the spring constant and mass by (5.10) but we are still missing something because we have only one constant of integration again. There are two ways to include another constant which lead to two different, but mathematically identical, expressions for the solution. The simplest method is to add a constant to the argument of the cosine function, ϕ, to get $x(t) = A\cos(\omega t + \phi)$. This constant has no effect on how the function behaves when differentiated:

$$x(t) = A\cos(\omega t + \phi) \implies \frac{dx}{dt} = -A\omega \sin(\omega t + \phi)$$

$$\implies \frac{d^2x}{dt^2} = -A\omega^2 \cos(\omega t + \phi) = -\omega^2 x$$
$$(5.13)$$

This more general function works in the equation of motion for the value of ω given in (5.10) and we now have two constants of integration: A and ϕ as required. In addition we are guaranteed that our solution for $x(t)$ will always give a real number that we can measure so we have finally managed to solve the equation of motion and come up with a solution for the position of the mass as a function of time.

$$x(t) = A\cos(\omega t + \phi) \qquad (5.14)$$

If we use our trig identities to expand this out we can also see the alternative approach we could have used to find the missing constant of integration:

$$x(t) = A\cos(\omega t + \phi) = A\cos\phi\cos(\omega t) + A\sin\phi\sin(\omega t)$$
$$x(t) = A'\cos(\omega t) + B'\sin(\omega t) \tag{5.15}$$

where $A'(= A\cos\phi)$ and $B'(= A\sin\phi)$ are constants. If instead of adding a constant ϕ to the argument, we had noted that $\sin(\omega t)$ was a valid solution too this would have lead directly to equation (5.15) which, as we showed above, is mathematically identical to the more usual form of the solution shown in (5.14). Similarly if we note that $\cos\theta = \sin(\theta + \pi/2)$ we can also write down our solution to the oscillator as:

$$x(t) = A\sin(\omega t + \phi') \tag{5.16}$$

where $\phi' = \phi + \pi/2$.

5.2.2 Physical Quantities

We now have a mathematical formula which describes the position of the mass for a mass-spring system as a function of the time, t as given by (5.14). However, we started this discussion by defining several quantities which describe the motion of an oscillator and we now need to figure out how to determine these from the solution we just derived.

Looking at (5.14) we know that the cosine function can only vary between +1 and -1 and so the displacement will vary between $+A$ and $-A$. Since the amplitude is defined as the largest distance from the equilibrium position this means that the amplitude of the motion is simply A as shown in figure 5.5.

Next let's consider the period of the motion. We can do this by asking at what times the displacement of the mass will be at its maximum value, A. If we put the requirement that $x(t) = A$ into our solution we get:

$$A\cos(\omega t + \phi) = A \implies \cos(\omega t + \phi) = 1 \implies (\omega t + \phi) = 2n\pi \tag{5.17}$$

where n is an integer. Let's consider the times for the first two solutions where $n = 0$ and $n = 1$

$$\omega t_0 + \phi = 0 \text{ and } \omega t_1 + \phi = 2\pi \tag{5.18}$$

The times t_0 and t_1 are two adjacent times at which the mass has a displacement of $+A$ with zero velocity and so the difference between these times is simply the period of the motion.

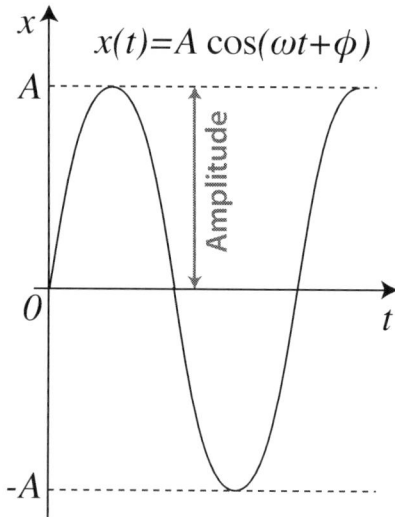

Fig. 5.5: Displacement versus time for a simple harmonic oscillator with an initial phase, $\phi = -\pi/2$ showing that the amplitude of the motion is A.

Hence as shown in figure 5.6 the period, T, is given by:

$$T = t_1 - t_0 = \frac{2\pi}{\omega} \qquad (5.19)$$

This relates the period of the motion to the quantity ω but to make it clear what this quantity ω is we need to replace the period of the motion with the frequency. Using our definition from (5.1) the equation above becomes:

$$T = \frac{1}{f} = \frac{2\pi}{\omega} \implies \omega = 2\pi f \qquad (5.20)$$

which is just our definition for the angular frequency from (5.2). Hence ω is the angular frequency of the oscillator and we can use this to derive both the period and the frequency of the system in terms of its physical parameters. We already know the angular frequency, ω, in terms of the spring constant and the mass from (5.10) and so using the relationships between the angular frequency and the period, T, and the frequency, f, (5.2) we find that for a mass-spring system:

$$f = \frac{1}{2\pi}\sqrt{\frac{k}{m}} \quad \text{and} \quad T = 2\pi\sqrt{\frac{m}{k}} \qquad (5.21)$$

These relationships demonstrate an important feature of *all* simple harmonic motion: the frequency is completely independent of the amplitude. At first this may seem very counter-intuitive since a larger amplitude requires that the mass move a lot further in order to complete a cycle which would normally be expected to take longer.

For our mass-spring system the larger the amplitude the larger the displacement from the equilibrium. Hooke's law tells us that the force of the spring increases linearly with extension and so larger amplitudes produce larger restoring forces on the mass. This, in turn, creates a larger acceleration which results in a higher velocity. The dynamics of the system are such that the increase in average speed exactly offsets the increase in the distance moved so that the period remains constant regardless of the amplitude.

Having a period which is independent of the amplitude is a feature common to all simple harmonic oscillators and is one of the reasons behind their use in timekeeping devices, such as pendulum clocks or torsional oscillators in mechanical watches. Indeed the constant period of the swinging chandelier in Pisa Cathedral is reputed to be the property which first drew Galileo's attention to studying oscillators.

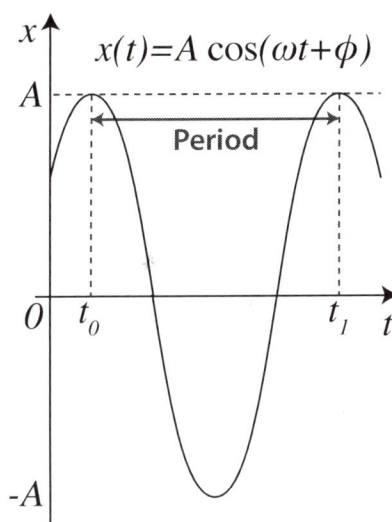

Fig. 5.6: Displacement versus time for a simple harmonic oscillator with an initial phase ϕ. The times t_0 and t_1 are the first two points where the oscillator has its maximum positive displacement. The period, T, is simply the difference between them as shown.

This now leaves just one unexplained physical quantity in our solution (5.14), ϕ, and to explain this we need to introduce a new concept: phase.

5.2.3 Phase

The phase of an oscillator is an angle which describes where an oscillator is in its cycle. It almost always measured in radians and so during a single period of the motion will change in value by exactly 2π. Since this is true for all oscillators regardless of amplitude or frequency phase is very useful as a means to compare where two oscillators are in their respective cycles.

If we look at the solution for simple harmonic motion given in (5.14) the phase of the oscillator is the argument of the cosine term i.e. $\omega t + \phi$. To understand how this represents an angle we need to compare simple harmonic oscillations to uniform circular motion i.e. motion in a circle at a constant speed.

Consider a particle moving with a constant speed, v, in a circle as shown in figure 5.7. At time t the particle will be at a certain point on the circle, as shown, and we can draw a line from the origin to the particle. This line is called a phasor (not to be confused with a phaser!) and the angle θ between this line and the positive x axis is called the phase.

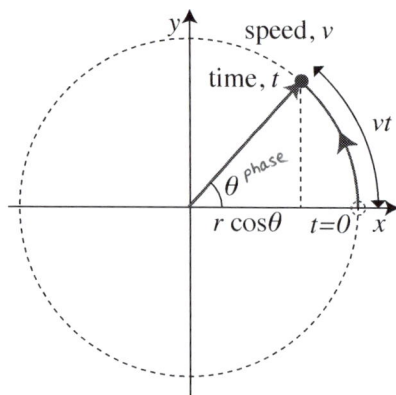

Fig. 5.7: A particle moving at a constant speed v in a circle will travel along an arc of the circle with length vt in a time t as shown here. The angle subtended by the arc at the centre of the circle, θ, gives the phase of the motion and the line from the original to the particle's location is a phasor.

Now consider the motion of the particle. Since it is travelling at a constant speed, v, the arc length through which it will have moved after a time t is just vt. Next, we want to calculate the phase of the particle at a time so we need to relate the arc length to the angle. The definition of an angle in radians is the ratio of the arc length, s, divided by the radius, r, of the circle, $\theta = s/r$, and so the phase angle is:

$$\theta = \frac{vt}{r} + \phi = \frac{v}{r}t + \phi \tag{5.22}$$

where ϕ is the phase of the system when $t = 0$ and so is determined by where the particle starts its motion. Now we remember the relationship between the angular and linear velocities of the particle:

$$\omega = \frac{v}{r} \qquad \omega = \frac{2\pi}{t} \tag{5.23}$$

which gives an expression for the phase angle of:

$$\theta = \omega t + \phi \tag{5.24}$$

where here ω is the constant angular velocity of the particle and so this equation shows that the phase will linearly increase with

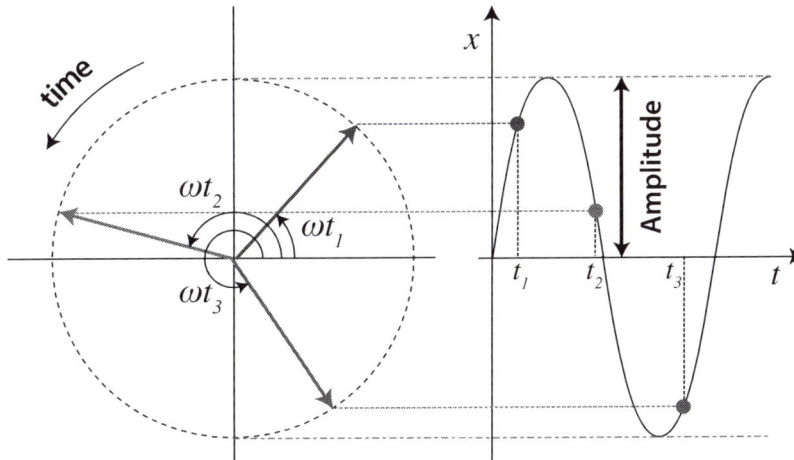

Fig. 5.8: Projection of one component of uniform circular motion showing how it corresponds to simple harmonic motion. The position of the particle is shown at three different times, t_1, t_2 and t_3.

time. Next, we consider just the x component of the particle's motion. Using simple trigonometry we have:

$$x = r \cos \theta = r \cos (\omega t + \phi) \qquad (5.25)$$

This has *exactly* the same form as our solution to the simple harmonic oscillator provided that the amplitude of the oscillator equals the radius of the circle and the angular frequency of the oscillator equals the angular velocity of the particle! In other words, the x component of a particle undergoing uniform circular motion is exactly the same as the displacement of a mass undergoing simple harmonic motion and the phase of the motion is simply the argument of the cosine function.

Furthermore we can do the same for the y component of motion:

$$y = r \sin \theta = r \cos \left(\omega t + \phi - \tfrac{\pi}{2}\right) \qquad (5.26)$$

where we have used the relationship that $\sin \theta = \cos(\theta - \pi/2)$. This shows that circular motion can be broken down into simple harmonic motion in both the x and y directions with a phase difference of $\pi/2$ between the two components. Figure 5.8 shows how the projection of one component of uniform circular motion leads to a sine-wave indicative of simple harmonic motion.

Interestingly if we change the phase difference, $\Delta\phi$, between the x and y components the path that the particle traces out changes from a circle ($\Delta\phi = \pm\pi/2$) to an ellipse and eventually to a line ($\Delta\phi = 0$ or π). For the circular case the two phase differences correspond to particles moving in opposite directions, clockwise vs. anticlockwise, and for the line, they correspond to two different gradients, +1 or -1. The range of curves obtained

$n_x = 1$, $n_y = 1$
$\phi = 0$

$n_x = 1$, $n_y = 1$
$\phi = \pi/4$

$n_x = 1$, $n_y = 2$
$\phi = 0$

$n_x = 1$, $n_y = 3$
$\phi = 0$

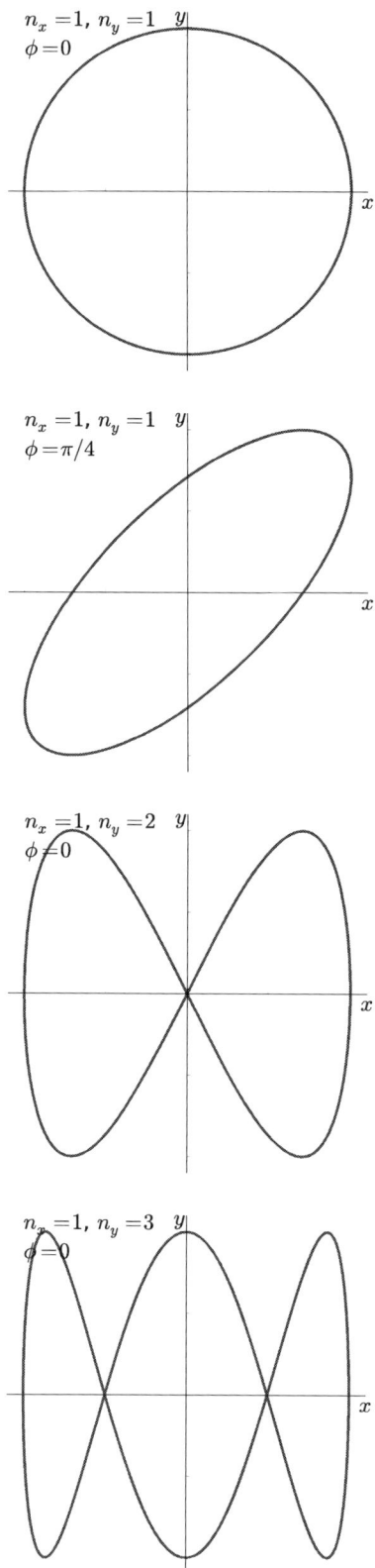

Fig. 5.9: Lissajous figures where $x = \cos(n_x t)$ and $y = \sin(n_y t + \phi)$ for differing values of n_x, n_y and ϕ.

by varying the phase difference between the x and y components are the simplest examples of a class of curves called *Lissajous figures* which were first investigated in detail in 1857 by Jules Antoine Lissajous and are now often demonstrated on oscilloscopes. Some examples are shown in figure 5.9.

We can now explain the physical significance of the quantity ϕ in our solution to the simple harmonic oscillator as the initial phase of the system. At time $t = 0$ the displacement of the mass is $x = A\cos\phi$ and so the initial phase determines the initial position of the mass.

Finally, it should be noted that the most important use of phase is when calculating phase differences between oscillators. The absolute value of phase depends purely on where the convention used defines the zero phase to be. For example, it is possible to rewrite the solution to the simple harmonic oscillator as a sine function:

$$x(t) = A\cos(\omega t + \phi) = A\sin(\omega t + \phi - \pi/2) = A\sin(\omega t + \phi') \quad (5.27)$$

where we have used the identity $\sin\theta = \cos(\theta - \pi/2)$ which gives a new initial phase, $\phi' = \phi - \pi/2$. In the comparison to circular motion, this is the same as choosing the zero phase to be along the positive y axis. Hence in terms of our oscillator using cosine sets zero phase at the maximum positive displacement and using sine sets it at the equilibrium position when the mass has a negative velocity.

In fact, we can actually choose any point in the cycle as our zero phase and measure all phases relative to that choice of zero, in exactly the same way that we can choose an origin and measure all positions relative to that zero. This is why displacements, which are a change in position, are often more useful than a position based on an arbitrary zero point. In the same manner, a difference in phase is often far more useful than knowing an absolute phase based on some arbitrary choice of zero.

5.2.4 Velocity and Acceleration

We now have a solution for the displacement of a simple harmonic oscillator but when dealing with dynamics we need to know velocities and accelerations as well. Starting with our solution for the displacement of the mass given in (5.14) we can convert this into a velocity simply by differentiating with respect to time:

$$v = \frac{dx}{dt} = -\omega A \sin(\omega t + \phi) \qquad (5.28)$$

Now if we want to compare the phase of the velocity to that of the displacement we need to rewrite this as a cosine function since, as we discussed in the previous section, the zero phase for sine is not the same as the zero phase for cosine. Using the identity $\sin\theta = \cos(\theta - \pi/2)$ our expression for the velocity becomes:

$$v = -\omega A \cos(\omega t + \phi - \pi/2) \qquad (5.29)$$

We can now compare this to the displacement as shown in figure 5.10. The first thing to note is that, like the displacement, the velocity will oscillate between positive and negative values with exactly the same angular frequency ω but with a different amplitude of ωA. There is also a difference in the phase. Naïvely looking at the equation we might conclude that the phase of the velocity is $\pi/2$ behind the displacement but this is wrong because it neglects the negative sign at the front of the expression. Inverting the sign of a cosine function is exactly the same as adding a factor of π to the argument. Hence we can rewrite our expression again to get rid of the negative sign:

$$v = -\omega A \cos(\omega t + \phi - \pi/2) = +\omega A \cos(\omega t + \phi + \pi/2) \qquad (5.30)$$

This shows that the phase of the velocity is actually $\pi/2$ *ahead* of the displacement. What this means is that when the displacement is at its minimum value, which occurs when the mass it at the equilibrium position, the velocity is at either its maximum or minimum value. This is what we would expect since for oscillations to happen the mass cannot stop at the equilibrium position and must move through it after which, regardless of direction, a force will act in the opposite direction to the velocity.

Returning to (5.28) we can now calculate the acceleration of the mass by differentiating again:

$$a = \frac{d^2x}{dt^2} = -\omega^2 A \cos(\omega t + \phi) \qquad (5.31)$$

Again we find that the acceleration oscillates back and forth with the same angular frequency, ω, and an amplitude, $\omega^2 A$, which is different to both the velocity and displacement. Comparing the phase is easier because we already have a cosine function but we still need to take care of the negative sign by introducing a phase change of π this gives:

$$a = \omega^2 A \cos(\omega t + \phi + \pi) \qquad (5.32)$$

Hence we can see that the acceleration is exactly out of phase with the displacement since there is a phase difference of π between them. This means that when the displacement is at its

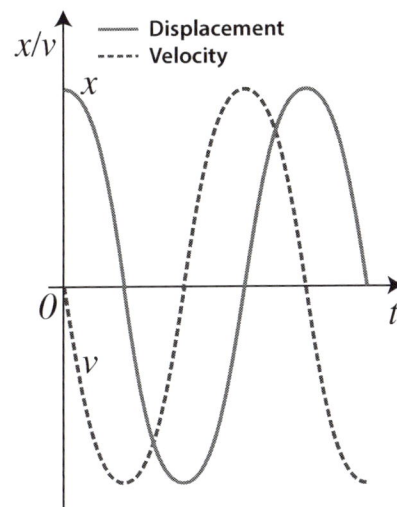

Fig. 5.10: Plot of displacement and velocity versus time for a simple harmonic oscillator where $\omega = 1$ so that both have the same amplitude in their respective units. When the displacement is a maximum the velocity is zero and when the displacement is zero the velocity has a maximum which is a result of the velocity having a phase $\pi/2$ ahead of the displacement.

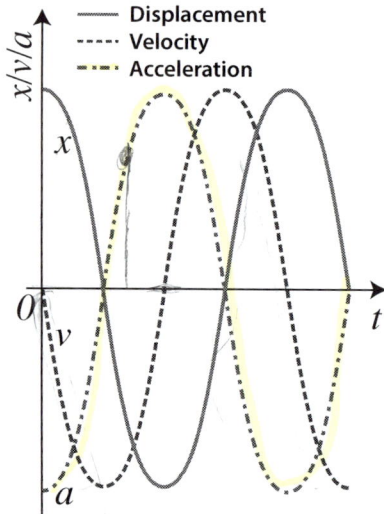

Fig. 5.11: Plot of displacement, velocity and acceleration versus time for a simple harmonic oscillator where $\omega = 1$ so that all have the same amplitude in their respective units. It is clear that the acceleration is exactly out of phase ($\Delta\phi = \pi$) with the displacement.

maximum positive value the acceleration is at its maximum negative value which is why the mass will return to the equilibrium position.

We can now compare all these quantities and figure 5.11 shows the displacement, velocity and acceleration of the mass on a linear plot where the x axis is time. To aid the comparison of the phase the amplitudes are drawn as being the same for all quantities but this only applies to the special case where $\omega = 1\,\text{rad}\,\text{s}^{-1}$. Figure 5.12 shows the displacement, velocity and acceleration phasors. Again the special case where $\omega = 1\,\text{rad}\,\text{s}^{-1}$ is chosen so that the phase differences are easy to see. All three phasors rotate with exactly the same frequency and so the phase difference between all three is fixed.

5.2.5   Boundary Conditions

When solving any differential equation we are effectively integrating and this gives rise of constants of integration which can only be determined by what are called the boundary conditions. In the case of a simple harmonic oscillator our variable is time, t, and so our boundary condition will be determined by the state of the system at a particular time.

Since we have a second order differential equation we have two constants of integration which are the amplitude and initial phase. This means that we need two boundary conditions in order to determine both quantities. Typically these are the ini-

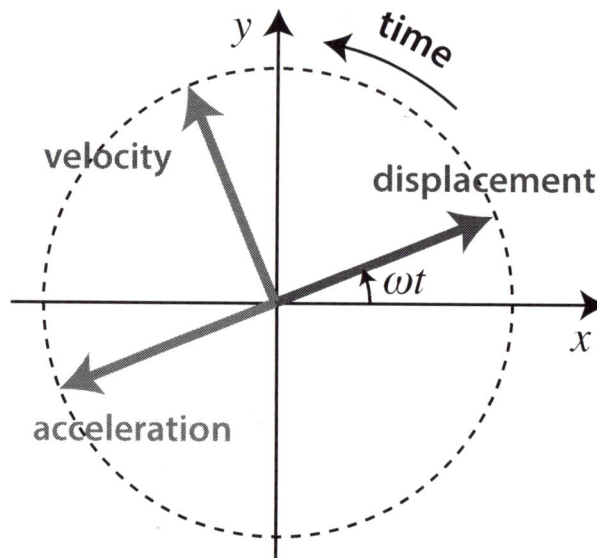

Fig. 5.12: Figure showing the phasors for the displacement, velocity and acceleration all normalized to the same length. The velocity is ahead of the displacement by a phase of $\pi/2$ and the acceleration is out of phase by exactly π. The actual values of the quantities represented is just the x component of the phasor.

tial position and velocity of the system but any two pieces of information at known times is sufficient e.g. the position of the system at two different times or the position and acceleration at a fixed time etc.

To determine the amplitude and initial phase we apply the conditions given to our solutions for the position, velocity or acceleration as appropriate. For example if the initial position of the system is x_0 and the initial velocity is v_0 then when $t = 0$ we can use our solutions for the position and velocity to obtain two simultaneous equations:

$$x_0 = A \cos \phi \tag{5.33}$$

$$v_0 = -A\omega \sin \phi \implies -\frac{v_0}{\omega} = A \sin \phi \tag{5.34}$$

To solve these we first square and add both equations to get:

$$x_0^2 + \left(\frac{v_0}{\omega}\right)^2 = A^2(\cos^2 \phi + \sin^2 \phi) = A^2 \tag{5.35}$$

and then we can also divide (5.34) by (5.33) to get:

$$-\frac{v_0}{\omega x_0} = \frac{A \sin \phi}{A \cos \phi} = \tan \phi \tag{5.36}$$

The equations (5.35) and (5.36) can then be simply rearranged to get our solutions for the amplitude and initial phase

$$A = \sqrt{x_0^2 + \left(\frac{v_0}{\omega}\right)^2} \tag{5.37}$$

$$\phi = \tan^{-1}\left(-\frac{v_0}{\omega x_0}\right) \tag{5.38}$$

This type of simultaneous equations occurs often with oscillations and so it is important to remember the trick required to solve them.

5.3 The Complex Solution

In section 5.2.1, we skipped over the complex solution we first obtained when trying to solve the equation of motion for a simple harmonic oscillator. Here we will revisit that first solution and, use some of the more advanced concepts related to complex numbers covered in chapter 1, to show that the real solution to the equation has the form we derived by the less rigorous inspection method. We will also see how the general complex solution appears on an argand diagram.

We start by returning to the general, complex solution we first obtained for the simple harmonic oscillator (5.11):

$$x(t) = Ae^{i\omega t} + Be^{-i\omega t} \tag{5.39}$$

where A and B are complex constants. We can simplify this by rewriting A and B in polar coordinates using the relationship given in (1.38):

$$x(t) = A'e^{i(\omega t + \phi)} + B'e^{-i(\omega t + \theta)} \tag{5.40}$$

where now A', B', ϕ an θ are all real constants and $A = A'e^{i\phi}$ and $B = B'e^{-i\theta}$. Now we need to find values for these constants such that the solution for $x(t)$ is always real. For the sum of two complex numbers to be real the second number must be the conjugate of the first:

$$z + \bar{z} = (x + yi) + (x - yi) = 2x \tag{5.41}$$

This means that we need the second term in (5.40) to be the complex conjugate of the first term so that the sum is real. To do this let's first consider what happens when we have a negative power for an imaginary exponential. Using the Euler identity (1.31) along with remembering that cosine is an even function and sine is an odd function we get:

$$e^{-i\omega t} = \cos(-\omega t) + i\sin(-\omega t) = \cos(\omega t) - i\sin(\omega t) = \overline{e^{i\omega t}} \tag{5.42}$$

Hence to get complex conjugate of an imaginary exponential all that is needed is to invert the sign. Now it we look at (5.40) for this to be real the complex conjugate of the first term must equal the second term and so we have the requirement that:

$$A'e^{-i(\omega t + \phi)} = B'e^{-i(\omega t + \theta)} \tag{5.43}$$

and so $B' = A'$ and $\theta = \phi$ so that the argument and modulus of both numbers is the same. Putting this into (5.40) and expanding it out using Euler's identity we get:

$$x(t) = A'e^{i(\omega t + \phi)} + A'e^{-i(\omega t + \phi)}$$
$$= 2A'\cos(\omega t + \phi) \tag{5.44}$$

All that remains is to define another real constant, $A = 2A'$ and we have the real solution to the simple harmonic oscillator:

$$x(t) = A\cos(\omega t + \phi) \tag{5.45}$$

This is the same result that we had before (5.14) but here we are derived it in a more mathematically rigorous fashion showing that this is the real part of the general solution to the equation of motion.

Having now established that the correct form for the real solution is (5.45) we can also introduce a new way of writing this down using a complex representation. Using the Euler identity we can rewrite (5.45) as:

$$x(t) = A\cos(\omega t + \phi) = \mathfrak{R}\left\{Ae^{i(\omega t+\phi)}\right\} \qquad (5.46)$$

Unfortunately, physicists are often a little sloppy with their mathematical notation so commonly the real operator, $\mathfrak{R}$, is omitted from (5.46) to give a solution which is written as:

$$x(t) = Ae^{i(\omega t+\phi)} \qquad (5.47)$$

This implied real operator can be very dangerous if great care is not taken. While some operations such as differentiating and integrating keep the real and imaginary parts separate others, such as multiplication, do not. Any operation which causes the imaginary part of this solution to affect the real part of a quantity will give a wrong result. This is because the imaginary part is not actually there as would be clear if the real operator was explicitly written down. So be aware of this and take great care when using this notation! The reason that such dangerous notation is used is that it can enormously simplify the maths frequently required to deal with oscillators by removing all the arcane trigonometric relationships required to deal with sines and cosines and replacing them with the far simpler rules for exponents.

The expression in (5.47) can be simplified even further by defining a new complex number, Z, where $Z = Ae^{i\phi}$. Putting this into the complex solution (5.47) we get:

$$x(t) = Ze^{i\omega t} \qquad (5.48)$$

where Z is called the complex amplitude and encodes both the real amplitude, A, and initial phase, ϕ, into a single variable where the real amplitude is the modulus of Z and the initial phase is the argument of Z.

Finally, it is very instructive to consider how our complex solution (5.47) looks on the complex plane. Looking at this expression we have a constant modulus, A, and a varying argument, $\omega t + \phi$. Hence this solution will trace out a perfect circle of radius A on the complex plane of an argand diagram as shown in figure 5.13. This is almost identical to the diagram shown in figure 5.8 where we showed that the x component of motion was equivalent to a simple harmonic oscillator. Here that same x component is the real part of the solution and so our complex solution for the simple harmonic oscillator is really just uniform circular motion in a complex plane where only the real part is relevant for describing the physics.

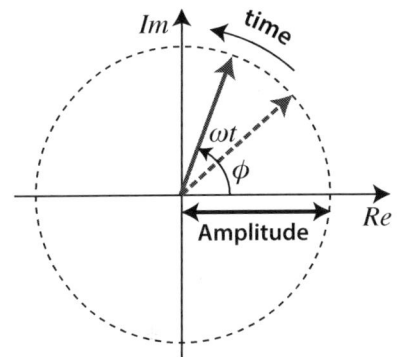

Fig. 5.13: The complex solution for a simple harmonic oscillator on the complex plane showing that it represents a particle undergoing uniform circular motion. Only the real component of the motion describes the physics.

5.4 Mass-Spring Systems

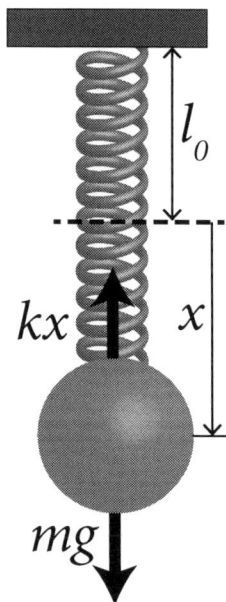

Fig. 5.14: Mass m hanging from a spring with constant k which is extended a distance x from its natural length, l_0. Such systems can also oscillate.

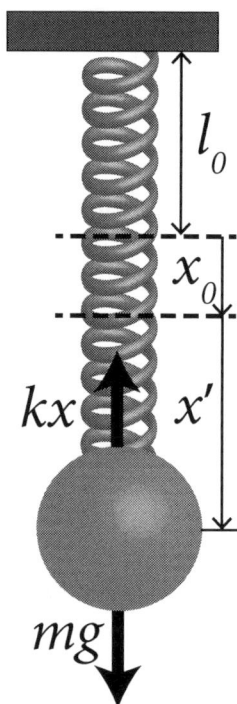

Fig. 5.15: This is the same mass-spring system shown in figure 5.14 but showing a different coordinate system where the displacement x' is from the equilibrium position and is not the extension of the spring.

In the previous section we introduced a horizontal mass-spring system and showed that this would undergo simple harmonic motion with an angular frequency of:

$$\omega = \sqrt{\frac{k}{m}} \qquad (5.49)$$

However, since frictionless surfaces are hard to come by the more common type of mass-spring system is one that is vertical as shown in figure 5.14. Resolving the forces acting on the mass and using Newton's second law we have:

$$\ddot{x} = \frac{dx^2}{d^2 t} = \vec{a}$$

$$m\ddot{x} = mg - kx \qquad (5.50)$$

where $\ddot{x}$ is Newton's dot notation for the second order derivative of x with respect to time. Unfortunately, the equation contains a constant term, mg, and so does not appear to be the usual simple harmonic motion equation that we have seen for the horizontal system. This is because we did not do *exactly* what we did before: we did not measure the displacement of the mass from the equilibrium position of the system! To fix this we need to define a new position variable, x', which is zero at the equilibrium position:

$$x' = x - x_0 \implies x = x' + x_0 \qquad (5.51)$$

where x' is the displacement from the equilibrium position, x is the total extension of the spring and x_0 is the extension of the spring at the equilibrium position, as shown in figure 5.15. Now at the equilibrium position the tension in the spring equals the weight of the object and so:

$$kx_0 = mg \qquad (5.52)$$

Now since x_0 is a constant (5.51) implies that $\ddot{x} = \ddot{x}'$ and so we can rewrite this equation in terms of our new position variable:

$$m\ddot{x}' = mg - kx' - kx_0 \qquad (5.53)$$

which looks even worse until you realize that (5.52) means that the weight and kx_0 terms cancel. The result is:

$$\ddot{x}' = -\frac{k}{m}x' \qquad (5.54)$$

which is exactly the same as our horizontal mass-spring system result! Hence a vertical mass-spring system will oscillate around the equilibrium position with the exact same characteristic frequency as the horizontal system. In general the displacement of a harmonic oscillator should always be measured from the equilibrium position. If there is a need to use a different co-ordinate system then the best approach is to solve the problem using the displacement from equilibrium and then transform the displacement to the new coordinate system afterwards.

5.5 Pendulums

A pendulum consists of a weight that is freely suspended from a pivot so that it can swing freely in a gravitational field. Since Galileo's discovery of the regular motion of pendulums they remained the most accurate means of measuring time until the 1930s and in 1851 Foucault also used a pendulum to demonstrate the rotation of the earth. There are two types of pendulums: the simple pendulum consisting of a point mass on the end of a string and the compound pendulum which consists of a rigid body which can freely pivot about a point.

5.5.1 The Simple Pendulum

A simple pendulum consists of a light, inextensible spring with a point mass, called a bob, attached to the end of it. When the bob is displaced horizontally from the equilibrium position, as shown in figure 5.16, the tension on the string and gravity act to restore the bob to the equilibrium position.

To describe the pendulum's motion consider the angular displacement, θ, from the equilibrium position as shown in figure 5.16. In this position the bob has moved through an arc of length s away from the equilibrium position. We now apply Newton's second law in the tangential direction which is always perpendicular to the string so the tension will not appear. In this direction applying Newton's second law to the mass gives:

$$F = ma \implies -mg\sin\theta = m\ddot{s} \implies \ddot{s} = -g\sin\theta \quad (5.55)$$

Now using the definition of an angle in radians, which is the arc length divided by the radius, we can differentiate this twice with respect to time remembering that the radius of the circle, l in this case, is a constant:

$$\theta = \frac{s}{l} \implies s = l\theta \implies \ddot{s} = l\ddot{\theta} \quad (5.56)$$

Substituting this value for $\ddot{s}$ into the equation from Newton's second law, (5.55), gives:

$$l\ddot{\theta} = -g\sin\theta \implies \ddot{\theta} = -\frac{g}{l}\sin\theta \quad (5.57)$$

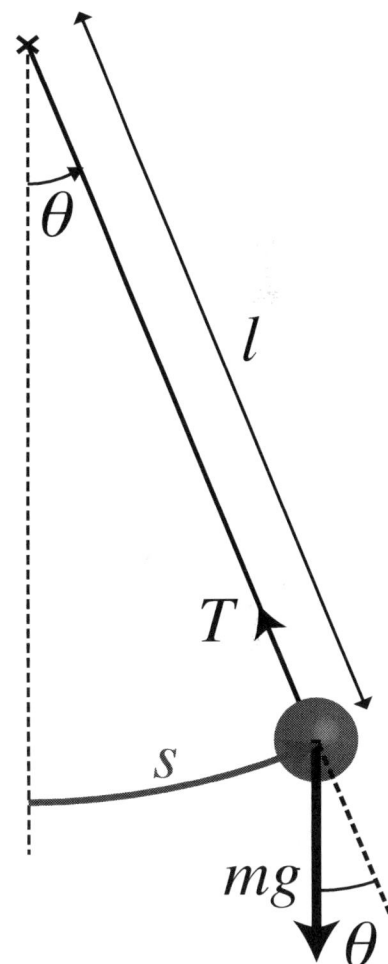

Fig. 5.16: A simple pendulum consisting of a small bob of mass m attached to the end of a light, inextensible string and displaced and angle θ from equilibrium.

This looks close to our simple harmonic equation but not quite the same because we have $\sin\theta$ and not just θ. However, if we limit ourselves to small amplitude oscillations of the pendulum the approximation that $\sin\theta \approx \theta$ from equation (1.4) applies and so the equation of motion (5.57) becomes:

$$\ddot{\theta} = -\frac{g}{l}\theta \qquad (5.58)$$

This is a simple harmonic oscillator in θ with an angular frequency of

$$\omega = \sqrt{\frac{g}{l}} \qquad (5.59)$$

and so the period and frequency of the pendulum for small amplitude oscillations are simply

small amp. occilations

$$T = 2\pi\sqrt{\frac{l}{g}} \quad \text{and} \quad f = \frac{1}{2\pi}\sqrt{\frac{g}{l}} \qquad (5.60)$$

This shows that for small amplitude oscillations the pendulum's period is independent of both the mass of the bob and the amplitude: it depends only on its length and the strength of gravity.

The Big Picture

Solving the motion of a pendulum for large amplitudes is very challenging and involves elliptical integrals. These can be approximated by a power series to yield an infinite power series for the pendulum's period of:

$$T = 2\pi\sqrt{\frac{l}{g}}\left(1 + \frac{1}{16}\theta_0^2 + \frac{11}{3072}\theta_0^4 + \frac{173}{737280}\theta_0^6 + \cdots\right) \qquad (5.61)$$

where θ_0 is the amplitude of the motion in radians. This clearly describes a complex oscillation and for the remainder of this text we will concern ourselves exclusively with the small angle approximation for pendulums.

For a 1 m pendulum on the surface of Earth with an amplitude of $10°$, the deviation between the true period and the small angle approximation is under 0.2% which is less that the deviation due to the variation of the gravitational field over the surface of the Earth.

5.5.2 The Compound Pendulum

In general, a pendulum does not have to be a bob on the end of a light string. Any rigid body which is pivoted at a single point that is not its centre of gravity may undergo oscillations and such a system is called a compound pendulum. One such system is shown in figure 5.17 which shows a rigid body that has a moment of inertia, I, about the pivot's axis of rotation. If the centre of gravity of this system is a distance l below the pivot point and the line between it and the pivot forms an angle θ with the vertical then the rotational form of Newton's second law gives:

$$I\ddot{\theta} = -mgl\sin\theta \tag{5.62}$$

Rearranging this and again using the small angle approximation we get:

$$\ddot{\theta} = -\frac{mgl}{I}\theta \tag{5.63}$$

for small amplitude oscillations. Now to check that this is correct let's treat the simple pendulum as a special case of the compound pendulum. For this situation, the bob is a point mass which is pivoted about a point a distance l from its centre of mass and so we have $I = ml^2$. It is then simple to see that the mass and one length in (5.63) cancel and it reduces to be identical to (5.58) which is consistent with a simple pendulum.

Returning to (5.63) we can see that by comparison to the simple harmonic quation of motion, $\ddot{x} = -\omega^2 x$, the angular frequency for any, general compound pendulum is just:

$$\omega = \sqrt{\frac{mgl}{I}} \tag{5.64}$$

and from this we can obtain the period and frequency of the motion using the relationships given in (5.1) and (5.2) which give:

$$T = 2\pi\sqrt{\frac{I}{mgl}} \quad \text{and} \quad f = \frac{1}{2\pi}\sqrt{\frac{mgl}{I}} \tag{5.65}$$

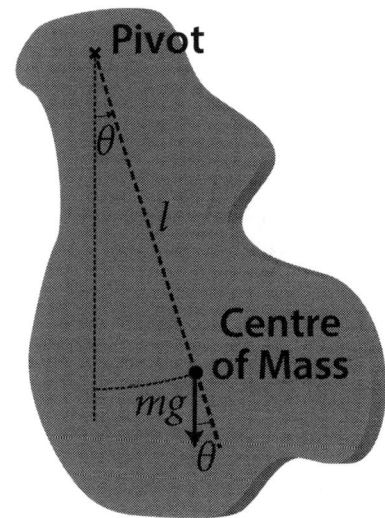

Fig. 5.17: A compound pendulum consisting of a rigid body of mass m suspended from a pivot. The moment of inertia about the pivot is I and the centre of gravity is a distance l from the pivot.

5.6 Energy

To understand how the energy of a simple harmonic oscillator behaves we will study two slightly different mechanical systems. First, to reduce complexity, we will consider the horizontal mass-spring system which contains two types of mechanical energy and then we will increase the complexity by considering the vertical mass-spring system which will include a third type of mechanical energy.

5.6.1 Horizontal Mass-Spring

The horizontal mass-spring system contains two types of mechanical energy: elastic potential energy and kinetic energy. The elastic potential is obtained by integrating the elastic force given by Hooke's law with respect to the displacement:

$$U = \int_0^x kx \, dx = \frac{1}{2}kx^2 \qquad (5.66)$$

Substituting in our expression for the displacement from (5.14) this gives:

$$U = \frac{1}{2}kA^2 \cos^2(\omega t + \phi) \qquad (5.67)$$

and so we can see that the potential energy of the system oscillates as well with the maximum corresponding to a maximum in the displacement as expected.

Using the velocity that we obtained in (5.28) we can also obtain the kinetic energy of the object undergoing oscillations using the classical kinetic energy formula:

$$T = \frac{1}{2}mv^2 = \frac{1}{2}m\omega^2 A^2 \sin^2(\omega t + \phi) \qquad (5.68)$$

So, like the potential energy, this oscillates with the same characteristic frequency but because it is a sine function, rather than a cosine, it remains out of phase with the potential energy. Now that we have seen that the energies both oscillate what about the total energy? Conservation of energy tells us that this must remain constant so let's examine the sum of the kinetic and potential:

$$E = T + U = \frac{1}{2}kA^2 \cos^2(\omega t + \phi) + \frac{1}{2}m\omega^2 A^2 \sin^2(\omega t + \phi) \quad (5.69)$$

Now for the mass-spring system we have been considering the angular frequency, ω, is simply $\sqrt{k/m}$. If we substitute this value into (5.69) then we get:

$$\begin{aligned} E &= \frac{1}{2}kA^2 \cos^2(\omega t + \phi) + \frac{1}{2}m\frac{k}{m}A^2 \sin^2(\omega t + \phi) \\ &= \frac{1}{2}kA^2 \left[\cos^2(\omega t + \phi) + \sin^2(\omega t + \phi)\right] \\ &= \frac{1}{2}kA^2 \end{aligned} \qquad (5.70)$$

So the energy is indeed constant and is equal to the potential energy at the maximum displacement from equilibrium. The relative oscillation of the energy between both the potential and kinetic is shown in figure 5.18.

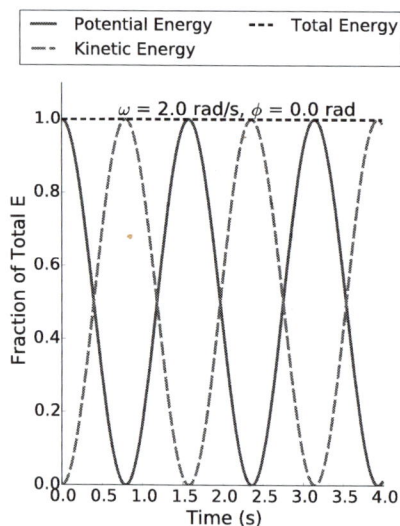

Fig. 5.18: Energy of an oscillator as a function of time. Both the potential and kinetic energies oscillate out of phase so that the total energy is constant.

5.6.2 Vertical Mass-Spring

For a vertical mass-spring system the energy calculation is made more complex by the need to consider the gravitational potential energy as well. When we attempted a dynamical solution to the vertical mass-spring we learnt that we needed to write the equations in terms of the displacement from the equilibrium position and so when considering the potential energy of the system we will want to use the potential energy relative to the equilibrium position.

Consider the vertical mass-spring system shown in figure 5.19 where the mass is displaced a distance x from the equilibrium position and the extension of the spring at the equilibrium position is x_0 so that the total extension of the spring is simply $(x_0 + x)$. We can now use this to calculate the total potential energy of the system at this point which will be the sum of the elastic and gravitational potential energies:

$$U = \underbrace{\frac{1}{2}k(x_0 + x)^2}_{\text{Elastic}} - \underbrace{mgx}_{\text{Gravitational}} \qquad (5.71)$$

Now at the equilibrium position we have zero gravitational potential energy and so the total the potential energy will be just:

$$U = \frac{1}{2}kx_0^2 \qquad (5.72)$$

and so we can use this to calculate the change in potential energy relative to the equilibrium position which gives:

$$\Delta U = \frac{1}{2}k(x_0 + x)^2 - mgx - \frac{1}{2}kx_0^2 = \frac{1}{2}kx^2 + kx_0x - mgx \quad (5.73)$$

Now at the equilibrium position the elastic force from Hooke's law must precisely balance the weight of the mass due to gravity and so we have:

$$kx_0 = mg \qquad (5.74)$$

Substituting this value for kx_0 into equation (5.73) we get:

$$\Delta U = \frac{1}{2}kx^2 \qquad (5.75)$$

which is the potential energy of the oscillator measured relative to the equilibrium position. Since this is a simple harmonic oscillator we already know the position as a function of time from (5.14) and so we have a expression for the potential energy which is:

$$\Delta U = \frac{1}{2}kA^2 \cos^2(\omega t + \phi) \qquad (5.76)$$

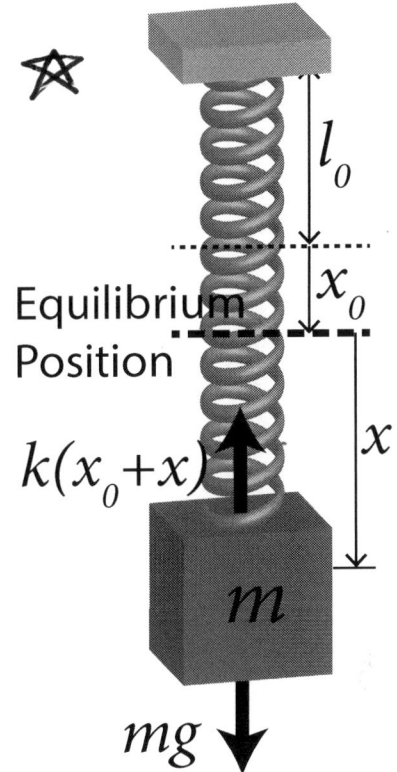

Fig. 5.19: Block on the end of a vertical spring showing the displacement from equilibrium, x, and the extension of the spring at equilibrium, x_0. The potential energy of the system is calculated relative to the potential energy at equilibrium.

which is identical to the equation (5.67) that we obtained for the horizontal mass-spring system. This shows that the total potential energy for both horizontal and vertical mass-spring systems is exactly the same provided that we measure it relative to the equilibrium position.

Since energy is conserved and no external forces act on the system the total energy of the system is constant. Hence when the mass is at the equilibrium position and the potential energy is zero the kinetic energy of the mass will be a maximum and equal to the maximum potential energy. This means that just as with the horizontal mass-spring system the energy will oscillate back and forth between potential and kinetic energy. The only difference is that the potential energy in the vertical case is split between elastic and gravitational potential.

The result that we have obtained here for potential energy is actually a general result for all simple harmonic oscillators: the potential energy of the oscillator is proportional to the square of the displacement from the equilibrium. For example, for a simple pendulum, the potential energy is proportional to the square of the angular displacement θ for small amplitude oscillations which can be easily shown using a power series expansion for $\cos\theta$.

The oscillation of the energy between potential and kinetic energy is common to all mechanical harmonic oscillators and the general principle of energy oscillating between two different forms is, in fact, common to all harmonic oscillators. For example in an alternating current circuit containing a capacitor and an inductor the energy oscillates between electrical and magnetic fields.

5.7 General Simple Harmonic Motion

We have now seen several examples of simple harmonic motion: the mass-spring system, the simple pendulum and the compound pendulum. In each of these cases we applied Newton's second law, either in its standard linear form or in a rotational form, to derive an equation of motion and in all cases we ended up with a differential equation of the form:

$$\ddot{z} = -\omega^2 z \qquad (5.77)$$

where $z(t)$ is the displacement of the system from equilibrium and, in the examples we have covered, is either a linear displacement or an angular displacement. Furthermore we have solved this equation to find that the displacement at a time t is given by:

$$z(t) = A\cos(\omega t + \phi) \qquad (5.78)$$

where A is the amplitude, ϕ the initial phase and ω is the angular frequency.

This provides a general way to now solve any simple harmonic oscillator. All that is required is to apply Newton's second law, in an appropriate form, to the problem and then to rearrange the result to get an equation of the form given in (5.77). Once this is achieved then the angular frequency can be simply determined by comparison and the solution written down using (5.78) with the amplitude and initial phase being determined by the initial conditions of the system. If this is not possible, and the equation of motion cannot be written in the required form, then the system is not undergoing simple harmonic motion.

Example 5.1

A block of mass m rests on a frictionless surface and is attached to three springs each with different spring constants which can be extended or compressed. The spring with constants k_1 and k_2 are connected to one side of the block and the spring with constant k_3 is connected to the other side as shown in figure 5.20. In the equilibrium position, all the springs have zero extension or compression. The block is now displaced slightly from the equilibrium position and released. What is the frequency of oscillation?

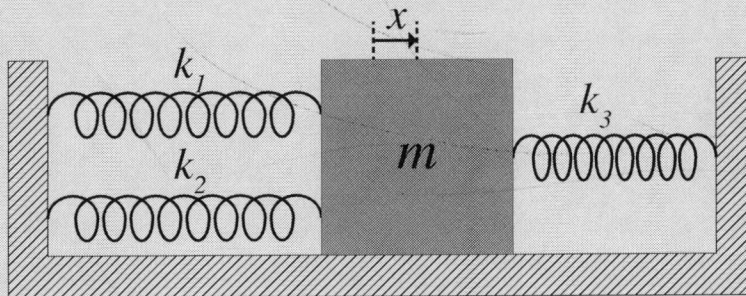

Fig. 5.20: A block of mass m rests on a frictionless surface and is connected to three springs with the spring constants shown. In the equilibrium position all the springs are at their natural length.

Solution:
We start by applying Newton's second law to the system shown in figure 5.20. If the displacement from the equilibrium position is x then all the springs will be compressed or extended by x. Taking the positive x direction as being to the right of the page this gives:

$$-k_1 x - k_2 x - k_3 x = m\ddot{x} \qquad (5.79)$$

Rearranging this gives:

$$\ddot{x} = -\frac{k_1 + k_2 + k_3}{m}x \tag{5.80}$$

Now we compare this equation directly with our general equation for simple harmonic motion given in (5.14):

$$\begin{aligned} \ddot{x} &= -\frac{k_1+k_2+k_3}{m}x \\ &\updownarrow \\ \ddot{x} &= -\omega^2 x \end{aligned} \tag{5.81}$$

Both equations have exactly the same form and so we can immediately see that the system we have here will undergo simple harmonic motion. Furthermore by comparing the coefficients for x we have the result:

$$\omega^2 = \frac{k_1 + k_2 + k_3}{m} \implies \omega = \sqrt{\frac{k_1 + k_2 + k_3}{m}} \tag{5.82}$$

This is the angular frequency of the system and it is now a simple matter of dividing by 2π to get the frequency of the oscillations we were asked to find:

$$f = \frac{\omega}{2\pi} = \frac{1}{2\pi}\sqrt{\frac{k_1 + k_2 + k_3}{m}} \tag{5.83}$$

Problems

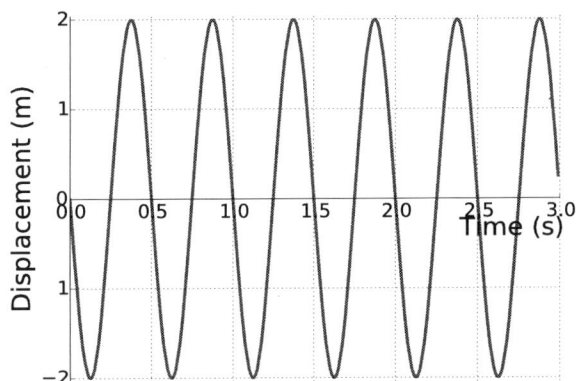

Fig. P5.1: *Plot of displacement, x, versus time, t, for a mass attached to a spring.*

Q5.1: Figure P5.1 shows the displacement of a mass attached to a spring as a function of time.
(a) What are the amplitude, frequency and initial phase of the motion when written using a cosine function?
(b) If the spring has a constant of $400\,\mathrm{N\,m^{-1}}$ what is the mass?

Q5.2: A mass of $4\,\mathrm{kg}$ is suspended from the bottom of a spring. The mass is pulled down a distance of $20\,\mathrm{cm}$ from the equilibrium position and then released after which it oscillates with a frequency of $0.5\,\mathrm{Hz}$. What is the speed of the mass as it passes through the equilibrium position?

Q5.3: A mass attached to the end of a spring is found to oscillate such that its maximum velocity is $0.6\,\mathrm{m\,s^{-1}}$ and its maximum acceleration is $1.5\,\mathrm{m\,s^{-2}}$. What are the angular frequency and amplitude of the oscillator?

Q5.4: A mass of $10\,\mathrm{kg}$ is on the end of a spring with a constant $98.7\,\mathrm{N\,m^{-1}}$. The displacement, velocity and acceleration of the mass are measured over time. At one particular instant the acceleration of the mass is found to have it maximum positive value how much time passes after this instant before the first time that:

(a) The velocity of the mass is at its largest positive value?
(b) The displacement of the mass is at its largest positive value?
(c) The acceleration of the mass is zero?

Q5.5: A simple pendulum of length $5\,\mathrm{m}$ in a gravitational field of $9.81\,\mathrm{m\,s^{-2}}$ starts at an initial displacement of $2\,\mathrm{mrad}$ and an initial angular velocity of $2\,\mathrm{mrad\,s^{-1}}$. Derive an expression for the angular displacement of the pendulum as a function of time, t.

Q5.6: A mass m is attached to a vertical spring with a constant k and has a second, identical mass m attached by light string to the first mass and hanging beneath it. The system is is at rest when the string snaps. If the gravitational field is g calculate the following quantities:
(a) the amplitude of the resulting motion
(b) the maximum velocity of the mass which remains attached to the string

Q5.7: An astronaut on the International Space Station needs to know the mass of a sample of material as part of an experiment. Unfortunately, being in freefall, it is not possible to use simple scales. Instead he attaches a known mass of $2\,\mathrm{kg}$ to a spring which is attached to a surface and notes that when it is disturbed from rest the period of oscillation is $2.5\,\mathrm{s}$. He then replaces the known mass with the sample and observes that the period is now $1.8\,\mathrm{s}$. What is the mass of the sample?

Q5.8: Two identical springs are hung vertically next to each other. The first spring has a mass m attached to it and the second spring has a mass of $2m$ attached to it. Both masses are pulled down from their equilibrium positions at released at rest at exactly the same time. What is the magnitude of the phase difference between the displacement of the two masses after the first mass has reached its equilibrium position for the first time?

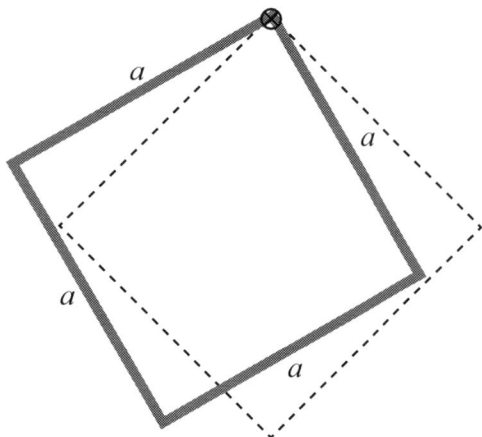

Fig. P5.2: *A square frame made from four identical, thin rods each of length a can rotate freely about an axis at one corner of the frame.*

Q5.9: A square frame consists of four identical, thin rods each of length a. The frame is pivoted at a corner so that it can rotate freely in the plane of the frame as shown in figure P5.2. The frame is rotated slightly away from the equilibrium position and allowed to oscillate in a gravitational field g. What is the period of the resulting motion? The moment of inertia of a rod of length l and mass m about an axis through its centre is $\frac{1}{12}ml^2$.

Q5.10: A flat object with an irregular shape can be suspended from two different points, A and B, such that it is free to pivot about an axis perpendicular to the plane of the object. First, the pivot is attached to the point A and when in equilibrium a vertical line is drawn on the object such that it passes through the pivot point. The same procedure is followed with the pivot attached at point B and the lines are found to intersect a distance a from the point A and a distance of $2a$ from the point B. While suspended from the point A the object undergoes small amplitude oscillations of period T when disturbed slightly from rest. What period will the small amplitude oscillations be if the object is pivoted from the point B?

Q5.11: A mass of 1 kg hangs at rest on the end of a vertical spring with a constant of $100\,\mathrm{N\,m^{-1}}$. The mass is hit with an object so that it gains a kinetic energy of 2 J and moves in the positive x

direction.
(a) What is the amplitude of the resulting oscillation?
(b) What is the expression for the kinetic energy of the mass as a function of time, t?

Q5.12: A mechanical, simple harmonic oscillator has an amplitude A and an angular frequency ω.
(a) How far from equilibrium, in terms of A, is the system when the potential and kinetic energies are equal given that there is zero potential energy at the equilibrium position?
(b) The system is at a position where the potential and kinetic energies are equal. How much time passes before this condition is true again?

Q5.13: A simple pendulum of length 0.5 m is placed in a lift and, whilst the lift is accelerating, the frequency of oscillation is found to be 0.05 Hz lower than when the lift is at rest. If the acceleration due to gravity is $9.81\,\mathrm{m\,s^{-2}}$ what is the acceleration of the lift and state whether it is accelerating upwards or downwards.

Q5.14: In the possibly not so distant future, a spacecraft is extracting minerals from an asteroid. In an attempt to determine the mineral content of one perfectly spherical asteroid with a uniform density ρ a small hole is drilled right through the centre of the asteroid and out of the other side. After the drilling process is complete a small piece of rock falls into the now empty hole. Given that the rock does not touch the sides and that the amplitude is small enough that it does not leave the hole show that the rock undergoes simple harmonic motion and calculate the period.

Q5.15: A perfectly frictionless wire is curved into a U-shape so that its height above the ground, y, is given by:

$$y = as^2$$

where a is a constant and s is the distance along the wire from the point where it touches the ground. The wire passes through a small, metal ring which has a mass m. The ring starts at $s = s_0$ and is released at rest. The gravitational field is g.

(a) Derive the expression which gives the displacement along the wire of the ring as a function of time t since release.

(b) A second, identical ring is now threaded through the wire. Starting at the lowest point of the wire $(y = s = 0)$ the first ring is moved along the right side of the wire a distance s_1 and the second ring is moved along the left side of the wire a distance $-s_2$. Both rings are then released at rest and each slide along the wire towards each other. At what height above the ground will the rings collide?

(c) The wire above is now reshaped so that it forms a parabola where the height above the ground, y, is given by $y = ax^2$ where x is the horizontal displacement from the equilibrium position. A physicist argues that because the potential energy of the ring a distance x from the equilibrium is mgx^2 this means that it will behave similarly to a horizontal mass-spring system which has a potential energy of $\frac{1}{2}kx^2$ and so it too must produce simple harmonic motion if only the horizontal component of the motion is considered. Is this assertion correct? Explain.

CHAPTER 6

Damped and Driven Oscillations

In the previous chapter, we discussed an ideal picture of an oscillator and came up with a solution where an oscillating system would continue to oscillate forever with a constant amplitude. However, our experience with oscillators in the real world contradicts this prediction. If you start a simple pendulum swinging over time its amplitude will get smaller and it will eventually stop moving. The reason for this is that energy is lost doing work against friction-like forces such as, in the case of a pendulum, air resistance.

To overcome these losses an oscillator can be driven by an oscillating driving force. For example, a clock will drive its pendulum typically either by a falling weight or via elastic energy stored in a string. Driven oscillators can exhibit quite surprising behaviour and when constructing buildings and bridges great care has to be taken to ensure that environmental factors do not drive oscillations of the structure since this can ultimately lead to failure and collapse!

In this chapter, we will extend our discussion of oscillations to include oscillators which have forces that oppose the motion of the system and, once we have solved for the different types of motion which result, we will consider these same systems with the addition of a periodic driving force.

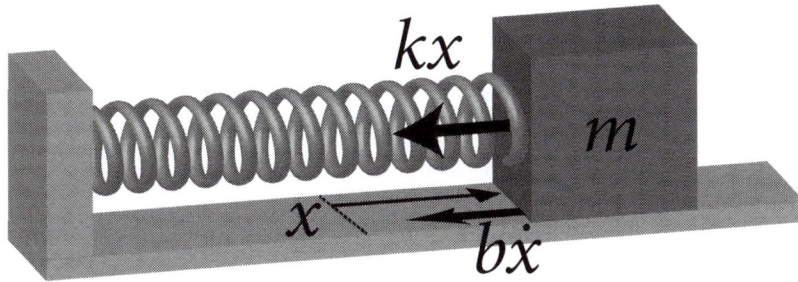

Fig. 6.1: A damped mass-spring system showing a retarding force which acts in the opposite direction to the velocity and has a magnitude proportional to it.

6.1 Damped Oscillations

Any real oscillator will lose energy as it oscillates due to friction, air resistance etc. Such systems are said to be *damped* because over time the amplitude of their oscillations will dampen down to zero. To model such a system let's return to our horizontal mass-spring system and consider a new, damping force that is proportional to the velocity and acts in the opposite direction to it. If the constant of proportionality is b then the damping force, $F_D = -b\dot{x}$. Figure 6.1 shows the forces acting on the mass and so Newton's second law gives an equation of motion:

$$m\ddot{x} = -kx - b\dot{x} \qquad (6.1)$$

Now let's define two new quantities:

$$\omega_0 = \sqrt{\frac{k}{m}} \quad \text{and} \quad \zeta = \frac{b}{2\sqrt{mk}} = \frac{b}{2m\omega_0} \qquad (6.2)$$

Clearly ω_0 is just the angular frequency of the undamped system but ζ is something new called the damping ratio. Rewriting (6.1) in terms of these and bringing all the terms onto the same side we get:

$$\ddot{x} + 2\zeta\omega_0\dot{x} + \omega_0^2 x = 0 \qquad (6.3)$$

Looking at this we can see that our existing simple harmonic solution will not work because we have a first order differential term, $\dot{x}$, and if we differentiate a cosine function we get a sine function. So let's revert to the solution we tried first for the simple harmonic oscillator: Ae^{qt}. Substituting this into (6.3) we get:

$$q^2 Ae^{qt} + 2\zeta\omega_0 q Ae^{qt} + \omega_0^2 Ae^{qt} = 0 \qquad (6.4)$$

We can then cancel the Ae^{qt} terms since this cannot be zero except in the most trivial of solutions where the object remains stationary at the equilibrium position. This gives:

$$q^2 + 2\zeta\omega_0 q + \omega_0^2 = 0 \qquad (6.5)$$

which is simply a quadratic equation in q that has solutions:

$$q = \frac{-2\zeta\omega_0 \pm \sqrt{4\zeta^2\omega_0^2 - 4\omega_0^2}}{2} = \omega_0\left(-\zeta \pm \sqrt{\zeta^2 - 1}\right) \quad (6.6)$$

This solution presents three possible scenarios. If $\zeta > 1$ then the square root will give a real number and we will have two real solutions for q, if $\zeta = 1$ there is only one real solution for q and if $\zeta < 1$ then the square root is of a negative number which will give an imaginary number as we saw before with the un-damped oscillator. These three solutions correspond to three different types of damping: over damped, critically damped and under damped respectively which we will discuss in the following sections.

6.1.1 Over Damped

This is the simplest case to consider. For over-damped systems where $\zeta > 1$ there are two real solutions for q which are:

$$q_+ = \omega_0\left(-\zeta + \sqrt{\zeta^2 - 1}\right) \quad \text{and} \quad q_- = \omega_0\left(-\zeta - \sqrt{\zeta^2 - 1}\right) \quad (6.7)$$

Constructing a general solution to our equation from these we have:

$$x(t) = Ae^{q_+ t} + Be^{q_- t} = e^{-\omega_0 \zeta t}\left(Ae^{\omega_0\sqrt{\zeta^2-1}\,t} + Be^{-\omega_0\sqrt{\zeta^2-1}\,t}\right) \quad (6.8)$$

where A and B are real constants since we require a real number for the displacement. However, since $\sqrt{\zeta^2 - 1} < \zeta$ this means that both q_+ and q_- are negative and so both will exponentially decay to zero. The result is that the displacement of the system exponentially decays to zero. The damping is so large that the system cannot oscillate and, if displaced from equilibrium, will simply return to that position and remain there. This behaviour is shown in figure 6.2 which shows the displacement versus time plot for an over damped system.

6.1.2 Under Damped

For the under damped case we have $\zeta < 1$ and so the solutions for q will be complex numbers having both a real and an imaginary part because they contain the square root of a negative number. Taking out a factor of i from inside the square root to

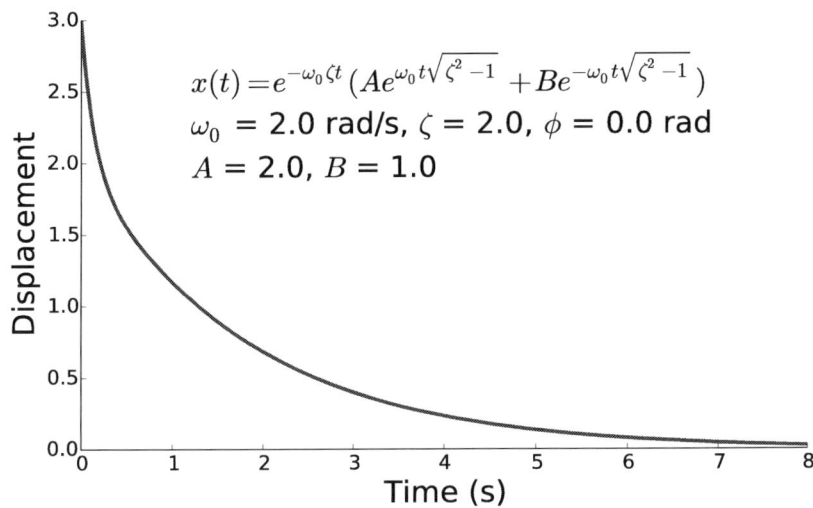

Fig. 6.2: Displacement of an over damped harmonic oscillator as a function of time. The system simply moves towards the equilibrium position without oscillating.

make the contents positive we have:

$$q_+ = \omega_0 \left(-\zeta + i\sqrt{1 - \zeta^2} \right) \qquad (6.9)$$

$$q_- = \omega_0 \left(-\zeta - i\sqrt{1 - \zeta^2} \right) \qquad (6.10)$$

Using these we can write our general solution as:

$$
\begin{aligned}
x(t) &= Ae^{q_+ t} + Be^{q_- t} \\
&= e^{-\omega_0 \zeta t} \left(Ae^{i\omega_0 \sqrt{1-\zeta^2} t} + Be^{-i\omega_0 \sqrt{1-\zeta^2} t} \right)
\end{aligned}
\qquad (6.11)
$$

This is the product of a completely real term, the negative exponential $e^{-\omega_0 \zeta t}$, and a part real, part imaginary, complex term consisting of the sum of two imaginary exponentials. However, we have seen this exact imaginary exponential term before in (5.11) when we tried to use an exponential to find the solution to the simple harmonic oscillator. Back there we argued that this was the general mathematical solution to the equation but that we were only interested in the real solution because we needed a real number for the displacement and the exact same argument applies here.

For the undamped oscillator, we showed, by inspection in section 5.2.1 and by derivation in section 5.3, that the real solution which was hidden inside the more general complex solution was a cosine function. Since we have the identical expression here the same thing is true and when we require the solution to be real we will end up with a cosine function again but this time with a frequency of $\omega_0 \sqrt{1 - \zeta^2}$. Hence our underdamped

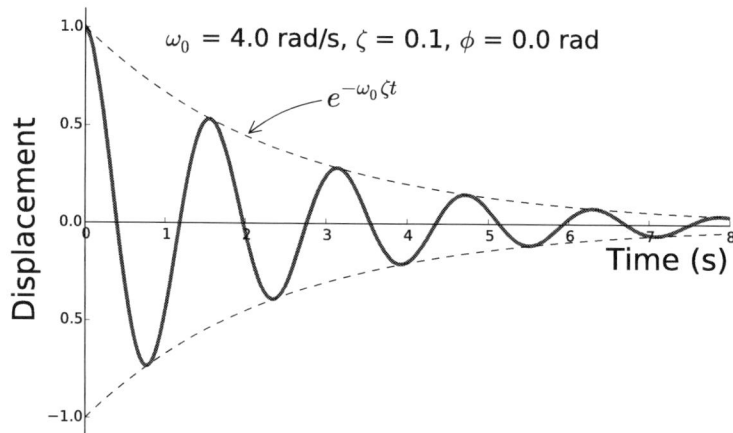

Fig. 6.3: Displacement of a under damped harmonic oscillator as a function of time. The system still oscillates but the amplitude of those oscillations exponentially decays with time.

solution is simply:

$$x(t) = Ae^{-\omega_0 \zeta t} \cos\left(\omega_0 \sqrt{1 - \zeta^2}t + \phi\right) \qquad (6.12)$$

where A and ϕ are real constants. This results in an oscillation but one with an exponentially decreasing amplitude due to the $e^{-\omega_0 \zeta t}$ term. In addition, the frequency of the oscillation is different from the free oscillator by a factor of $\sqrt{1 - \zeta^2}$. The resulting motion is shown in figure 6.3 which shows how the oscillation occurs inside an exponentially decaying envelope. To check that this solution is consistent we can also check the limit of no damping where $\zeta = 0$. In this case, the real exponential becomes a constant and the frequency of oscillation becomes ω_0 which is exactly consistent with a non-damped oscillator.

6.1.3 Critically Damped

The case of a critically damped oscillator corresponds to the damping ratio, $\zeta = 1$. In this situation there is only one solution for q which is:

$$q = -\omega_0 \zeta \qquad (6.13)$$

This would suggest that our general solution is simply $Ae^{-\omega_0 t}$. However, looking at this solution there is only one constant of integration and, as we saw in (5.12) when we were trying a solution to the undamped oscillator, this means that we are missing a solution because when we integrate twice we *must* have two constants of integration.

Clearly adding a constant to the exponential argument will not work in this case since that is the same as multiplying by a constant so let's try adding a factor of t instead to give a function of the form $Ate^{-\omega_0 t}$. Differentiating this gives:

$$x = Ate^{-\omega_0 t} \tag{6.14}$$

$$\dot{x} = -\omega_0 Ate^{-\omega_0 t} + Ae^{-\omega_0 t} \tag{6.15}$$

$$\ddot{x} = \omega_0^2 Ate^{-\omega_0 t} - 2\omega_0 Ae^{-\omega_0 t} \tag{6.16}$$

Substituting these into our differential equation that is now $\ddot{x} + 2\omega_0\dot{x} + \omega_0^2 x = 0$ gives:

$$\omega_0^2 Ate^{-\omega_0 t} - 2\omega_0 Ae^{-\omega_0 t} - 2\omega_0^2 Ate^{-\omega_0 t} + 2\omega_0 Ae^{-\omega_0 t} + \omega_0^2 Ate^{-\omega_0 t} = 0 \tag{6.17}$$

and so this is also a solution too! However, it is important to note that this only works in the critically damped case because it relies on $\zeta = 1$ to cancel the different types of terms created by the differentiation. The result is that our general solution in the critically damped case is now:

$$x(t) = (A + Bt)e^{-\omega_0 t} \tag{6.18}$$

where A and B are constants. This special case corresponds to no oscillations, only exponential decay of the position. Its form is shown in figure 6.4 and similar in appearance to the over damping case. One thing to note is that a critically damped oscillator is the fastest to return to the equilibrium position. In the over damped case the damping is so large that it slows the return to equilibrium and in the under damped case the system oscillates about the equilibrium with a slowly decaying amplitude.

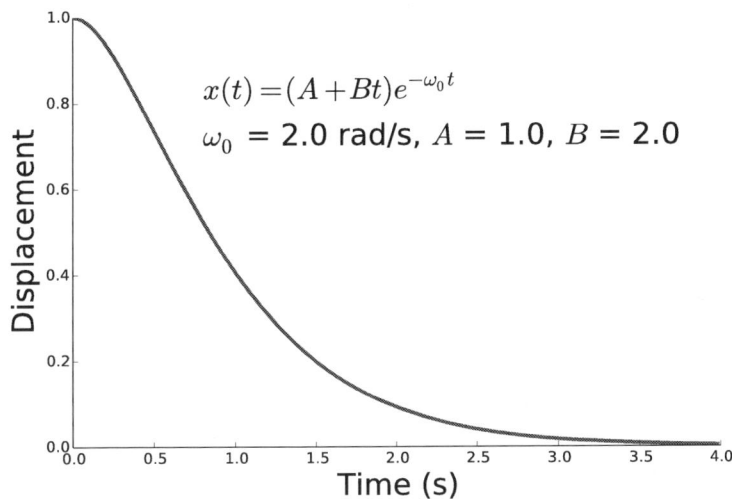

$$x(t) = (A + Bt)e^{-\omega_0 t}$$

$\omega_0 = 2.0$ rad/s, $A = 1.0$, $B = 2.0$

Fig. 6.4: Displacement of a over damped harmonic oscillator as a function of time. The system simply moves towards the equilibrium position without oscillating and does so faster than an over damped oscillator.

6.2 General Damped Oscillator

We have now seen how to solve the case of a damped mass-spring system. However, we want to be able to solve the equation of motion for any damped harmonic oscillator. The key to doing this is to first apply Newton's laws of motion to the system to derive the equation of motion. If the displacement of the system from equilibrium is z then the result should be an equation which has the form:

$$\ddot{z} + a\dot{z} + b = 0 \tag{6.19}$$

where a and b are constants that will depend on the physical parameters of the system. If it is impossible to write the equation of motion in this form then the system is not a damped, harmonic oscillator. Next we compare this with the equation we derived for the mass-spring system, (6.3).

$$
\begin{array}{ccccc}
\ddot{x} & + & 2\zeta\omega_0\dot{x} & + & \omega_0^2 x & = 0 \\
\uparrow & & \uparrow & & \uparrow & \\
\ddot{z} & + & a\dot{z} & + & b & = 0
\end{array}
\tag{6.20}
$$

Since we know a and b we can use these to determine the natural, undamped frequency of motion ω_0 and the damping ratio, ζ. To make each term in our new equation match the terms in the equation we have already solved we require:

$$\omega_0 = \sqrt{b} \quad \text{and} \quad \zeta = \frac{a}{2\sqrt{b}} \tag{6.21}$$

Once we have the natural, undamped frequency and the damping ratio we can immediately determine the damping regime which applies: under damped, over damped or critically damped. Then the solution for the appropriate damping case is used to determine the displacement of the system as a function of time.

To see how this works let's return to our ocean buoy example and consider what happens when we add in the drag due to the viscosity of the water.

Example 6.1

A horizontal mass-spring system consisting of a small, spherical mass m with radius r resting on a frictionless surface is placed underwater as shown in figure 6.5. The water generates a drag force which is given by Stokes' law: $F_D = 6\pi\eta r v$ where r is the radius of the sphere, v is the velocity of the sphere and η is the coefficient of viscosity for water which here is constant. The drag force acts so as to always oppose the motion of the sphere. The spring has a constant k and can be considered to be unaffected by the water and the sphere is dense enough that it does not float. What is the condition on the radius of the sphere for the system to be critically damped?

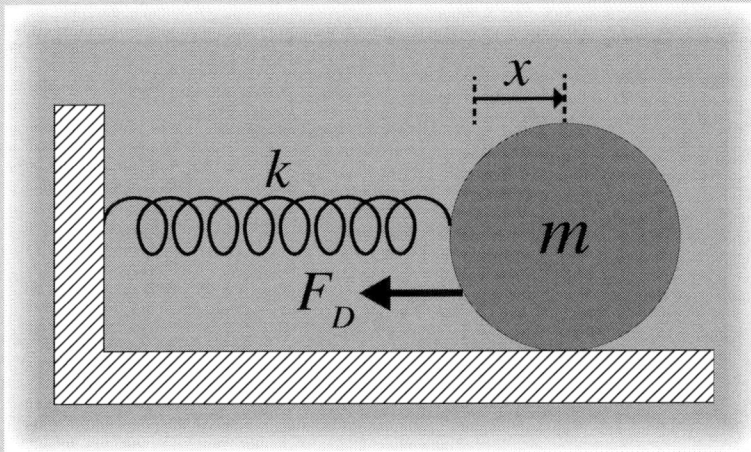

Fig. 6.5: A sphere of mass m and radius r is attached to a spring and rests on a frictionless surface. The entire system is placed underwater which causes a drag force, F_D to act on the sphere when it is moving.

Solution:
The first step is to draw a diagram with the forces acting on the sphere. This is shown in figure 6.5 where the displacement x is measured relative to the equilibrium position and there are three forces acting: weight, spring and the drag. Applying Newton's second law to the sphere with the positive x direction being to the right gives:

$$-kx - 6\pi\eta r\dot{x} = m\ddot{x}$$

which we can rearrange to give:

$$\ddot{x} + \frac{6\pi\eta r}{m}\dot{x} + \frac{k}{m}x = 0$$

Now we just compare this with the damped harmonic oscillator equation of motion (6.3) and we can see that we have all the same terms and so the system is clearly a damped harmonic oscillator. To determine the natural frequency of the system and the damping ratio all we need to do is compare the coefficients of the x and $\dot{x}$ terms:

$$\ddot{x} \quad + \quad \frac{6\pi\eta r}{m}\dot{x} \quad + \quad \frac{k}{m}x \quad = 0$$
$$\updownarrow \qquad\qquad \updownarrow$$
$$\ddot{x} \quad + \quad 2\zeta\omega_0\dot{x} \quad + \quad \omega_0^2 x \quad = 0 \qquad (6.22)$$

First look at the x coefficients this gives:

$$\omega_0^2 = \frac{k}{m} \implies \omega_0 = \sqrt{\frac{k}{m}}$$

and since ω_0 is supposed to be the natural frequency of the undamped system this is exactly what we would expect for a mass-spring system. Next we look at the $\dot{x}$ coefficients to find the damping ratio:

$$2\zeta\omega_0 = \frac{6\pi\eta r}{m} \implies \zeta = \frac{3\pi\eta r}{m\omega_0}$$

Substituting in the value of ω_0 we just obtained above we get an expression for the damping ratio of:

$$\zeta = \frac{3\pi\eta r}{\sqrt{mk}}$$

Now for critical damping we require that $\zeta = 1$ and so we have:

$$\frac{3\pi\eta r}{\sqrt{mk}} = 1 \implies r = \frac{\sqrt{mk}}{3\pi\eta}$$

6.3 Driven Oscillators

So far we have confined our discussion to systems which are disturbed from equilibrium and then left to oscillate by themselves. However, one possibility that needs to be considered is how an oscillator will respond to being driven. Driving an oscillator can take many different forms. One of the most common that many of us have encountered when a child is "pumping" a swing to increase the amplitude. This works by altering the moment of inertia about the axis of rotation at certain points in the cycle. However, here we will consider a simpler system: a damped mass-spring system driven by an external, oscillating force that is given by the real part of $F_0 e^{i\omega t}$, i.e. $F_0 \cos(\omega t)$, as shown in figure 6.6.

The equation of motion for this system, from Newton's second law is simply:

$$m\ddot{x} = -kx - b\dot{x} + F_0 \cos(\omega t) \tag{6.23}$$

Fig. 6.6: Damped mass-spring system showing spring's tension force, the damping force opposed to the velocity and the external driving force, $F_0 \cos(\omega t)$.

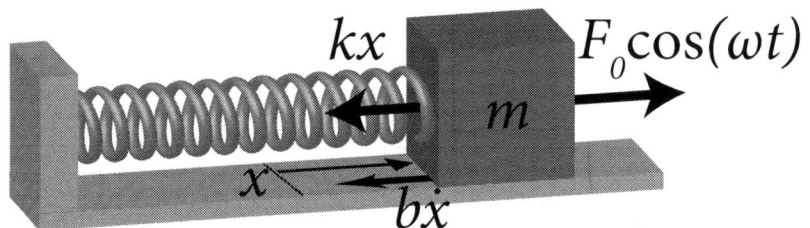

As before when dealing with a damped oscillator we can define two new quantities:

$$\omega_0 = \sqrt{\frac{k}{m}} \quad \text{and} \quad \zeta = \frac{b}{2\sqrt{mk}} = \frac{b}{2m\omega_0} \tag{6.24}$$

where ω_0 is the natural frequency of vibration of the undamped, free system and ζ is the damping ratio. Using these we can rewrite (6.23) to give:

$$\ddot{x} + 2\zeta\omega_0\dot{x} + \omega_0^2 x = \frac{F_0}{m}\cos(\omega t) \tag{6.25}$$

This looks similar to our damped equation of motion (6.3) but with the additional driving force term on the righthand side. The easiest way to solve this is to actually write down our driving force in imaginary exponential form:

$$\ddot{x} + 2\zeta\omega_0\dot{x} + \omega_0^2 x = \frac{F_0}{m}e^{i\omega t} \tag{6.26}$$

which immediately suggests a solution of the form $Ze^{i\omega t}$ where Z is the complex amplitude which contains both the real amplitude and a phase. This approach results in a complex equation which we separate into two simultaneous equations be requiring that the real and imaginary parts are both equal and the resulting algebra is far simpler than the trigonometric approach. To illustrate this we will first solve equation (6.25) using pure trigonometry and then compare this to the imaginary exponential method.

6.3.1 Trigonometric Solution

The additional term on the right hand side of (6.25) suggests a solution for x that is of the form $A\cos(\omega t - \phi)$ where ϕ is the phase difference between the driving force and the position of the mass. In this case all the differentials of x will be sine and cosines of the same ωt term and so the sum can be equal to the driving force expression for a suitable value of the phase difference, ϕ. First let's calculate the differentials:

$$x = A\cos(\omega t - \phi) \tag{6.27}$$
$$\dot{x} = -\omega A\sin(\omega t - \phi)$$
$$\ddot{x} = -\omega^2 A\cos(\omega t - \phi)$$

where A and ϕ are real constants. The choice to use a phase offset of $-\phi$ instead of $+\phi$ is purely convention: the results will be the same either way with just a sign flip for the final derived value of ϕ. To see whether we have guessed correctly that this is a solution we need to substitute these into our equation of motion (6.25) and then find values for A and ϕ:

$$-\omega^2 A\cos(\omega t-\phi)-2\zeta\omega_0\omega A\sin(\omega t-\phi)+\omega_0^2 A\cos(\omega t-\phi) = \frac{F_0}{m}\cos(\omega t)$$
(6.28)

To solve this we first need to expand out the trig terms using the trig addition identities, specifically:

$$\cos(\omega t - \phi) = \cos(\omega t)\cos\phi + \sin(\omega t)\sin\phi \qquad (6.29)$$
$$\sin(\omega t - \phi) = \sin(\omega t)\cos\phi - \cos(\omega t)\sin\phi \qquad (6.30)$$

When we put these into (6.28) we get the following:

$$-\omega^2 A\cos(\omega t)\cos\phi - \omega^2 A\sin(\omega t)\sin\phi$$
$$-2\zeta\omega_0\omega A\sin(\omega t)\cos\phi + 2\zeta\omega_0\omega A\cos(\omega t)\sin\phi$$
$$+\omega_0^2 A\cos(\omega t)\cos\phi + \omega_0^2 A\sin(\omega t)\sin\phi = \frac{F_0}{m}\cos(\omega t)$$
(6.31)

Looking at (6.31) we have a combination of both $\sin(\omega t)$ and $\cos(\omega t)$ terms. However, this equation must be true for all values of t and since the sine and cosine terms each vary differently with t the sine and cosine terms must both, independently cancel out with each other because a $\sin(\omega t)$ term cannot cancel out a $\cos(\omega t)$ term for all possible values of t. This means that we can split the equation into two by collecting the sine terms and the cosine terms separately which gives:

$$-\omega^2 A\cos(\omega t)\cos\phi + 2\zeta\omega_0\omega A\cos(\omega t)\sin\phi$$
$$+\omega_0^2 A\cos(\omega t)\cos\phi = \frac{F_0}{m}\cos(\omega t) \quad (6.32)$$
$$-\omega^2 A\sin(\omega t)\sin\phi - 2\zeta\omega_0\omega A\sin(\omega t)\cos\phi$$
$$+\omega_0^2 A\sin(\omega t)\sin\phi = 0 \qquad (6.33)$$

Dividing through (6.32) and (6.33) by $\cos(\omega t)$ and $\sin(\omega t)$ respectively we get:

$$-\omega^2 A\cos\phi + 2\zeta\omega_0\omega A\sin\phi + \omega_0^2 A\cos\phi = \frac{F_0}{m} \qquad (6.34)$$
$$-\omega^2 A\sin\phi - 2\zeta\omega_0\omega A\cos\phi + \omega_0^2 A\sin\phi = 0 \qquad (6.35)$$

Starting with equation (6.35) we now divide this through by $A\cos\phi$ to get an equation which we can solve to find ϕ:

$$-\omega^2\tan\phi - 2\zeta\omega_0\omega + \omega_0^2\tan\phi = 0$$
$$\implies \phi = \tan^{-1}\left(\frac{2\omega_0\omega\zeta}{\omega_0^2 - \omega^2}\right) \qquad (6.36)$$

Now we need to return to (6.34) and use our expression for $\tan \phi$ to solve for a value of A, the amplitude of the motion. We start by taking out a factor of $A \cos \phi$ on the lefthand side:

$$A \cos \phi \left(\omega_0^2 - \omega^2 + 2\zeta \omega_0 \omega \tan \phi \right) = \frac{F_0}{m} \qquad (6.37)$$

Next we need to derive a trigonometric identity to relate cosine to tangent:

$$\sin^2 \phi + \cos^2 \phi = 1 \quad \Longrightarrow \quad \cos^2 \phi = \frac{1}{1 + \tan^2 \phi} \qquad (6.38)$$

and now we can use this, along with our result for $\tan \phi$ from 6.36, to substitute into (6.37) in order to get an equation for A:

$$A \frac{\omega_0^2 - \omega^2}{\sqrt{(\omega_0^2 - \omega^2)^2 + 4\omega_0^2 \omega^2 \zeta^2}} \left(\omega_0^2 - \omega^2 + \frac{4\omega_0^2 \omega^2 \zeta^2}{\omega_0^2 - \omega^2} \right) = \frac{F_0}{m} \qquad (6.39)$$

which we can solve to get:

$$A = \frac{F_0}{m} \frac{1}{\sqrt{(\omega_0^2 - \omega^2)^2 + 4\omega_0^2 \omega^2 \zeta^2}} \qquad (6.40)$$

6.3.2 Complex Solution

For the solution using imaginary exponentials, we return to our equation of motion written using an imaginary exponential for the driving force (6.26). As we previously noted this suggests a solution of the form $Ze^{i\omega t}$ but here we will rewrite our complex amplitude Z to get a suggested solution of the form $Ae^{i(\omega t - \phi)}$ where A is the amplitude and ϕ is the phase difference between the response of the system and the driving force.

Now we start by calculating the derivatives:

$$x = Ae^{i(\omega t - \phi)} \qquad (6.41)$$
$$\dot{x} = i\omega Ae^{i(\omega t - \phi)}$$
$$\ddot{x} = -\omega^2 Ae^{i(\omega t - \phi)}$$

Next, we substitute these into our equation of motion (6.25) and then find values for A and ϕ:

$$-\omega^2 Ae^{i(\omega t - \phi)} + 2\zeta \omega_0 . i\omega Ae^{i(\omega t - \phi)} + \omega_0^2 . Ae^{i(\omega t - \phi)} = \frac{F_0}{m} e^{i\omega t} \qquad (6.42)$$

Every term contains $e^{i\omega t}$ and so we can cancel this from the equation and rearrange to get:

$$\left(-\omega^2 + 2i\omega_0\omega\zeta + \omega_0^2\right)Ae^{-i\phi} = \frac{F_0}{m}$$

$$\left(-\omega^2 + 2i\omega_0\omega\zeta + \omega_0^2\right)A = \frac{F_0}{m}e^{i\phi}$$

$$\left(-\omega^2 + 2i\omega_0\omega\zeta + \omega_0^2\right)A = \frac{F_0}{m}\left(\cos\phi + i\sin\phi\right)$$

(6.43)

Looking at (6.43) we have a complex number of both sides of the equation. For this to be equal both the real and imaginary parts must each be equal separately. Hence, since we only have real constants present we can split this single equation into two separate equations by equating the real and imaginary parts independently to get:

$$A(\omega_0^2 - \omega^2) = \frac{F_0}{m}\cos\phi \qquad (6.44)$$

$$2\omega_0\omega\zeta A = \frac{F_0}{m}\sin\phi \qquad (6.45)$$

To eliminate ϕ we need to square and add both (6.44) and (6.45) since $\sin^2\phi + \cos^2\phi = 1$. This gives:

$$A^2(\omega_0^2 - \omega^2)^2 + 4\omega_0^2\omega^2\zeta^2A^2 = \frac{F_0^2}{m^2}$$

$$\implies A = \frac{F_0}{m}\frac{1}{\sqrt{(\omega_0^2 - \omega^2)^2 + 4\omega_0^2\omega^2\zeta^2}} \qquad (6.46)$$

To find ϕ we divide (6.45) by (6.44) to get the tangent of ϕ and then just take the inverse:

$$\phi = \tan^{-1}\left(\frac{2\omega_0\omega\zeta}{\omega_0^2 - \omega^2}\right) \qquad (6.47)$$

So clearly the function, $x = Ae^{i(\omega t + \phi)}$, is a valid solution to our equation of motion for a driven oscillator (6.25) for the values of A and ϕ given in (6.46) and (6.47) respectively which match precisely the values obtained via the trigonometric approach shown in section 6.3.1. However, the approach we used here was much shorter, had a far clearer rationale for the choice of initial function and avoided the need to remember multiple trig formulae. Despite the name using complex numbers can actually result in maths which is far less complex!

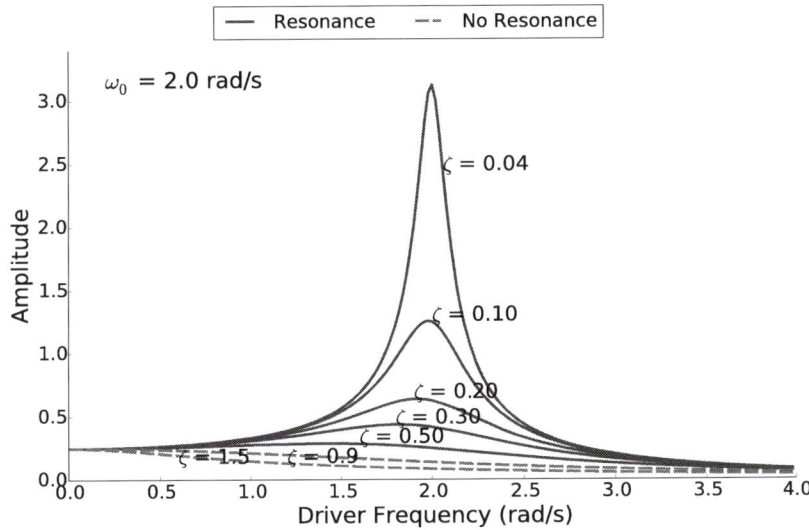

Resonance occurs
when driver frequency
close to natural frequency.

Fig. 6.7: Amplitude for a driven oscillator with a natural frequency of $2\,\mathrm{rad\,s^{-1}}$ as a function of driver frequency. Various damping ratios are shown with those lines which are dashed not exhibiting any resonance.

to find max resonance,
find $\dfrac{dA}{d\omega} = 0$

6.3.3 Resonance

Now that we have a solution for our driven oscillator we need to look at the result. The first thing to notice is that the amplitude of the motion varies as a function of the driving frequency, ω, and has a maximum value. To find this maximum value we first differentiate the amplitude given by (6.40) with respect to the angular frequency, ω:

$$\frac{\mathrm{d}A}{\mathrm{d}\omega} = \frac{-2\left[\omega(\omega^2 - \omega_0^2) + 2\omega\omega_0^2\zeta^2\right]}{\left[(\omega_0^2 - \omega^2)^2 + 4\omega_0^2\omega^2\zeta^2\right]^{3/2}} \qquad (6.48)$$

At the maximum value of the amplitude the rate of change of amplitude with respect to angular frequency will be zero. Hence, if the value of the driving angular frequency which gives the maximum amplitude is ω_r then we have:

$$\frac{\mathrm{d}A}{\mathrm{d}\omega} = 0 \qquad (6.49)$$

$$-2\left[\omega_r(\omega_r^2 - \omega_0^2) + 2\omega_r\omega_0^2\zeta^2\right] = 0$$

$$\omega_r^2 - \omega_0^2 + 2\omega_0\zeta^2 = 0$$

$$\omega_r^2 = \omega_0^2(1 - 2\zeta^2)$$

$$\omega_r = \omega_0\sqrt{1 - 2\zeta^2} \qquad (6.50)$$

when ζ is small $\omega_r \approx \omega_0$

ζ must be $< \dfrac{1}{\sqrt{2}}$

This is called the resonant frequency and when the system is driven at this frequency the amplitude of the system in response to the driving force is at its maximum value. This effect, where there is a maximum amplitude response of the system,

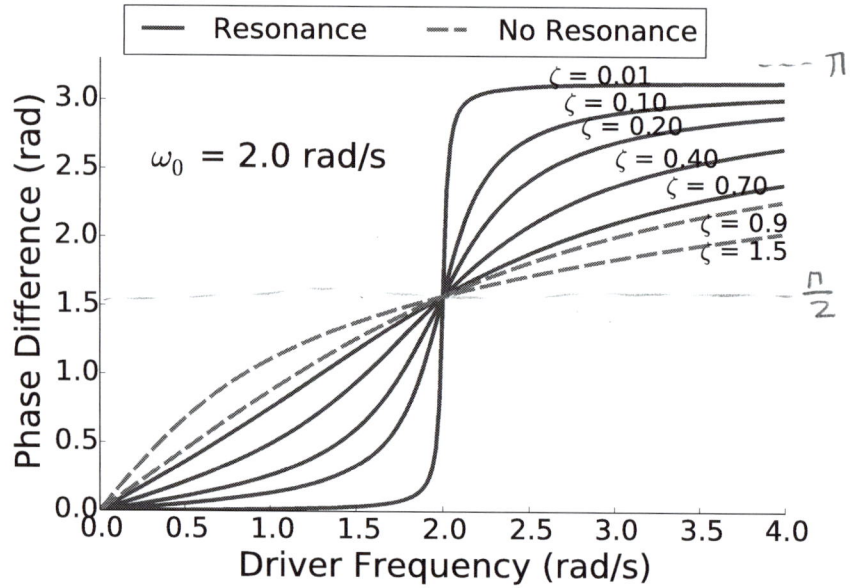

Fig. 6.8: Phase difference between the driving force and the displacement for various damping ratios and a fixed natural frequency of 2 rad/s. All systems have a phase lag of $\pi/2$ at the natural, undamped oscillation frequency. Those which also have a resonance also have a point of inflection at this frequency with less damping corresponding to a sharper phase switch as the drive frequency passes through the natural frequency.

$$\phi = \tan^{-1}\left(\frac{2\zeta\omega_0\omega}{\omega_0^2 - \omega^2}\right)$$

phase difference between driving force and displacement

rapid Δphase for minimal damping, slower change for more damping

① very low driver frequency ↳ ~ in phase

② $\omega = \omega_0$ $\tan^{-1}("\infty")$ $\theta = \frac{\pi}{2}$

③ $\omega \to \infty$ $\tan^{-1}(-01)$ $\theta = \pi$

is called resonance. Looking at the expression for the resonant frequency given in equation (6.50) there will only be a real solution if $\zeta < 1/\sqrt{2}$ and so only systems which are very under damped will exhibit resonance (recall that the condition for a system to be under damped is just $\zeta < 1$, the requirement for resonance is lower). Figure 6.7 shows how the amplitude varies with frequency for different damping ratios. The smaller the damping the greater the amplitude at resonance and the closer the resonance is to the natural frequency of oscillation. Note that at all times the system moves with the frequency of the driver and not at the natural frequency. However, when the driving force has a frequency close to the natural frequency, the amplitude of the motion will be larger.

Next, we consider the phase lag of the motion. Since ϕ is not always zero there will be a phase difference between the driving force and the displacement of the mass and this phase difference depends on the relative values of the natural and drive frequencies. Figure 6.8 shows how the phase difference varies with the drive frequency. It varies between 0 for $\omega = 0$ to π in the limit that $\omega \to \infty$ and when $\omega = \omega_0$ the phase difference is exactly $\pi/2$.

6.4 Transient Solution

In the previous section we solved the equation of motion of a driven, damped harmonic oscillator but we did not get the entire solution. We know this because we derived exact solutions for both the phase and amplitude of the motion without any unknown constants arising. However, since we solved a second order differential equation we must end up with two constants of integration. We last saw this happen when solving the case of a critically damped oscillator when we initially ended up with only a single constant of integration which showed that we were missing a possible solution to the equation. Here the problem is the same: clearly we have not got the entire solution to the driven, damped harmonic oscillator.

To see where we missed something let's return to our equation of motion for the driven oscillator, (6.25):

$$\ddot{x} + 2\zeta\omega_0\dot{x} + \omega_0^2 x = \frac{F_0}{m}\cos(\omega t) \qquad (6.51)$$

Now if we call the solution which we derived in section 6.3 $S(t)$ when we put $S(t)$ into the left hand side of the equation we will end up with $\frac{F_0}{m}\cos(\omega t)$. However, let us now consider the equation of motion for the undriven, damped harmonic oscillator. In this case, we have an equation of motion (6.3):

$$\ddot{x} + 2\zeta\omega_0\dot{x} + \omega_0^2 x = 0 \qquad (6.52)$$

We derived the different solutions to this equation of motion in section 6.1 so let's call the solution $T(t)$ where this function will be the under, over or critically damped solution depending on the value of the damping ratio, ζ. Clearly if we put $T(t)$ into the above equation we will end up with zero.

So far we have just reviewed the solutions we obtained for the driven and damped oscillators. Now we want to consider what happens when we add them together. Consider a function of the form:

$$X(t) = S(t) + T(t) \qquad (6.53)$$

If we put this into our equation of motion for the driven oscillator, (6.51), and remember that $S(t)$ is a solution of (6.51) and $T(t)$ is a solution of (6.52) then we get:

$$\ddot{X} + 2\zeta\omega_0\dot{X} + \omega_0^2 X = \underbrace{\ddot{S} + 2\zeta\omega_0\dot{S} + \omega_0^2 S}_{=\frac{F_0}{m}\cos(\omega t)} + \underbrace{\ddot{T} + 2\zeta\omega_0\dot{T} + \omega_0^2 T}_{=0}$$

$$= \frac{F_0}{m}\cos(\omega t) \qquad (6.54)$$

Hence $X(t)$ is also a solution to the driven oscillator equation of motion. This is where our constants of integration are hiding. If we examine the solutions to the damped harmonic oscillator we see that each one has two constants which have to be determined by the initial conditions of the oscillator being considered.

Having now found a complete solution to the driven, harmonic oscillator let's consider the characteristics of this new combined solution. If we look at the function $T(t)$ it will always exponentially decay with time regardless of whether it is the under, over or critically damped solution. This means that a long time after the start of the motion the value of $T(t)$ will be vanishingly small and the solution will be entirely dominated by $S(t)$. Hence we call $S(t)$ the steady-state solution and $T(t)$ the transient solution. In general, both are required to describe the motion of a driven oscillator but, a long time after there has been any change to the motion the steady-state solution will be the only component remaining.

Problems

Q6.1: A simple pendulum with a length of 1 m and a damping ratio, $\zeta = 0.04$, is set in motion. How long does it take the amplitude of the oscillations to decay to half of their initial amplitude? [$g=9.81\,\mathrm{m\,s^{-2}}$]

Q6.2: A student calculates that under ideal conditions, without any damping, a particular mass-spring system should oscillate at a frequency of 2 Hz. However when an experiment is performed it is determined that the system oscillates with a frequency of 1.5 Hz and with an exponentially decreasing amplitude.
(a) What is the damping ratio of the experimental system?
(b) How long does it take for the system's amplitude to decay to half of its initial value?

Q6.3: A damped mass-spring system, with mass m and spring constant k, has the damping tuned such that when the mass is disturbed from equilibrium it returns to the equilibrium position and remains there in the fastest possible time. Initially the mass is displaced a distance a from the equilibrium position and released at rest. Starting with the general equation of motion for a damped harmonic oscillator derive an expression for the displacement from the equilibrium position, x, of the mass as a function of time after the mass is released, t using only the quantities provided.

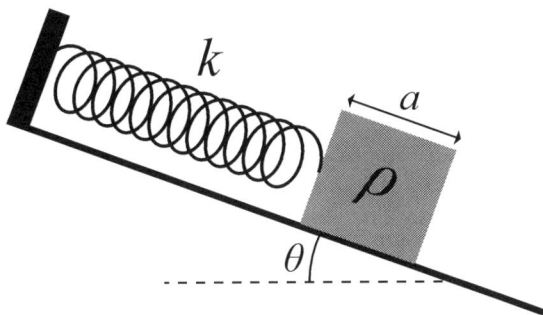

Fig. P6.1: *Cube with sides of length a shown resting on a slope inclined at an angle θ. The spring exerts a force on the cube which is parallel to the slope.*

Q6.4: A cube with sides of length a and a density ρ is attached to the end of a light spring with a constant k. The cube is then placed on a surface which is sloped at an angle θ to the horizontal and the other end of the spring is fixed such that the spring exerts a force on the cube which is parallel to the direction of the maximum slope as shown in figure P6.1. The gravitational field is g.
(a) If the surface is frictionless derive an expression for the angular frequency of the oscillation in terms of the quantities given.
(b) Unable to find a frictionless surface a physicist instead lubricates the surface of the slope with a uniform layer of oil which has a viscosity η. The cube rests on the oil layer so that its surface is parallel to the sloping surface. Given that the oil layer maintains a constant uniform thickness as the cube oscillates how thick does the layer of oil need to be to ensure that the system is critically damped?

Q6.5: The shock absorbers and suspension of a car mean that it can be treated as a damped mass-spring system with a mass of 1,500 kg, a spring constant of 20,000 N m^{-1} and a damping coefficient of 2,000 kg s^{-1}. When the car drives over a pothole it starts to bounce up and down with some amplitude.
(a) How long will it take for the amplitude of the oscillations to halve?
(b) The strength of the shock absorbers can be adjusted to increase or decrease the damping coefficient. What value of the damping coefficient will cause the car to return, and remain, at its equilibrium position in the fastest possible time?

Q6.6: A mass of 2 kg is attached to the end of a vertical spring with a constant of 200 N m^{-1}. The system is placed under water and is found to oscillate with a decaying amplitude and an angular frequency of 9.8 Hz. While still submerged a small, oscillating driving force is applied to the top of the spring with an angular frequency, ω, which can be varied, while the amplitude remains constant. Assuming that plenty of time is ·

allowed to pass after any change in the driving frequency what value of ω will give the largest amplitude oscillation of the mass?

Q6.7: A damped, vertical mass-spring system is connected to a small motor which drives the system at a variable frequency ω and constant amplitude. It is observed that at an angular frequency of $10\,\mathrm{rad\,s^{-1}}$ the phase difference between the motor and the mass is exactly $\pi/2$. It is also found that when the motor is turned off and the system is disturbed from equilibrium it takes 1 s for the initial amplitude to reduce to half of its initial value.

(a) What is the value of the damping ratio?

(b) The motor is now turned on and scanned through a range of different angular frequencies. In each case, the motor is set to a specific frequency and then the system is left for a while before the amplitude of the mass' oscillation is measured. At what angular frequency will the amplitude of the mass be largest?

(c) Why it is important to wait after changing the frequency of the motor before measuring the amplitude of the system?

tem is:

$$\ddot{x} + \frac{\eta a^2}{mh}\dot{x} + \frac{k}{m}x = 0$$

(b) Derive expressions for both the natural, angular frequency, ω_0, of the undamped system and the damping ratio, ζ, in terms of the quantities given above.

(c) For a cube with a total mass of $3\,\mathrm{kg}$ and sides of length $10\,\mathrm{cm}$ it is found that it returns to the equilibrium point in the fastest possible time without oscillating when the thickness of the oil layer is $100\,\mathrm{\mu m}$ and the spring constant is $300\,\mathrm{N\,m^{-1}}$. What is the dynamic viscosity of the oil?

(d) A system with the values of parameters given above is now attached to a motor which drives the system at an angular frequency which can be varied. What is the minimum thickness of oil required for this system to produce a resonance?

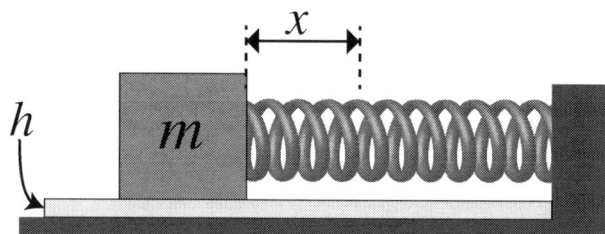

Fig. P6.2: *A cube of mass m and sides of length a is attached to a spring with constant k and rests on a thin layer of oil.*

Q6.8: A cube of mass, m, is attached to a spring with a spring constant of k. The block rests on a flat surface but, in order to reduce the friction, a thin layer of liquid oil with constant thickness h is placed between the block and the surface as shown in figure P6.2. The block is then displaced from the equilibrium position and released. The dynamic viscosity of the oil is η and the length of one side of the cube is a.

(a) Show that the equation of motion for the sys-

CHAPTER 7

Waves

A wave is a disturbance that propagates through a medium of connected oscillators. Waves can transmit energy, momentum and information from one location to another. Mechanical waves require a physical medium, such as a fluid or solid, where the components of the medium vibrate. These are the types of wave which we will discuss here. However, waves, such as electromagnetic waves, do not require a physical medium but instead are an excitation of a fundamental field of nature which is why light can travel through a vacuum where there is no physical medium. More recently the discovery of gravitational waves by the LIGO observatory in 2016 shows that even space-time itself can oscillate and transmit a wave although the energies required to make a wave large enough to detect require colliding black holes and neutron stars!

7.1 Wave Properties

The medium of a mechanical wave can be thought of as a system which has a tiny oscillator at each point of space. Each of these oscillates with different phases to create the wave. As a result, waves have properties which we have already seen for oscillators. However, unlike a single oscillator, waves extend through space on one or more dimensions and this gives rise to properties that single oscillators do not have. In addition parameters, like velocity, can have different meanings when describing waves instead of single oscillators.

7.1.1 Amplitude

This property matches precisely the definition for a single oscillator. The amplitude of a wave is simply the maximum displacement of the wave medium from the equilibrium position. However, since the wave occupies a region of space the amplitude is a function of both position and time.

Just as with oscillators, the units of the amplitude depends on the type of the wave. For mechanical waves, it is a length and so measured in metres. However, for electromagnetic or gravitational waves the units will be the units of the associated field strength.

7.1.2 Frequency

Just like the single oscillator, the frequency of a wave is the number of complete oscillations per second that a single point on the wave will make in a second. The SI units are hertz (Hz) which is an inverse second (s^{-1}).

Unlike an oscillator, waves are very rarely described as having a period and are almost exclusively described using frequency. Often this will require large SI prefixes, for example, microwaves have a frequency in the GHz range (10^9 Hz) and the near infrared waves used to penetrate your clothes and scan your body at the airport are in the THz range (10^{12} Hz).

Waves also have an *angular frequency*, ω, which is defined in the same way as that for an oscillator:

$$\text{Angular Frequency, } \omega = 2\pi f \qquad (7.1)$$

where f is the frequency. It has SI units of radians per second (rad/s).

7.1.3 Wavelength and Wavenumber

The shortest distance between two points on the wave which are vibrating with the same phase is called the wavelength which is conventionally denoted with the greek lower case letter lambda, λ. This is a length measurement and so has SI units of metres. Often wavelength is defined as the distance between two adjacent crests. This is consistent with the definition but any two points with the same phase will suffice.

In addition to the wavelength another commonly occurring quantity is the wavenumber which is defined as:

$$\text{Wave Number, } k = \frac{2\pi}{\lambda} \qquad (7.2)$$

This is to the wavelength what angular frequency is to the frequency. It has SI units of radians per metre (rad/m) although often the radians are dropped (since they are dimensionless) and the units used are simply inverse metres (m^{-1}).

7.1.4 Velocities

The simplest type of wave velocity is called the phase velocity. It is defined as the velocity of a point of constant phase, for example, the speed that a wave crest moves with. Consider the wave shown in figure 7.1. In the time it takes the crest to move the horizontal distance between two crests, i.e. the wavelength, the medium at the destination has had sufficient time to complete one cycle of its motion.

The time taken for the crest to travel a distance of λ is simply:

$$\text{Time, } t = \frac{\lambda}{c} \qquad (7.3)$$

where c is the phase velocity of the wave. The time taken to complete one cycle of the wave oscillation is just:

$$\text{Time, } t = \frac{1}{f} \qquad (7.4)$$

where f is the frequency of the wave. Since the times shown in both (7.3) and (7.4) must be the same we get what is sometimes referred to as the wave equation by non-physicists:

$$c = f\lambda \qquad (7.5)$$

However, as we shall see in section 7.4, what physicists call the wave equation is something rather more complicated and whose solutions are functions which describe waves.

In addition to the phase velocity waves may also have a *group velocity* which is the speed at which a change in amplitude will propagate. For an ideal wave medium both the phase and group velocities are the same but for waves in real media, this is not the case. We will discuss some of the phenomena which this causes for light waves later in section 11.2.

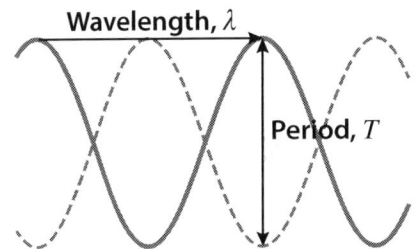

Fig. 7.1: Diagram of a wave. In the time it takes for the crest to move along the horizontal arrow shown the oscillator at the end point must complete one entire cycle of motion.

Transverse Wave

Longitudinal Wave

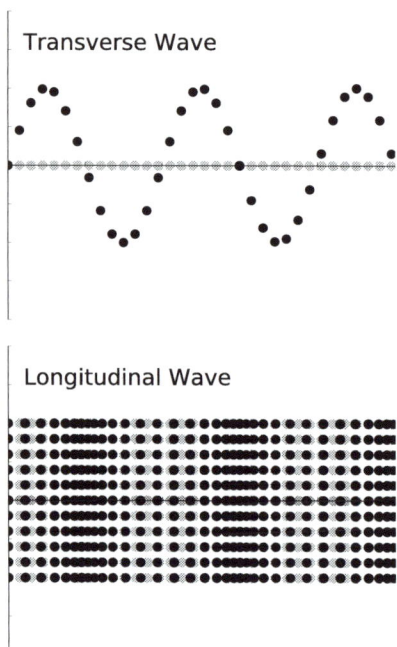

Fig. 7.2: The two basic types of mechanical wave shown as a snapshot of two waves each moving from left to right. Transverse waves (top) displace the medium perpendicular to the direction of motion of the wave while longitudinal waves (bottom) displace the medium in the direction of motion.

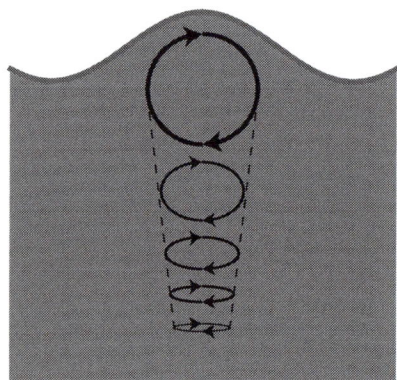

Fig. 7.3: Motion of water molecules in a surface water wave which moves from left to right. The circular motion is a combination of transverse and longitudinal components and the transverse, vertical component is suppressed with increasing depth.

7.2 Types of Waves

Mechanical waves have two basic types depending on the direction of the displacement of the medium with respect to the wave velocity. If the particle displacement is perpendicular to the wave this is called a *transverse wave*. Alternatively, if the displacement is parallel to the direction of the wave this is called a *longitudinal wave*. Both these wave types are shown in figure 7.2.

An example of a transverse wave is a vibrating string whereas sound waves are longitudinal waves. This is because fluids, such as air, do not support shear stresses which means that the motion of one particle does not cause a neighbour particle to move in the same direction unless there is a collision involved. The result is that transverse waves in fluids are typically quickly attenuated whereas longitudinal pressure waves are not. This is how we know that the Earth's core is liquid: it only transmits longitudinal waves, not transverse waves.

Some waves are both transverse and longitudinal at the same time. A very common example of this is a surface water wave. In these waves, the particles undergo a circular motion and so have both a transverse and longitudinal displacement as shown in figure 7.3. For this reason, water waves are quite complex to describe mathematically and so despite these being an incredibly common type of wave we will not be discussing them in detail.

7.3 Wave Functions

With an oscillator, we had a function which described its displacement as a function of time. However, with a wave, we need a function which will describe the displacement of the wave's medium as a function of both time and position. To keep things as simple as possible we will stick to one-dimensional waves and which limits the function to two variables: position (x) and time (t).

When we studied the oscillator we started by deriving an equation of motion for the system and then we solved this equation to find the function which described the displacement of the oscillator as a function of time. The same approach can be taken with waves but is more mathematically challenging because the presence of two variables means that we will have a partial differential equation of motion to solve. The full mathematical details behind deriving and then solving this equation are given in appendix A where section A.1 describes how the equation is derived and section A.2 shows how it is solved. However here we will take a simpler, less mathematically rigorous, approach by considering a wave as a system of connected oscillators and deriving a function to describe it before then showing that this function is the solution of a more general equation.

For a single oscillator we had a solution for the displacement which was a simple cosine function. Hence at a fixed point in the medium where we can treat it as a simple oscillator the displacement will be given by (5.14):

$$\psi(t) = A\cos(\omega t + \xi) \tag{7.6}$$

where ψ is the displacement and ξ is the initial phase. If we take a picture of a wave at time $t = 0$ we can see that each point of the wave has a different initial phase which depends on its position, x (see figure 7.4). Hence in this case ξ is not a constant but a function of the position in the medium i.e. $\xi(x)$. Using the definition of the wavelength we know that after a displacement of $+\lambda$ in x we have a phase change of -2π for a wave travelling in the positive x direction: the phase at higher values of x will be behind that at lower values because at lower values of x the medium will have undergone more oscillations. If we assume that the phase changes linearly with position we have a relationship between position and initial phase which is:

$$\xi(x) = -\frac{2\pi}{\lambda}x - \phi = -kx - \phi \tag{7.7}$$

where ϕ is a constant and k is the wavenumber defined in (7.2). From (7.7) it is easy to see that $\xi(x + n\lambda) = \xi(x) - 2n\pi$ for any integer n as we require. All that remains is to substitute this back into (7.6) and we have:

$$\Psi(x, t) = A\cos(\omega t - kx - \phi) \tag{7.8}$$

We can now flip the sign because cosine is an even function, $\cos(x) = \cos(-x)$ which gives:

$$\Psi_\rightarrow(x, t) = A\cos(kx - \omega t + \phi) \tag{7.9}$$

for a wave moving in the positive x direction where A is the amplitude and ϕ is constant which gives the initial phase at $x = 0$ and $t = 0$.

If we consider a wave propagating in the negative x direction then after a displacement of $+\lambda$ in x we have a phase change

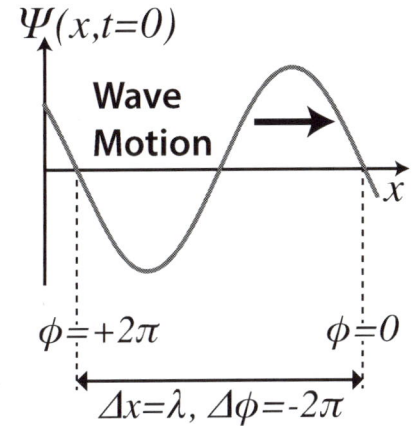

Fig. 7.4: Picture of a one-dimensional wave at time $t = 0$ showing that the initial phase of the wave, $\xi(x)$, varies as a function of position. For a wave moving left to right, as shown, a change in position of $+\lambda$ corresponds to a phase change of -2π.

of $+2\pi$ because now higher values of x are ahead in phase compared to lower values. This means that the sign of the x term above will be positive which implies that the kx and ωt terms will have the same sign to get:

$$\Psi_{\leftarrow}(x, t) = A\cos(kx + \omega t + \phi) \tag{7.10}$$

for a wave moving in the negative x direction.

These functions, which describe the displacement due to a wave, are called wave functions. Here we will stick to this method of writing them down but it is important to note that this is a convention that is not unique. Firstly, just as with the oscillator, the functions can be written as a sine function instead because the only difference between sine and cosine is a constant phase offset which can be accommodated by changing the value of ϕ by $\pi/2$. Secondly, as we saw in (7.8) only the relative signs of the kx and ωt terms are important. Hence we could rewrite the wave function for a wave moving in the positive x direction as:

$$\Psi_{\rightarrow}(x, t) = A\sin(\omega t - kx + \phi') \tag{7.11}$$

The choice of convention to use is purely arbitrary. The important distinction is that waves where the kx and ωt terms have the same sign propagate in the negative x direction and those where the terms have opposite signs propagate in the positive x direction.

Lastly if instead of starting with the cosine form of the solution for the oscillator we had used the complex solution (5.47) from section 5.3 we would have an imaginary exponential function:

$$\Psi_{\rightarrow}(x, t) = Ae^{i(kx - \omega t + \phi)} \quad \text{and} \quad \Psi_{\leftarrow}(x, t) = Ae^{i(kx + \omega t + \phi)} \tag{7.12}$$

As with the complex oscillator solution there is an implied real operator in both the functions above because the function's value is the physical displacement of the wave medium as a function of both position, x, and time, t. This form of representing a wave is very common in physics and is actually essential when dealing with quantum mechanical waves as we will see in section 12.6.

7.4 The Wave Equation

Now that we know about partial derivatives we need to use them to find a differential equation whose solutions are wave functions. For the oscillator we had a second order differential equation so we will start with second order differentials. However for a wave we have two variables so we will first calculate the second order partial differentials for x and t. Starting with either (7.9) or (7.12) this gives:

$$\frac{\partial^2 \Psi}{\partial x^2} = -k^2 \Psi \quad \text{and} \quad \frac{\partial^2 \Psi}{\partial t^2} = -\omega^2 \Psi \qquad (7.13)$$

Combining these we get:

$$\frac{1}{k^2}\frac{\partial^2 \Psi}{\partial x^2} = \frac{1}{\omega^2}\frac{\partial^2 \Psi}{\partial t^2} \implies \frac{\partial^2 \Psi}{\partial x^2} = \frac{k^2}{\omega^2}\frac{\partial^2 \Psi}{\partial t^2} \qquad (7.14)$$

Now we need to remember our definition for k and ω:

$$k = \frac{2\pi}{\lambda} \quad \text{and} \quad \omega = 2\pi f \implies \frac{k}{\omega} = \frac{1}{f\lambda} = \frac{1}{c} \qquad (7.15)$$

where c is the phase velocity. Putting this back into our partial differential equation (7.14) we get:

$$\frac{\partial^2 \Psi}{\partial x^2} = \frac{1}{c^2}\frac{\partial^2 \Psi}{\partial t^2} \qquad (7.16)$$

This is the wave equation and the functions, $\Psi(x,t)$, which are solutions of this equation are waves. Although we derived this by working backwards from a sinusoidal wave function this is a general result true for any propagating wave and it can be easily shown that the wave function for a wave travelling in the negative x direction, (7.10), is also a solution.

Appendix A covers the completely general derivation of (7.16). Solving the equation to derive the allowed wave functions is also shown in this appendix because it requires slightly more knowledge of partial derivates than was covered in chapter 1. However, like the oscillator, the completely general solution to the equation are four imaginary exponential functions which, just as we found with the oscillator, pair up when we require that the function be real for all x and t to give two cosine functions (7.9) and (7.10): one for each possible direction of motion of the wave.

7.5 Waves on a String

Now that we have a mathematical treatment for waves we can start to look at real, physical waves. The simplest example of this is a wave on a string. Consider a string with tension T when undisturbed and mass per unit length ρ and take a small length, δx of the string as shown in figure 7.5.

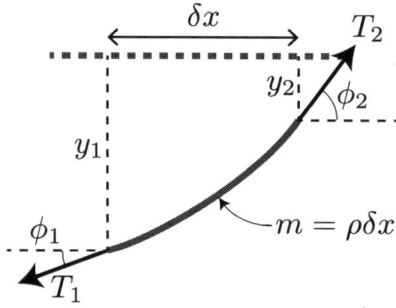

Fig. 7.5: Element of a string, with mass per unit length ρ, which is displaced from the equilibrium position by a wave. For small amplitude displacement the tension in the string remains a constant T.

To keep things simple we will make the assumption that we are only dealing with small amplitude, transverse waves on the string. In this case both the angles ϕ_1 and ϕ_2 can be considered small. Applying Newton's second law to the element of the string shown in figure 7.5 in the x direction we get:

$$T_{1x} = T_1 \cos \phi_1 \approx T \qquad (7.17)$$
$$T_{2x} = T_2 \cos \phi_2 \approx T \qquad (7.18)$$

where we have used the small angle approximation for cosines (1.6) and also assumed that the tension in the string is constant which will be the case for small amplitude waves. Next we apply Newton's second law in the y direction to get:

$$T_2 \sin \phi_2 - T_1 \sin \phi_1 = \rho \delta x \frac{\partial^2 y}{\partial t^2} \qquad (7.19)$$

where the acceleration of the element in the y direction is a partial derivative because the x value is constant and y is a function of both x and t. Now divide (7.19) through by T and substitute for T using equations (7.17) and (7.18) to get:

$$\frac{\rho}{T} \delta x \frac{\partial^2 y}{\partial t^2} = \frac{T_2 \sin \phi_2}{T_2 \cos \phi_2} - \frac{T_1 \sin \phi_1}{T_1 \cos \phi_1} \qquad (7.20)$$
$$= \tan \phi_2 - \tan \phi_1 \qquad (7.21)$$

Now the tangents of ϕ_1 and ϕ_2 are related to the gradient of the string at a fixed instant in time:

$$\tan \phi_1 = \left.\frac{\partial y}{\partial x}\right|^{x} \quad \text{and} \quad \tan \phi_2 = \left.\frac{\partial y}{\partial x}\right|^{x+\delta x} \qquad (7.22)$$

where the vertical bar notation is used to specify the value of x at which the partial derivative should be evaluated. Putting these into (7.21) gives:

$$\frac{\rho}{T} \delta x \frac{\partial^2 y}{\partial t^2} = \left.\frac{\partial y}{\partial x}\right|^{x+\delta x} - \left.\frac{\partial y}{\partial x}\right|^{x} \qquad (7.23)$$
$$\frac{\rho}{T} \frac{\partial^2 y}{\partial t^2} = \frac{1}{\delta x}\left(\left.\frac{\partial y}{\partial x}\right|^{x+\delta x} - \left.\frac{\partial y}{\partial x}\right|^{x}\right) \qquad (7.24)$$

The righthand side of the equation is just the change in the value $\partial y / \partial x$ between x and $x + \delta x$. Hence if we define $f(x, t) = \partial y / \partial x$ then we can rewrite the righthand side of (7.24) as:

$$\frac{1}{\delta x}\left(\left.\frac{\partial y}{\partial x}\right|^{x+\delta x} - \left.\frac{\partial y}{\partial x}\right|^{x}\right) = \frac{1}{\delta x}[f(x + \delta x, t) - f(x, t)] \qquad (7.25)$$

In the limit that $\delta x \to 0$ this simply becomes a partial differential of f and all we then need to do is substitute back in our definition of f:

$$\lim_{\delta x \to 0} \frac{1}{\delta x}[f(x + \delta x, t) - f(x,t)] = \frac{\partial f}{\partial x} = \frac{\partial}{\partial x}\left(\frac{\partial y}{\partial x}\right) = \frac{\partial^2 y}{\partial x^2} \tag{7.26}$$

Putting this back into (7.24) gives:

$$\frac{\rho}{T}\frac{\partial^2 y}{\partial t^2} = \frac{\partial^2 y}{\partial x^2} \tag{7.27}$$

This is the wave equation for waves on the string. To determine the phase velocity of the wave we simply make a direct comparison with the general wave equation (7.16):

$$\text{String Equation}: \quad \frac{\partial^2 \Psi}{\partial x^2} = \frac{1}{c^2}\frac{\partial^2 \Psi}{\partial t^2}$$
$$\updownarrow \quad \updownarrow \quad \updownarrow$$
$$\text{Wave Equation}: \quad \frac{\partial^2 y}{\partial x^2} = \frac{\rho}{T}\frac{\partial^2 y}{\partial t^2} \tag{7.28}$$

Hence we can see that the phase velocity of the wave on a string is:

$$\text{Phase Velocity, } c = \sqrt{\frac{T}{\rho}} \tag{7.29}$$

This shows that the waves on a string move faster as the tension is increased and slower as the string's mass per unit length is increased.

7.6 Acoustic Waves

Acoustic waves are longitudinal waves which travel through a medium causing it to undergo adiabatic compression and expansion where the term 'adiabatic' means without the transfer of heat or mass to the surroundings. The sound waves in air which we hear are the most common example of an acoustic wave but so too are the waves emitted by earthquakes or even the plasma pressure waves observed on the sun.

Since the wave is compressing and expanding the bulk volume of a medium we need to switch from using mass and force to using pressure and density as defined in sections 3.2 and 3.1 respectively. These are related through a property of the material called the bulk modulus which we discussed in section 2.3.3.

Using these quantities we can apply Newton's second law to an element of the medium and derive the wave equation for acoustic waves. Like any wave, we can describe an acoustic wave in terms of the displacement of medium. However, for acoustic waves in a fluid, we also have the option of expressing them as a pressure wave. The full derivation of the pressure wave equation quite complicated and requires a good degree of familiarity with partial derivatives. The full details are given in appendix section A.3.1 and the resulting equation in one dimension (A.47) is:

$$\frac{\partial^2 p}{\partial x^2} = \frac{\rho}{B}\frac{\partial^2 p}{\partial t^2} \tag{7.30}$$

where $p(x, t)$ is the deviation of the pressure from that of the undisturbed medium and B is the bulk modulus of the medium as defined in (2.15).

Similarly we can derive a wave equation in terms of the displacement of the medium from equilibrium, $\psi(x, t)$. The details of this derivation are also not trivial and the full details are given in appendix section A.3.3. The resulting equation is (A.59):

$$\frac{\partial^2 \psi}{\partial x^2} = \frac{\rho}{B}\frac{\partial^2 \psi}{\partial t^2} \tag{7.31}$$

Both (7.30) and (7.31) agree that the phase velocity for acoustic waves, c, is:

$$c = \sqrt{\frac{B}{\rho}} \tag{7.32}$$

This means that for denser media acoustic waves are slower and they are faster for media with higher bulk moduli. To check this result we can put in the numbers for air which has an adiabatic bulk modulus, $B = 142\,\text{kN m}^{-2}$, and a density $\rho = 1.2\,\text{kg m}^{-3}$. This gives an acoustic wave velocity in air of:

$$c = \sqrt{\frac{142 \times 10^3}{1.2}} = 344\,\text{m s}^{-1}$$

which agrees with the typically quoted value of $340\,\text{m s}^{-1}$ for the speed of sound in air at room temperature and pressure.

While both of these wave equations describe the same acoustic waves it is interesting to note that there is a phase difference between the pressure deviation and the displacement of the medium. Again from the derivations given in appendix A we have a relationship between the pressure deviation and the displacement of the medium which is (A.61):

$$p(x, t) = -B\frac{\partial \psi}{\partial x} \tag{7.33}$$

Since $\psi(x, t)$ is a cosine function, (7.9), differentiation gives a negative sine function which introduces phase difference of $\pi/2$

between the pressure and displacement in the same manner that the displacement and velocity for a simple harmonic oscillator are $\pi/2$ out of phase because the velocity is the differential of the displacement cosine function. This means that the points of maximum and minimum pressure correspond to the points of zero displacement which is not what you might have intuitively expected.

We can also use equation (7.33) to find the relationship between the pressure and displacement amplitudes. For a displacement wave we have a wave function:

$$\psi(x, t) = A_\psi \cos(kx - \omega t + \phi) \tag{7.34}$$

where A_ψ is the amplitude of the displacement wave. Now we can substitute this wave function into equation (7.33) to get:

$$p(x, t) = BkA_\psi \sin(kx - \omega t + \phi) \tag{7.35}$$

and so the amplitude of the pressure wave is simply:

$$A_p = BkA_\psi = c\rho\omega A_\psi \tag{7.36}$$

where we use (7.32) to replace the bulk modulus, B, and the relationship between wave speed, frequency and wavenumber, $c = \omega/k$, to rearrange the expression.

7.7 Wave Power

All waves represent a disturbance from the equilibrium conditions of a system and so must contain energy which, since the wave is moving, is carried by the wave to another location. For a one-dimensional wave, the power of the wave is the instantaneous rate at which the wave transports energy.

To calculate the rate at which a wave on a string transmits energy consider the small string element shown in figure 7.6. This shows an element of a small amplitude string wave. For a wave moving towards the right, as shown in figure 7.6, the force on the lefthand end of the element is doing work on the element which is why it continues to oscillate. Hence the power of the wave is just the power of this force.

The displacement of the string element is entirely transverse to the wave's direction of motion and so only the transverse y component of the tension is relevant for power transmission. The power of this force is just:

$$\text{Power}, P = T_y v \tag{7.37}$$

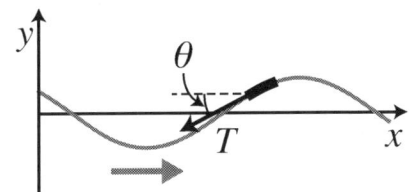

Fig. 7.6: Small element of a string which has a wave travelling to the right. The power transmission of the wave is the power of the force at the right-hand end of the element which does work on the part of the string to the right of the element.

where $v(x, t)$ is the velocity of the piece of the string at position x at time t and T_y is the y component of the tension. Now the velocity of the string at a fixed position, x, is just the rate of change of displacement, y, with respect to time t. Hence this is just the partial derivative of y with respect to t:

$$v(x, t) = \frac{\partial y}{\partial t} \tag{7.38}$$

As we already discussed in section 7.5, for small amplitude oscillations the x component of the string's tension is approximately equal to the tension in the string and so we have:

$$\frac{\partial y}{\partial x} = -\tan\theta = -\frac{T_y}{T_x} \approx -\frac{T_y}{T} \implies T_y = -T\frac{\partial y}{\partial x} \tag{7.39}$$

where the negative sign is required because the y component of the force has the opposite sign to the slope: the positive gradient shown in figure 7.6 gives a negative y component. Putting this into our expression for the power, (7.37), along with the expression for the velocity given in (7.38) gives:

$$\text{Power, } P = -T\frac{\partial y}{\partial x}\frac{\partial y}{\partial t} \tag{7.40}$$

This result is valid for any shape of wave on a string provided that it is small amplitude. However we are most interested in our sinusoidal wave so we need to evaluate it for this situation. Evaluating the derivatives we get:

$$y(x, t) = A\cos(kx - \omega t + \phi) \tag{7.41}$$

$$\frac{\partial y}{\partial t} = \omega A\sin(kx - \omega t + \phi)\,[= v(x, t)] \tag{7.42}$$

$$\frac{\partial y}{\partial x} = -kA\sin(kx - \omega t + \phi) \tag{7.43}$$

Substituting these values for the derivatives into our expression for the power given in (7.40) gives:

$$\text{Power, } P = T\omega kA^2\sin^2(kx - \omega t + \phi) \tag{7.44}$$

However for a wave on a string we know the phase velocity from (7.29) and so we have:

$$\omega k = \frac{\omega^2}{c} = \omega^2\sqrt{\frac{\rho}{T}} \tag{7.45}$$

Putting this into our expression for the power we get:

$$\text{Power, } P = \sqrt{T\rho}\,\omega^2 A^2\sin^2(kx - \omega t + \phi) \tag{7.46}$$

This is the instantaneous power transmission along the string. As expected it is never less than zero since this would represent energy flowing opposite to the direction of motion of the wave. In addition it is clear that the power is not constant and varies, dropping as low as zero when the particle velocity at that point is zero.

The variation of the wave power with time means that a useful quantity to calculate is the mean wave power. The only time dependence comes from the $\sin^2$ function which for any period containing a whole number of oscillations, averages to precisely $\frac{1}{2}$. Hence the average power of a sinusoidal wave on a string is:

$$\text{Power, } P = \frac{1}{2}\sqrt{T\rho}\ \omega^2 A^2 \qquad (7.47)$$

The result obtained here is particular to waves on a string although the method used to obtain the power can be used for other waves. While it is always true that the power depends on the square of the wave amplitude the dependence on the frequency is not always guaranteed. The SI unit of power is the watt (W) which corresponds to a joule per second (Js^{-1}) which is true for all waves regardless of type.

7.8 Wave Intensity

For a one-dimensional wave on a string, there is no possibility for the wave to spread out because there is only one possible direction it can move: along the string. However, for waves in two or more dimensions, this is not the case and waves will generally spread out in all possible directions and so the size of the wavefront will increase. To study this phenomenon lets consider the power transmission of acoustic waves.

Just as we did for a wave on a string we need to consider the power of the force acting on the near side of an element of the medium. Figure 7.7 shows such an element with a pressure acting on the side upon which the wave is incident i.e. the side nearest to the source of the wave. As this pressure deviation compresses the element it will do work on it and so have a power associated with it.

Looking at figure 7.7 the net force acting on the element is just the pressure deviation multiplied by the area. The power of this force is obtained by multiplying it by the velocity and so we have:

$$\text{Power} = Fv = p\delta y\delta z\frac{\partial\phi}{\partial t} \qquad (7.48)$$

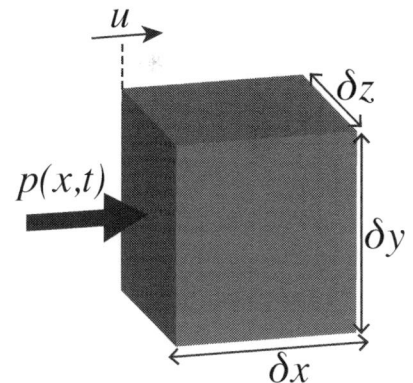

Fig. 7.7: Small element of a fluid showing the pressure deviation, $p(x,t)$, acting on the face of the element closest to the source of the acoustic waves. This pressure deviation results in a net force which will do work on the element.

To simplify things further we can use equation (7.31) to get everything in terms of the displacement, ϕ. This gives:

$$\text{Power} = -B\delta y \delta z \frac{\partial \phi}{\partial x} \frac{\partial \phi}{\partial t} \qquad (7.49)$$

which is very similar to the initial expression we obtained for the power of a wave on a spring. To evaluate this further we need to use the solution to the wave equation for the fluid displacement and then differentiate it. Since we will be multiplying these expressions together the 'implied real' notation is dangerous here so we will use the trigonometric expressions but for simplicity omit the constant phase:

$$\phi(x, t) = A\cos(kx - \omega t) \qquad (7.50)$$

$$\frac{\partial \phi}{\partial t} = \omega A \sin(kx - \omega t) \qquad (7.51)$$

$$\frac{\partial \phi}{\partial x} = -kA \sin(kx - \omega t) \qquad (7.52)$$

Putting these all into equation (7.49) we get:

$$\text{Power} = B\omega k \delta y \delta z A^2 \sin^2(kx - \omega t) \qquad (7.53)$$

As we did before with the string we can use this to calculate the mean power of the sound wave. The average of the $\sin^2$ function is just 1/2 and so the mean power is just:

$$\text{Mean Power} = \frac{1}{2} B\omega k \delta y \delta z A^2 \qquad (7.54)$$

However this is not a useful expression since to calculate the wave power we now have to take the limit as δy and δz go to zero and the integrate over their allowed values which depend on the characteristics of the wave source and medium.

This is a general problem with waves in more than one dimension and so, instead of power, we create a new quantity called intensity which is defined as the power per unit area of the wave. In the SI system, this is measured in units of watts per square metre (W m^{-2}). Combining this definition with the mean power from (7.54) the mean intensity of an acoustic wave is:

$$\text{Mean Intensity, } I = \frac{1}{2} B\omega k A^2 = \frac{1}{2}\sqrt{B\rho}\,\omega^2 A^2 \qquad (7.55)$$

This shows that the intensity of an acoustic wave is proportional to the square of the amplitude as well as the square of the frequency.

To understand how wave intensity changes with distance from a source consider a point source of acoustic waves which has a total power output of P. The waves at a distance, r, from the source as shown in figure 7.8, will all have been emitted from the source at the same time. Hence to conserve energy the total power of the wave on the spherical surface must be equal to the power of the source. This means that:

$$\text{Total Power, } P = \int_S I ds \qquad (7.56)$$

where S is the spherical surface we are integrating over and I is the intensity of the waves as a function of the position on that surface. However, by symmetry, the intensity must be the same at every point on the surface and so this integral simply becomes:

$$P = 4\pi r^2 I \qquad (7.57)$$

where I is the constant intensity of the waves at this surface and $4\pi r^2$ is just the area of the sphere. Rearranging this we get:

$$\text{Intensity, } I = \frac{P}{4\pi r^2} \qquad (7.58)$$

Hence the intensity of the acoustic waves falls off with the square of the distance from the source. This result is not just valid for acoustic waves but for any wave which propagates in three dimensions from a point source: our derivation only relied on conservation of energy which is common to all waves!

For waves constrained to two dimensions, such as surface waves, the wave will spread out as a ring and not as a sphere which changes how the intensity falls with distance. In this case, the wave intensity is measured in watts per unit length (W m^{-1}) and this is multiplied by the circumference of the ring the wavefront forms around the point source, as shown in figure 7.9. The result is that, for two-dimensional waves, the expression for the wave intensity a distance r from a point source with power P becomes:

$$\text{Intensity, } I = \frac{P}{2\pi r} \qquad (7.59)$$

The fact that the waves can only spread in two dimensions does not affect the relationship between intensity and amplitude and so when calculating the effect on the wave amplitude for waves in any dimension it is always safe to assume that the intensity is proportional to the square of the amplitude.

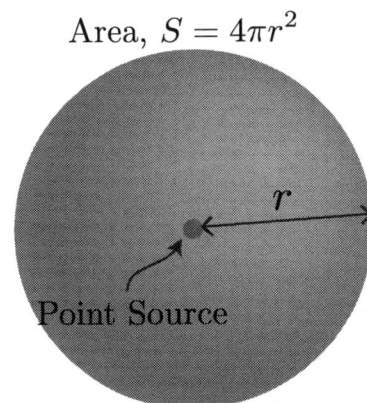

Fig. 7.8: Point source of acoustic waves emitting waves with a constant power, P. The spherical wavefront a distance, r, from the source is shown. The total power of the wave at this surface must equal the power of the source in order to conserve energy.

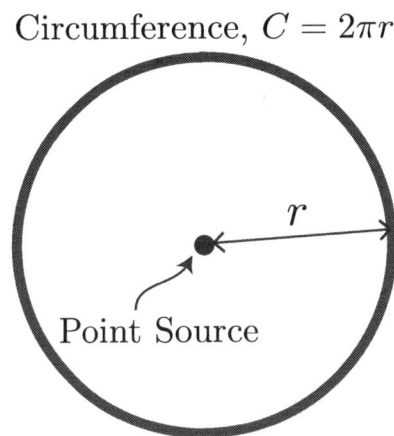

Fig. 7.9: Point source of surface waves emitting waves with a constant power, P. The circular wavefront at a distance, r, from the source is shown. The total power of the wave at this circle must equal the power of the source in order to conserve energy.

7.9 The Decibel Scale

The decibel scale is a non-SI logarithmic unit which expresses the ratio of two power or intensity values. It was originally in-

Fig. 7.10: Alexander Graham Bell (1847-1922) after whom the decibel is named. He invented the telephone in 1876 while a British citizen living in Canada although he later became a US citizen in 1882.

Moffett Studio circa 1914-19, Public Domain,

Library and Archives Canada / C-017335

vented to quantify the loss of signal along telegraph and telephone lines. Originally the unit used was "miles of standard cable" (MSC). 1 MSC was the power loss for a $5,000\,\mathrm{rad\,s^{-1}}$ frequency signal over 1.6 km (1 mile) of standard telephone cable. Bell Telephone Laboratories defined a new unit, called the "transmission unit" (TU) in 1924, which was equal to 1.056 MSC and then in 1928 renamed this unit as one tenth of a "bel" which was a new unit named after the inventor of the telephone, Alexander Graham Bell (figure 7.10), the British-Canadian inventor of the telephone.

As one tenth of a bel, the decibel (dB) is defined as:

$$\beta = (10\,\mathrm{dB}) \log_{10}\left(\frac{I}{I_0}\right) \tag{7.60}$$

where I_0 is the reference intensity and I is the measured intensity. The same equation also works for a ratio of any two measurements and commonly used decibel scales cover a large variety such as pressure, power, energy as well as intensity. This gives the decibel two unusual features: it can be used to measure a large variety of different physical measurements and it uses a logarithmic scale which means it can cover a huge range of magnitudes without the need to resort to unusual SI prefixes. The logarithmic nature also means that the units can be added where standard units are multiplied. For example, if we have two amplifiers one with a gain of 20 dB and one with a gain of 15 dB the total gain after going through both is just 35 dB whereas with a simple gain factor we would have had to multiply.

Table 7.1: Several commonly encountered decibel scales used for different purposes and each with its own reference.

Symbol	Description
dB SIL	Sound Intensity Level, measures sound intensity using a reference of $10^{-12}\,\mathrm{W\,m^{-2}}$.
dB SPL	Sound Pressure Level, measures sound intensity via pressure measurements using a reference of $20\,\mathrm{\mu N\,m^{-2}}$
dBm	Used in audio electronics and radio to measure an electronic sound signal relative to a power of 1 mW.
dBFS	Used in audio electronics to measure a signal amplitude relative to the maximum amplitude which a device can reproduce without clipping.
dBJ	Energy relative to $1\,\mathrm{J}=1\,\mathrm{W\,Hz^{-1}}$ and it is used to measure the power density of an emission spectrum, typically in radio.

The diverse types of measurement which the decibel system can be used for coupled with the need to define a reference measurement mean that there are a large variety of decibel scales in use and several are shown in table 7.1. One of the most common uses of the decibel scale is to measure the intensity of sound. The human ear is sensitive to sound over a 12 order of magnitude intensity range from the softest, audible sound at $10^{-12}\,\mathrm{W\,m^{-2}}$ up to $1\,\mathrm{W\,m^{-2}}$ which is the threshold of discomfort, actual pain takes approximately $100\,\mathrm{W\,m^{-2}}$.

This huge range of sound intensity is easily covered by a logarithmic scale and Sound Intensity Level (SIL) decibel scale is used to measure the intensity of sounds in the context of human hearing. To do this it defines the reference point, I_0, to be the softest, audible sound at an intensity of $10^{-12}\,\mathrm{W\,m^{-2}}$. Hence, using equation (7.60) we see that the sound intensity which causes pain is:

$$\beta = (10\,\mathrm{dB})\log_{10}\left(\frac{100}{10^{-12}}\right) = 140\,\mathrm{dB(SIL)}$$

and similarly the softest, audible sound has an intensity of $0\,\mathrm{dB(SIL)}$.

Problems

Q7.1: Airports now include special body scanners which can see through clothing to determine whether a passenger is carrying any weapons. [The speed of light in vacuo is $3 \times 10^8 \, \text{m s}^{-1}$]
(a) The scanners used in Canada and Europe use safe, non-ionizing terahertz electromagnetic waves. If the frequency of the radiation is 1 THz what is the wavelength?
(b) Some of the scanners in the USA which are slowly being replaced use the back-scattering of low intensity, ionizing X-rays. If the wavelength of the X-rays used is 5 nm what is the frequency of the radiation?

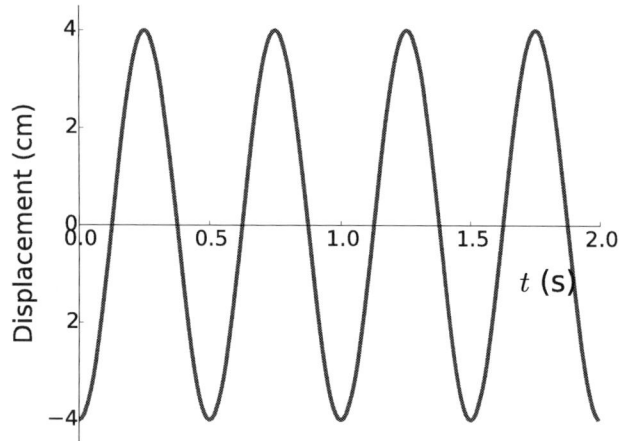

Fig. P7.1: *Plot of displacement vs. time at a fixed point on a wave.*

Q7.2: The displacement versus time for a fixed point on a wave is shown in figure P7.1.
(a) What is the amplitude of the wave?
(b) What is the frequency of the wave?
(c) Given that the wave shown is sinusoidal what is the maximum speed of the medium and at what point on the plot shown is the medium travelling at this maximum speed?

Q7.3: Figure P7.2 shows the displacement versus position for a wave. The initial displacement is shown by the dashed line and the displacement one second later is denoted by the solid line.

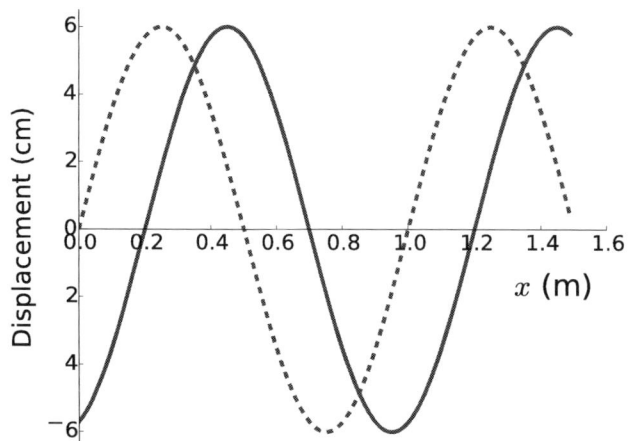

Fig. P7.2: *Plot of displacement vs. position for a wave. The dashed line shows the initial displacement and the solid line shows the displacement 1 s later.*

(a) What is the amplitude of this wave?
(b) What is the wavelength of this wave?
(c) What is the phase velocity of the wave?
(d) What is the frequency of the wave?

Q7.4: Two fishermen are sitting on the end of a pier such that the floats on the end of their fishing lines are positioned 2 m apart and bob up and down on water waves which are travelling parallel to the line between the floats. One of the fishermen notes that the two floats bob up and down exactly out of phase such that then one is at its highest the other is at its lowest and that each float takes 1 s to complete a cycle. What is the fastest possible velocity of the water waves?

Q7.5: A wave on a string is travelling with a phase velocity of $12 \, \text{m s}^{-1}$ and has an amplitude of 6 cm and a wavelength of 24 cm. The waves are travelling in the positive x direction and at time $t = 0$ the displacement of the string at $x = 0$ is at its lowest value.
(a) What are the wavenumber, frequency and period for this wave?
(b) Determine the expression which gives the displacement, y, of the string as a function of position, x, and time, t.
(c) What is the displacement of the string at the position where $x = 0.65$ m and $t = 1$ s and how

much time will pass before this point of the string reaches its maximum, positive displacement at this location?

Q7.6: A transverse wave travelling along a string is described by the wave function:

$$\Psi(x, t) = 0.5\cos(x - 5t)$$

(a) What is the phase velocity of the wave?
(b) Derive the expression for the velocity of the string as a function of position and time and determine the maximum velocity that any piece of the string can achieve.

Q7.7: A length, l, of rope with a mass per unit length of ρ hangs vertically below a rock climber with the end hanging freely just above the ground. To attract the climber's attention a person standing on the ground below grabs the end of the rope and gives it a brief shake. This sends a small, transverse wave pulse up the rope to the climber. If the gravitational field if g how long does it take the pulse to reach the climber?

Q7.8: In a thunderstorm, there is often a significant delay observed between seeing lightning and hearing the thunder which it generates.
(a) Assuming that light is approximately instantaneous what is the bulk modulus of air on Earth given that at the distance of 1.75 km there is a 5.1 s delay observed between lighting and its associated thunder? Would this calculation provide a higher or lower value than the true value if the actual speed of light were taken into account? The density of Earth's atmosphere is 1.2 kg m^{-3}.
(b) Lightning has also been observed in dust storms on Mars. Assuming that the time delay between observing the lightning and hearing the thunder is still 5.1 s how far away is the lightning strike from the observer given that the density of the martian atmosphere is 0.015 kg m^{-3} and that the bulk modulus is 864 N m^{-2}?

Q7.9: A whale makes a sound with a wavelength of 40 m underwater. What is the wavelength when this same sound wave passes into air given that B(water)=2.15 GN m^{-2}, B(air)=142 kN m^{-2}, ρ(water)=1,000 kg m^{-3} and ρ(air)=1.2 kg m^{-3} where B is the bulk modulus and ρ is the density.

Q7.10: A loudspeaker is tested both in air and in helium by generating a sound of constant frequency and displacement amplitude. Given that the bulk modulus of air and helium are approximately equal and that the densities of air and helium are 1.2 kg m^{-3} and 0.18 kg m^{-3} respectively what is the ratio of the pressure amplitude in air to that in helium?

Q7.11: Some biologists claim that dolphins can stun their prey with the loud clicking sounds they make underwater although recent papers have thrown some doubt on this. If the pressure amplitude of a dolphin click is 1,000 N m^{-2} and the frequency is 50 kHz what is the displacement amplitude felt by a small fish in the path of the wave given that B(water)=2.15 GN m^{-2} and ρ(water)=1,000 kg m^{-3}?

Q7.12: When an underwater earthquake occurs it triggers waves in both water and rock. Assuming that both these waves can be treated as acoustic waves in the bulk at what distance from the epicentre does an earthquake provide a 5 min warning of a tsunami? For water $B = 2.2$ GN m^{-2} and $\rho = 1,000$ kg m^{-3} and for typical rock $B = 50$ GN m^{-2} and $\rho = 2,800$ kg m^{-3}.

Q7.13: When a very massive star runs out of nuclear fuel to burn the core of the star collapses under gravity causing a massive explosion which can outshine the rest of the galaxy which contained the original star. One such supernova releases 1×10^{43} J of electromagnetic energy over a period of 40 days. At what distance, in light years, from the Earth would such an event have the same intensity as the sun given that sunlight's intensity is 1,413 W m^{-2}? [a light year is the distance which light travels in one year, 9.46×10^{15} m].

Q7.14: A steel wire can support two different types of wave. Small amplitude, transverse waves have a wave equation:

$$\frac{\partial^2 \Psi}{\partial x^2} = \frac{\mu}{T}\frac{\partial^2 \Psi}{\partial t^2}$$

where μ is the mass per unit length of the wire and T is the tension in the wire. However it is also possible to have a longitudinal wave which has the wave equation:

$$\frac{\partial^2 \Psi}{\partial x^2} = \frac{\rho}{Y} \frac{\partial^2 \Psi}{\partial t^2}$$

where ρ is the density of the wire and Y is the Young's modulus of the wire. In both cases $\Psi(x, t)$ is the displacement of the wire's material from the equilibrium position.

(a) Write down the algebraic expression for the phase velocity of a longitudinal wave on the wire, c_l, in terms of the quantities given above.

(b) A wire of length l_0 is stretched to produce and extension Δl and then clamped at both ends. Show that the ratio of the phase velocity of the transverse waves, c_t, to that of the longitudinal waves, c_l, is given by:

$$\frac{c_t}{c_l} = \sqrt{\frac{\Delta l}{l_0}}$$

CHAPTER 8

Wave Phenomena

The previous chapter dealt with the physical properties of waves and how they propagate through a medium. In this chapter, we will deal with a variety of different phenomena which arise both from the interactions of two waves which overlap, such as beats and interference patterns, as well as the effect of the motion of both wave sources and observer through the physical medium of the waves such as doppler shift and shockwaves.

However, to start we will consider the very simple principle which lies behind many of these phenomena that of linear superposition which arises from the linear nature of the wave equation itself.

8.1 Principle of Superposition

The principle of superposition states that the displacement at any point due to two or more waves being present there is just the sum of the displacements of all the individual waves. For example if we have two transverse waves on a string then the displacement of the string as a function of position and time is just:

$$y(x, t) = y_1(x, t) + y_2(x, t) \tag{8.1}$$

where $y_1(x, t)$ and $y_r(x, t)$ are the wave functions for the two waves. This principle, sometimes called *linear superposition* is a direct consequence of the linear nature of the wave equation. Since $y_1(x, t)$ and $y_2(x, t)$ must each be solutions of the wave equation by reason of them each being waves we have from (7.16):

$$\frac{\partial^2 y_1}{\partial x^2} = \frac{1}{c^2} \frac{\partial^2 y_1}{\partial t^2} \text{ and } \frac{\partial^2 y_2}{\partial x^2} = \frac{1}{c^2} \frac{\partial^2 y_2}{\partial t^2} \quad (8.2)$$

Adding these equations together gives:

$$\frac{\partial^2 y_1}{\partial x^2} + \frac{\partial^2 y_2}{\partial x^2} = \frac{1}{c^2} \frac{\partial^2 y_1}{\partial t^2} + \frac{1}{c^2} \frac{\partial^2 y_2}{\partial t^2} \quad (8.3)$$

$$\implies \frac{\partial^2(y_1 + y_2)}{\partial x^2} = \frac{1}{c^2} \frac{\partial^2(y_1 + y_2)}{\partial t^2} \quad (8.4)$$

Hence the linear nature of the wave equation means that the sum of any two solutions of the wave equation is itself a solution. For a medium where the wave equation is not linear this principle of linear superposition would not apply.

As you would expect this means that for two identical wave pulses travelling in the same direction at the same time the displacement of the pulse is exactly twice the displacement of one of the waves. However if the two waves are travelling in opposite directions they will temporarily add together to create a single, large pulse before each wave continues in its original direction as seen in figure 8.1.

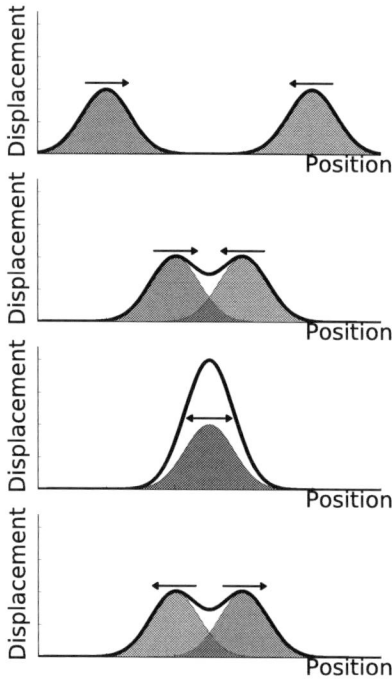

Fig. 8.1: Interaction of two waves travelling in opposite directions. The waves add together to give a single, large wave and then separate again.

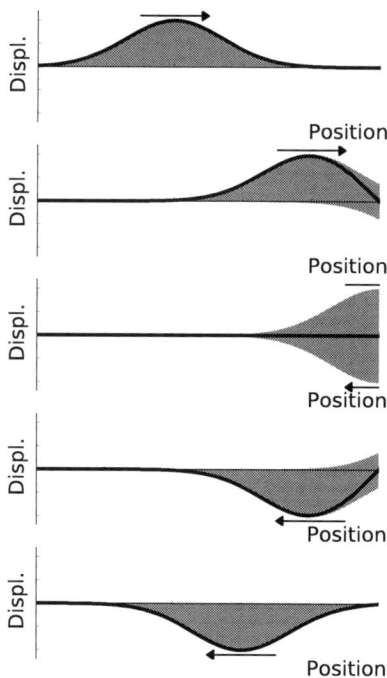

Fig. 8.2: Reflection of a wave at the clamped end of a string. The incident wave is reflected with a phase shift of π.

8.2 Boundary Conditions

When a wave reaches the boundary of the medium it is propagating through then, unless there is some mechanism to absorb the wave's energy, it must reflect off that boundary in order to conserve energy. This means that the displacement at the boundary will be the sum of the incoming wave and the reflected wave. However, it is often the case that certain conditions apply to the displacement of the wave's medium at a boundary.

Consider a wave on a string which is clamped at one end. When this wave reflects off the end of the string the displacement of the end of the string will be equal to the sum of the incoming wave and the outgoing, reflected wave. However, because the end is clamped, we have a boundary condition that the displacement of the end of the string must be zero. This means that when the pulse reflects off the end of the string the wave must be inverted so that the sum of the incident and reflected wave remains precisely zero at the clamped end as shown in figure 8.2.

If the end of the string is not clamped and is free to oscillate vertically then there is no requirement that the displacement must be zero. Instead, the boundary condition, in this case, is that the slope of the string at the end is zero, i.e. $\frac{\partial y}{\partial x} = 0$, because there can be no transverse force on the end of the string. This requires that the reflected wave is not inverted so that their slopes, rather than their displacements, cancel as shown in figure 8.3.

8.3 Interference

Interference is a phenomenon which occurs when two waves both occupy the same region of space. For waves with the same frequency and wavelength the critical factor in determining the result is the phase difference between the two waves. For example consider two waves with exactly the same phase as shown in figure 8.4. In this situation, the waves add together to give a single wave with an amplitude equal to the sum of the individual amplitudes. Mathematically we add the wave functions at a fixed point in space and assume that the waves both travel the same distance x_0 to get there then we have:

$$\begin{aligned}
\Psi(x_0, t) &= \Psi_1(x_0, t) + \Psi_2(x_0, t) \\
&= A_1 \cos(kx_0 - \omega t) + A_2 \cos(kx_0 - \omega t) \\
&= (A_1 + A_2) \cos(kx_0 - \omega t) \quad\quad (8.5)
\end{aligned}$$

Since the amplitudes add to each other this is known as *constructive interference*.

However if the waves are out of phase by an angle of π then instead of adding the waves will cancel and the remaining amplitude will be the difference of the two individual amplitudes as shown in figure 8.5. Mathematically we still add the wave functions at a fixed point in space but we include a phase difference of π which flips the sign of one of the functions

$$\begin{aligned}
\Psi(x_0, t) &= \Psi_1(x_0, t) + \Psi_2(x_0, t) \\
&= A_1 \cos(kx_0 - \omega t) + A_2 \cos(kx_0 - \omega t + \pi) \\
&= A_1 \cos(kx_0 - \omega t) - A_2 \cos(kx_0 - \omega t) \\
&= (A_1 - A_2) \cos(kx_0 - \omega t) \quad\quad (8.6)
\end{aligned}$$

Since the amplitudes cancel each other this is known as *destructive interference*.

In both the previous cases we have assumed that the waves start in phase and travel exactly the same distance, x_0, to get to the point where they interfere. However, suppose the sources of the waves are such that the distance travelled by the waves

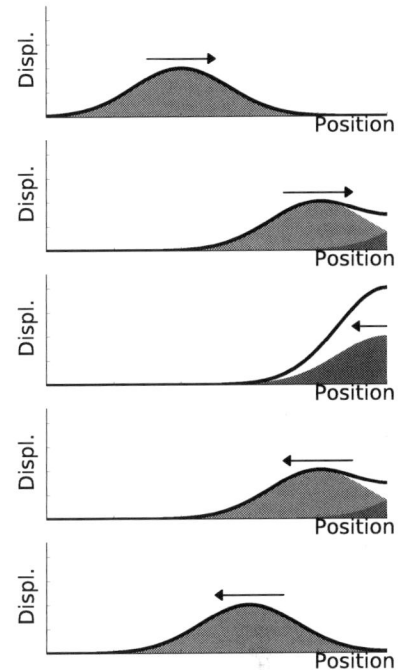

Fig. 8.3: Reflection of a wave at the free end of a string. The incident wave is reflected without any phase shift.

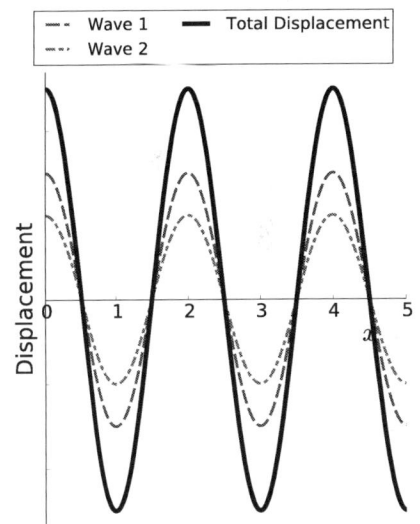

Fig. 8.4: Interference of two waves (dashed) which have the same phase but different amplitudes to give a single, larger amplitude wave (solid). This is constructive interference.

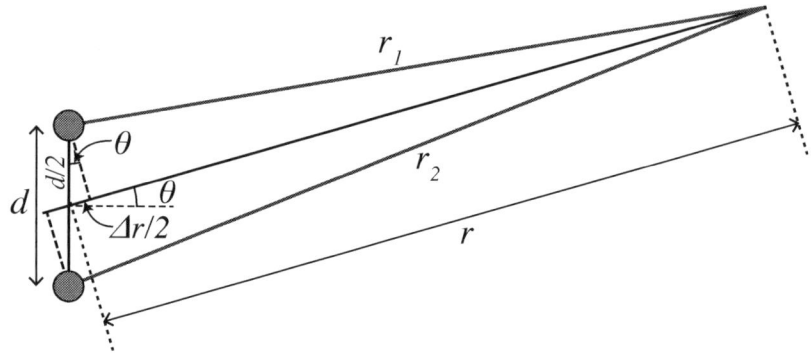

Fig. 8.6: Two wave sources showing the different path lengths between the sources and the point a distant r away.

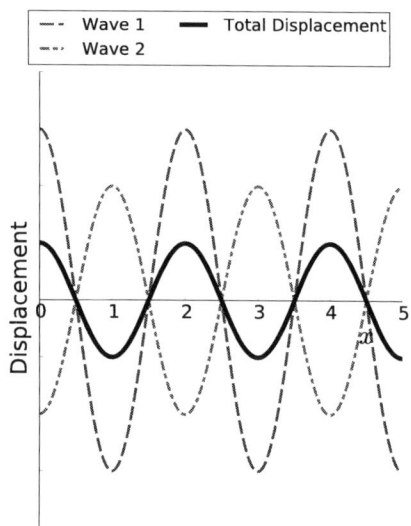

Fig. 8.5: Interference of two waves (dashed) which have opposite phases but different amplitudes to give a single, smaller amplitude wave (solid). This is destructive interference.

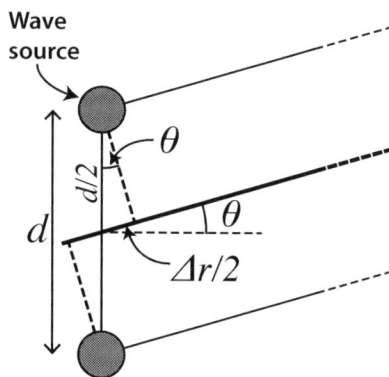

Fig. 8.7: Close-up of the two wave sources. The path length difference between the sources at the angle shown is Δr.

is not exactly the same. If this were the case then the phase change due to the distance travelled by the wave, which is kx_0 in equations (8.5) and (8.6), would not be the same. This means that even if the waves started out in phase there may be points where the waves will cancel out due to a difference in the path length between the two sources and the point being considered.

Looking at (8.6) it is clear that the phase change due to the path length is just the wavenumber, k, multiplied by the path length. Hence for two waves which start in phase the phase difference due to a change in the path length is just:

$$\Delta\phi = k\Delta l \tag{8.7}$$

where $\Delta\phi$ is the phase difference and Δl is the difference in path length. The fact that a difference in path length corresponds to a difference in phase is a critical concept in understanding inference phenomena.

To see this in action we need to consider waves propagating in two dimensions. Figure 8.6 shows a setup with two identical sources of waves a distance d apart. If we take a point a distance r from the centre of the two sources at an angle θ as shown in the figure then we can calculate the difference in the distance which the waves from the two sources have to travel. For large distances, the waves from the two sources will travel along almost parallel lines to reach the point and so the path length difference is just the projection of the line joining the sources onto the line from their centre to the now distant point as shown in figure 8.7.

Looking at the triangle shown in figure 8.7 simple trigonometry gives the relationship that:

$$\frac{\Delta r}{2} = \frac{d}{2}\sin\theta \implies \Delta r = d\sin\theta \tag{8.8}$$

where d is the separation of the sources and Δr is the difference in the path length between the two sources and the point where the waves interfere.

Using equation (8.7) this path length difference can be converted into a phase difference which gives:

$$\Delta\phi = k\Delta r = kd\sin\theta \tag{8.9}$$

where k is the wavenumber of waves from the two sources. For constructive interference this phase difference must be a multiple of 2π and so our condition for constructive interference, and hence a large wave amplitude, is:

$$kd\sin\theta = 2n\pi \implies \sin\theta = \frac{2n\pi}{kd} \tag{8.10}$$

where n is any integer. We can simplify this one step further by remembering the definition of the wavenumber, $k = \frac{2\pi}{\lambda}$:

$$\sin\theta = \frac{n\lambda}{d} \tag{8.11}$$

where λ is the wavelength. Similarly for destructive interference, where the wave amplitude will be a minimum, we require that the phase difference be an odd number of π i.e. $(2n+1)\pi$ where n is any integer. This leads to a condition for destructive interference of:

$$kd\sin\theta = (2n+1)\pi \implies \sin\theta = \frac{(n+\frac{1}{2})\lambda}{d} \tag{8.12}$$

The result is that at large distances from the sources at certain angles the two sources will constructively interfere to generate large amplitude waves and at angles between these the two sources will destructively interfere, reducing the wave amplitude, even to zero if the two sources have identical amplitudes. This is clearly shown in figure 8.8.

8.4 Beats

In the previous section, we considered what happens when waves with the same frequency and wavelength but different phases interfere. However, we can also have two waves with slightly different frequencies interfere and this gives rise to a phenomenon known as beats.

For two sources with different frequencies, the phase difference between the sources will change with time because each

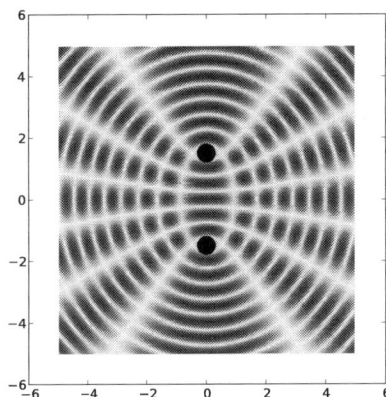

Fig. 8.8: Height map of waves from two sources interfering with each other. As is clear there are certain angles for which the waves completely cancel and others for which the waves add constructively.

source oscillates at a different frequency. This means that the effect will be one that varies in time and so path length differences will not be relevant since these provide a constant phase offset. Hence to understand this phenomenon we need to look at the interference between the time-dependent part of two waves with slightly different frequencies, ω_1 and ω_2.

To keep things simple we assume that both waves have the same amplitude and we choose a fixed point in space such that the phase of the two waves is initially the same and equal to zero so there is no constant phase offset. Next we define two new quantities corresponding to the average angular frequency, $\overline{\omega}$, and the difference in angular frequency, $\Delta\omega$, where:

$$\overline{\omega} = \tfrac{1}{2}(\omega_1 + \omega_2) \text{ and } \Delta\omega = \omega_1 - \omega_2 \qquad (8.13)$$

Using these definition we can rewrite our original frequencies to get:

$$\omega_1 = \overline{\omega} + \tfrac{1}{2}\Delta\omega \text{ and } \omega_2 = \overline{\omega} - \tfrac{1}{2}\Delta\omega \qquad (8.14)$$

Writing down the resulting displacement due to each wave at this fixed point in space we will get an expression which we can then expand in terms of the average and difference in frequencies using the trig identifies for the cosine of the sum, and difference, of two angles:

$$\psi(t) = A\cos(\omega_1 t) + A\cos(\omega_2 t) \qquad (8.15)$$
$$= A\cos(\overline{\omega}t + \tfrac{1}{2}\Delta\omega t) + A\cos(\overline{\omega}t - \tfrac{1}{2}\Delta\omega t)$$
$$= A\cos(\overline{\omega}t)\cos(\tfrac{1}{2}\Delta\omega t) - A\sin(\overline{\omega}t)\sin(\tfrac{1}{2}\Delta\omega t)$$
$$+ A\cos(\overline{\omega}t)\cos(\tfrac{1}{2}\Delta\omega t) + A\sin(\overline{\omega}t)\sin(\tfrac{1}{2}\Delta\omega t)$$
$$= 2A\cos(\overline{\omega}t)\cos(\tfrac{1}{2}\Delta\omega t) \qquad (8.16)$$

Now if we rewrite this expression in terms of ω_1 and ω_2 we get:

$$\psi(t) = 2A\cos\left(\frac{\omega_1 + \omega_2}{2}t\right)\cos\left(\frac{\omega_1 - \omega_2}{2}t\right) \qquad (8.17)$$

Now consider the case where the two frequencies are large but have a small difference. In this case the $\omega_1 - \omega_2$ term in (8.17) will give a very low frequency oscillation while the $\omega_1 + \omega_2$ will give a high frequency. The result is a high frequency wave which has an "amplitude" of $2A\cos\left[\frac{\omega_1 - \omega_2}{2}t\right]$ so that the amplitude oscillates slowly as shown in figure 8.9.

If two sound waves with almost the same frequency and amplitude are played next to each other we hear this oscillation in the amplitude as beats. Looking at (8.17) the frequency of the envelope wave is $(f_1 - f_2)/2$ where f_n is the frequency of the wave. However looking at figure 8.9 there is no difference

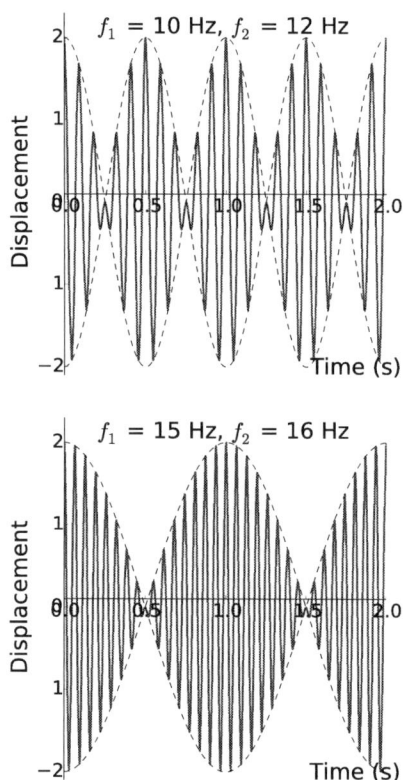

Fig. 8.9: Two examples of beats. The top plot shows a low frequency wave and a 2 Hz beat frequency while the lower plot shows a higher frequency wave but a lower beat frequency of 1Hz because there is less of a frequency difference between the two waves.

between having a positive or negative value for the amplitude since the high-frequency wave oscillates between positive and negative values. Hence each amplitude lobe in figure 8.9 corresponds to half a wavelength of the envelope function and so the beat frequency is actually:

$$\text{Beat Frequency, } f_b = |f_1 - f_2| \qquad (8.18)$$

which is twice the frequency of the envelope wave.

8.5 Standing Waves

Standing waves are an incredibly important phenomenon and are essential to understanding the physics of all musical instruments and, at a more fundamental level, they are also the physics of atoms and molecules. We will consider two types of standing wave here: standing waves on a string and standing sound waves in a pipe.

8.5.1 Standing Waves on Strings

Consider a string of length l which is clamped at both ends. As we already saw in section 8.2 this means that we have a boundary condition at each end of the string which requires that the displacement is zero ($y = 0$) when $x = 0$ and when $x = l$. Now if we consider our standard solution to the wave equation we have a wave function for a positive travelling wave which is:

$$y(x, t) = A\cos(kx - \omega t + \phi) \qquad (8.19)$$

However this will reflect off the end of the string and acquire a phase change of π which means we can just flip the sign of the reflected, negative travelling, wave to give:

$$y(x, t) = A\cos(kx - \omega t + \phi) - A\cos(kx + \omega t + \phi) \qquad (8.20)$$

where the sign of ωt flips because the reflected wave travels in the opposite direction. If we group the $kx + \phi$ terms and treat them as a single expression we can expand out the cosine functions using the trig identifies for the sum and difference of two angles:

$$y(x, t) = A\cos(kx + \phi)\cos(\omega t) + A\sin(kx + \phi)\sin(\omega t)$$
$$- A\cos(kx + \phi)\cos(\omega t) + A\sin(kx + \phi)\sin(\omega t)$$
$$= 2A\sin(kx + \phi)\sin(\omega t) \qquad (8.21)$$

At this point we can now apply our boundary conditions. The first is that when $x = 0$, $y = 0$ which will only be true for all values of time, t, if:

$$\sin(\phi) = 0 \implies \phi = 0 \tag{8.22}$$

This simplifies our wave function to:

$$y(x, t) = 2A \sin(kx) \sin(\omega t) \tag{8.23}$$

Lastly the second boundary condition requires that when $x = l$, $y = 0$. Again for this to be true for all values of time, t, we require that:

$$\sin(kl) = 0 \implies kl = 2n\pi \implies k = \frac{n\pi}{l} \implies \lambda = \frac{2l}{n} \tag{8.24}$$

where n is an integer. This result is quire remarkable: we have taken a continuous, newtonian mechanical system and come up with a result that only certain, quantized values of wavelength are possible! Only these particular values for the wavelength will satisfy the boundary conditions and so are allowed.

Figure 8.10 shows some of the vibration modes of the string for low values of n. It is important to remember that the $\sin(\omega t)$ term means that the string is still vibrating up and down with the same frequency everywhere but, as is evident in the figure, the amplitude at certain points on the string is zero. These points are called *nodes* and the points where the vibration has a maximum amplitude is called an *anti-node*. Each of these allowed wavelengths is called a *normal mode*, or *harmonic*, of the string. The longest wavelength, or lowest frequency, normal mode is referred to as the *fundamental mode* with higher frequency modes being referred to as the second, third etc. harmonics.

The nodes on the string will occur where $\sin(kx) = 0$. Putting in the definition of the wavenumber this gives the condition:

$$\sin\left(\frac{2\pi}{\lambda}x\right) = 0 \implies \frac{2\pi}{\lambda}x = m\pi \implies c = m\frac{\lambda}{2} \tag{8.25}$$

where m is an integer. This means that the nodes on the string will always be separated by $\lambda/2$. Since the anti-nodes are when $\sin(kx) = 1$ these will be separated from the nearest node by $\lambda/4$.

Starting with the wavelength given in (8.24) we can convert this into a frequency:

$$\lambda_n = \frac{2l}{n} \implies \frac{c}{f_n} = \frac{2l}{n} \implies f_n = \frac{nc}{2l} \tag{8.26}$$

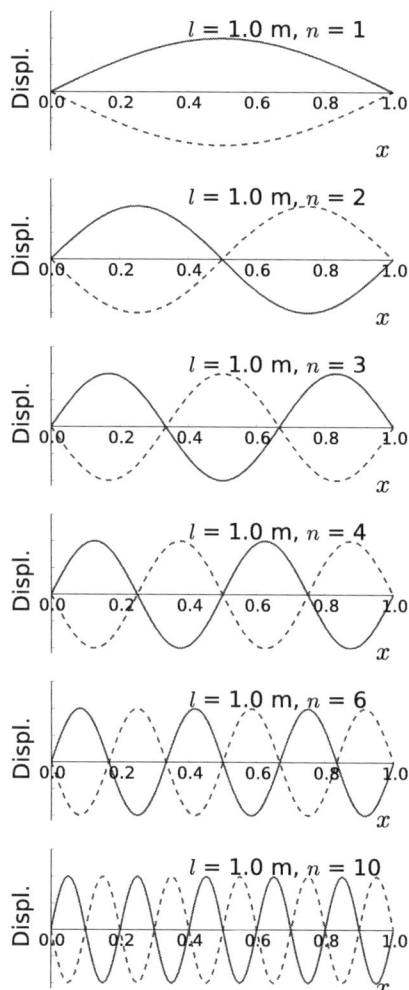

Fig. 8.10: Displacement of the string for various possible standing waves of the string starting with $n = 1$ for a string of length 1 m.

Hence the frequency is also quantized. This quantization of the allowed frequencies is how all stringed instruments work. In the case of a violin, the musician shortens the length of the string using their fingers so that the frequency of vibration corresponds to the frequency of the desired note. However, as we have seen for a fixed length of string, more than one frequency of vibration, or harmonic, is possible. When a violin bow is drawn across a string multiple harmonics will be excited and the unique mixture of harmonics, each of which will be amplified to a different extent by the sound box, give the violin its distinctive sound. The reason why larger instruments give lower notes is also clear from (8.26): a longer length of string corresponds to a lower frequency for the normal modes of vibration.

8.5.2 Standing Waves in Pipes

The phenomenon of standing waves occurs in more than just strings. Sound waves in a pipe can reflect off the ends of a pipe and generate standings waves. It should come as no surprise that this is how wind and brass instruments generate their notes.

Unlike the transverse waves on a string, sound waves are longitudinal waves. What this means is that if the end of the pipe is closed the displacement is constrained to be zero because the air molecules against the wall at the end cannot vibrate longitudinally and so the end must have a node. Conversely, at the open end of the pipe the air is free to vibrate without resistance and so an anti-node is formed. What this means is that the modes of vibration for the pipe will depend on whether its ends are open or closed.

A pipe with two open ends must have an anti-node at each end. The longest wavelength modes are shown in figure 8.11. Since each end must have an anti-node and the separation between adjacent anti-nodes is $\lambda/2$ we can conclude that the allowed wavelengths are: $2l$, l, $2l/3$, $l/2$, $2l/5$... Hence the expression for allowed wavelengths is just:

$$\lambda = \frac{2l}{n} \implies f = \frac{nc}{2l} \tag{8.27}$$

which is the same as it was for the string. The only difference here is that the pipe now has anti-nodes at each end of it.

Now consider a pipe which has one end open and one end closed. This requires that there be an anti-node at one end and

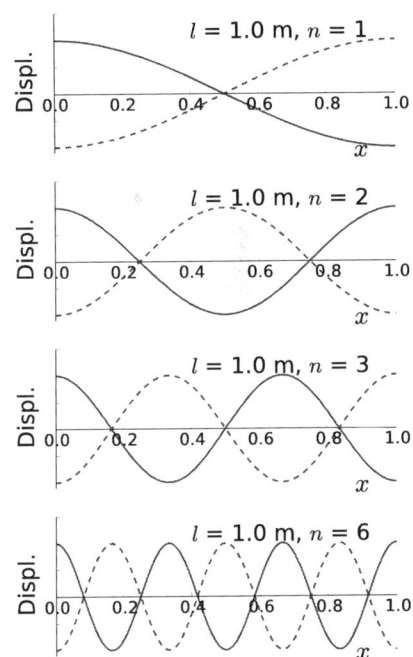

Fig. 8.11: Displacement of air in a sound wave in a 1 m long pipe with two open ends.

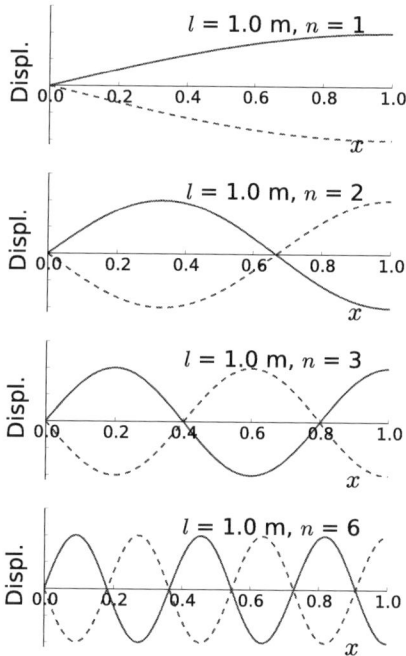

Fig. 8.12: Displacement of air in a sound wave in a 1 m long pipe with one open end and one closed end.

a node at the other. Figure 8.12 shows some of the longest wavelength modes allowed. Since the separation between a node and an anti-node is just $\lambda/4$ this means that the relationship between the length of the pipe and the wavelength is:

$$l = \frac{\lambda}{4}, \frac{3\lambda}{4}, \frac{5\lambda}{4}, \frac{7\lambda}{4} \cdots = (2n-1)\frac{\lambda}{4} \qquad (8.28)$$

where n is an integer. Rearranging this we obtain expressions for the allowed wavelengths and frequencies:

$$\lambda = \frac{4l}{2n-1} \implies f = \frac{(2n-1)c}{4l} \qquad (8.29)$$

Hence the allowed frequencies of a pipe with one open end are not the same as for one with two open ends. In particular the longest wavelength mode has $\lambda = 4l$ which is twice the maximum length of two open ends. Although we will not cover it explicitly here a pipe with two close ends behaves exactly the same as the previous string example since it requires nodes at each end. Hence the allowed modes will be identical to those for a pipe open at two ends and given in (8.27) since the separation between the two nodes at each end is a half-integer multiple of the wavelength just as it is for the separation between two anti-nodes.

8.6 Doppler Effect

When a source of waves is moving the wave crests in front of the source end up closer together and those behind it end up spaced further apart due to the motion of the source relative to the medium. This means that the time between crests in front of the source reduced and the time between crests arriving behind the source is increased with the result that the wavelength and frequency of the waves must have been changed. The same effect occurs if an observer is moving towards or away from a stationary source.

8.6.1 Non-relativistic Doppler Effect

To quantify the non-relativistic doppler effect consider figure 8.13 where both the source and the observer are moving towards each other and relative to the medium of the waves. The time between the source emitting two wave crests is $1/f_s$ where f_s is the source's frequency. During this time the source will move a distance v_s/f_s where v_s is the velocity of the source. Hence

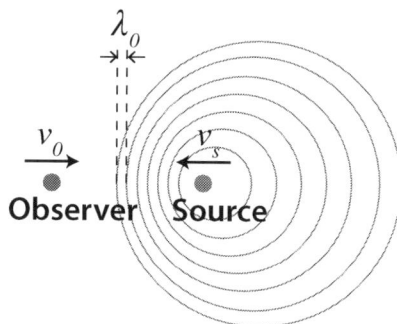

Fig. 8.13: Wave source moving towards an observer with a velocity v_s relative to the wave medium. The wave crest in front bunch together giving a shorted effect wavelength, λ_o, for the observer while those behind would have a longer wavelength.

the wavelength as seen by the observer will be:

$$\lambda_o = \lambda_s - \frac{v_s}{f_s} = \frac{c}{f_s} - \frac{v_s}{f_s} = \frac{c - v_s}{f_s} \qquad (8.30)$$

where c is the phase velocity of the wave relative to the medium and λ_s is the wavelength of the source. Now because the observer is moving towards the waves the velocity of the waves relative to the observer is:

$$c_o = c + v_o \qquad (8.31)$$

where v_o is the velocity of the observer relative to the medium. Hence the wave frequency which the observer will see is just:

$$f_o = \frac{c_o}{\lambda_o} = \left(\frac{c + v_o}{c - v_s}\right) f_s \qquad (8.32)$$

where v_o is positive for motion towards the source and v_s is positive for motion towards the observer and both are measured relative to the medium of the wave.

Note that here we have only considered velocities towards or away from each other for both the source and the observer. For the case of a transverse velocity neither the source nor observer are approaching or receding from each other. Hence the observer wavelength and wave velocity will be the same as the stationary case and there is no doppler shift.

8.6.2 Relativistic Doppler Effect

In the previous section, we considered the shift in frequency due to the motion of the source and the observer relative to the medium. However, this is not always possible: light waves propagate through a vacuum and one of the central tenets of relativity is that it is impossible to measure your velocity relative to the vacuum. Hence our doppler shift can only depend on the relative velocity of the source to the observer. In addition, there are time dilation effects to consider and, as we shall see, this introduces a transverse doppler effect.

To calculate the relativistic doppler effect consider the diagram shown in figure 8.14 where a wave source is moving with a speed v at an angle θ to the observer. This figure shows the motion of the source between the production of two wave fronts which are emitted with a separation of Δt_p in the observers frame of reference. This gives a time between the arrival of the two wavefronts at the observer of

$$T_o = \Delta t_p + \frac{v}{c}\Delta t_p \cos\theta = \Delta t_p(1 + \beta\cos\theta) \qquad (8.33)$$

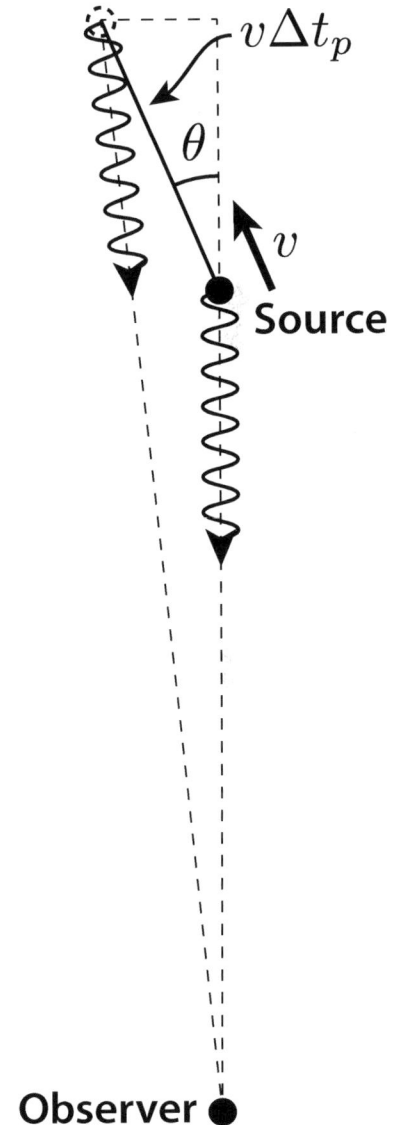

Fig. 8.14: This figure shows the displacement of a wave source between the emission of two wavefronts. If the time between emission is small then the increase in distance from the observer will be $\Delta t_p v \cos\theta$.

where $\beta = v/c$ and c is the speed of light in vacuo. Now since the source is moving relative to the observer we must consider the effects of time dilation on the production of the wave fronts by the source. If the time between the emission of two wavefronts in the source's frame of reference is T_s then:

$$\Delta t_p = \gamma T_s \quad \text{where} \quad \gamma = \frac{1}{\sqrt{1 - \beta^2}} \qquad (8.34)$$

Now we can substitute this expression for Δt_p into (8.33) to get

$$T_o = T_s \frac{(1 + \beta \cos \theta)}{\sqrt{1 - \beta^2}} \qquad (8.35)$$

However we usually are more interested in frequency rather than the wave period so remembering that these are related by $f = 1/T$ we have a relativistic doppler shift formula of

$$f_o = f_s \frac{\sqrt{1 - \beta^2}}{(1 + \beta \cos \theta)} \qquad (8.36)$$

where f_o is the observed frequency and f_s is the source frequency in its own reference frame. Hence the frequency shift only depends on the relative speed of the source with respect to the observer, v (which appears in β), as well as its direction, θ.

Now consider the case where the motion is directly away from the source and so $\theta = 0$. Under these conditions we have $\cos \theta = 1$ and so:

$$f_o = f_s \frac{\sqrt{1 - \beta^2}}{(1 + \beta)} = f_s \frac{\sqrt{(1 + \beta)(1 - \beta)}}{(1 + \beta)}$$

$$\implies \quad f_o = f_s \sqrt{\frac{1 - \beta}{1 + \beta}} = f_s \sqrt{\frac{c - v}{c + v}} \qquad (8.37)$$

For relative motion directly towards the observer we have $\theta = \pi$ and so $\cos \theta = -1$ which gives

$$f_o = f_s \sqrt{\frac{1 + \beta}{1 - \beta}} = f_s \sqrt{\frac{c + v}{c - v}} \qquad (8.38)$$

which appears significantly different for the non-relativistic doppler shift we calculated in (8.32) for the similar case where source and observer are moving towards one another.

For many cases where the doppler shift is encountered the relative velocity of the source with respect to the observer is a lot less than c and so β has a very small value. If we make the assumption that β is small enough that terms of order β^2 or higher are negligible then we can use the Taylor series expansion around $\beta = 0$ of our formula for motion towards the observer:

$$f_o = f_s(1+\beta)^{1/2}(1-\beta)^{-1/2} \approx f_s(1+\frac{1}{2}\beta+\ldots)(1+\frac{1}{2}\beta+\ldots) \quad (8.39)$$

Multiplying this out and again neglecting terms of order β^2 or higher we get

$$f_o \approx f_s(1+\beta) \quad (8.40)$$

and hence the change in frequency, defined as difference between the observed frequency and the source frequency ($f_o - f_s$), at velocities where $v \ll c$ is simply

$$\frac{\Delta f}{f_s} \approx \frac{v}{c} \quad (8.41)$$

when the relative velocity of the source is towards the observer. In the case where the same approximation holds ($v \ll c$) but the relative motion is away from the observer, the sign of the frequency change is flipped. It can be important to use this approximation when dealing with velocities well below c since the full formula contains the expressions $1 \pm \beta$ which may be so close to one that calculators, and if you are not careful even computers, may introduce large rounding errors.

Relativity also introduces a type of doppler shift which is not seen in the non-relativistic case. If we consider the situation where $\theta = \pi/2$ then the source is moving perpendicular to the observer and is neither increasing nor decreasing its distance to the observer. In the classical, non-relativistic case this results in no doppler shift since there is no motion towards, or away from, the source. However, for the relativistic case, we have $\cos\theta = 0$ and we get a doppler shift of

$$f_o = f_s\sqrt{1-\beta^2} \quad (8.42)$$

This is called the *transverse doppler shift* and is caused by the observer seeing the source undergoing time dilation due to the relative velocity. It is worth noting that at velocities well below c this effect is far smaller than the standard doppler effect. If we consider the case where $v \ll c$ then we can perform a Taylor expansion around $\beta = 0$ since $\beta = v/c$ and so will be very small. This gives:

$$f_o = f_s(1-\beta^2)^{1/2} \approx f_s\left(1-\frac{\beta^2}{2}-\ldots\right) \quad (8.43)$$

Hence the change in frequency between the source's frequency and the observed frequency is given by:

$$\frac{\Delta f}{f_s} \approx -\frac{1}{2}\left(\frac{v}{c}\right)^2 \qquad (8.44)$$

where $\Delta f = f_o - f_s$. This is always a reduction in frequency and since it is proportional to the square of v/c it gives a far smaller shift in frequency than the case for motion towards or away from the source where the shift is proportional to v/c.

Example 8.1

A car is driving directly towards a police speed radar trap at 80 km/h. If the frequency of waves which the radar gun emits is 10 GHz what is the difference in frequency of the reflected wave? Maintaining a constant speed the car draws level with the speed trap such that it is travelling perpendicular to the line of sight to the policeman. What is the new shift in frequency?

Solution:
First, we need to convert the speed of the car into a value for β given that c is 3×10^8 m/s:

$$\beta = \frac{v}{c} = \frac{80/3.6}{3 \times 10^8} = 7.407 \times 10^{-8}$$

With a value for β this small we clearly need to use the low-velocity approximation to avoid introducing large rounding errors in the calculation. Now we need to consider the effect of a reflection. The car will see an incoming frequency which is doppler shifted due to the motion of the policeman towards the car in the car's frame of reference. This is the frequency which the car will reflect i.e. the frequency which the car will emit in its own frame. However, the policeman sees this reflected wave source moving towards him and so the wave is again doppler shifted. Hence the change in frequency is twice that given in (8.41) because for a reflection there are two shifts and so:

$$\Delta f = 2f_s\frac{v}{c} = 2 \times 10 \times 10^9 \times 7.407 \times 10^{-8} = 1.48 \text{ kHz}$$

When the car draws level to the speed trap the only doppler effect will be due to the time dilation caused by the transverse motion of the car. Hence we use equation (8.44) for the transverse doppler effect remembering that the fact it is a reflection doubles the frequency change since the incoming wave before reflection is also shifted:

$$\Delta f = -f_s \frac{v^2}{c^2} = 10 \times 10^9 \times (7.407 \times 10^{-8})^2 = 54.9 \text{ } \mu\text{Hz}$$

This illustrates how small the transverse doppler effect is for typical, everyday velocities.

The relativistic doppler effect is an extremely important tool for astronomy. Edwin Hubble (1889-1953) discovered in 1929 that the visible spectrum of galaxies was shifted increasingly towards the red end of spectrum (so-called redshift) the further they are away from us. This discovery could only be explained by the expansion of space itself and lead to the current Big Bang model of cosmology. The doppler shift of starlight has also been used to measure the orbital velocity of stars around the centre of distant galaxies which has provided clear evidence of Dark Matter - a new type of matter which is not made of atoms and which physicists are still trying to understand.

Fig. 8.15: Edwin Powell Hubble (1889-1953) who, in 1929, used the doppler shift of distant galaxies to show that the universe was expanding.

Photo: Johan Hagemeyer, 1931 (public domain)

8.7 Shockwaves

The Doppler effect describes what happens to the frequency and wavelength of waves from a source which is moving through the wave medium at a speed less than that of the phase velocity of the waves. However, it is possible to move faster than the propagation speed of waves although how easy this depends on the type of wave at the medium being considered. For example, water waves typically travel at a few metres per second and you can drag your figure through water faster than the waves it produces whilst exceeding the speed of sound in air, approximately 340 m s^{-1}, requires considerable effort.

To see what happens lets first consider what happens when a source is moving at the same speed relative to the medium as the waves it produces. In this case, the source will emit a spherical wavefront and move at the same speed as the wavefront in the source's direction of motion. This means that when the next wave front is emitted it will add to the previous wavefront directly in front of the source which leads to a large amplitude wave arising directly in front of the source as shown in figure 8.16. For an aircraft approaching the speed of sound this build up of a large amplitude wave directly in front of the plane leads to a sudden increase in drag and other effects which prevented early jet aircraft from flying at supersonic speeds which lead to the phenomenon being called the sound barrier.

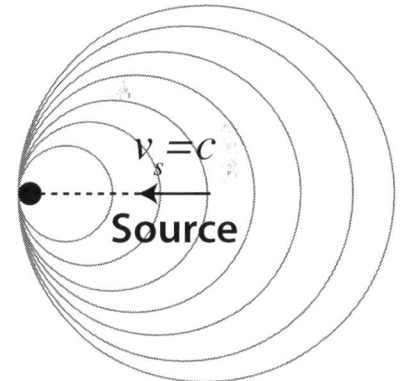

Fig. 8.16: A source travelling through a medium at the same speed as the phase velocity of the waves it emits will cause a large amplitude wave to build up directly in front of the source as shown here. For sound waves in air this led to a huge increase in drag on aircraft approaching the speed of sound, a phenomenon known as the sound barrier.

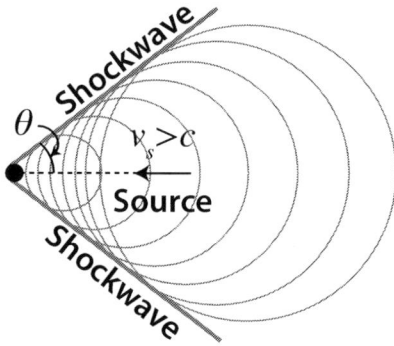

Fig. 8.17: When a source moves through a medium faster than the phase velocity of the waves it emits it creates a cone-shaped shockwave behind it from the constructive interference of the waves as shown here.

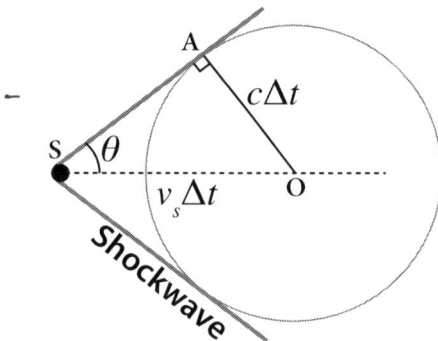

Fig. 8.18: Diagram showing how the opening angle of the shockwave, θ, is related to the speed of the source, v_s, and the phase velocity of the waves, c, it emits.

Once the speed of the waves in the medium is exceeded the large amplitude wave in front of the source dissipates to form a cone shaped shockwave which opens up behind the source as shown in figure 8.17. The shockwave is formed by the constructive interference of the waves emitted by the source as it travels through the medium. The faster the source the narrower the shockwave cone formed. To derive the relationship between the opening angle of the cone and the speed of the wave source we need to consider the distance travelled by both the source in a time Δt as well as the distance travelled by the waves it emits at the start of that same period. This is shown in figure 8.18.

Looking at the triangle SAO in figure 8.18 we can see that the sine of the angle θ is:

$$\sin\theta = \frac{c\Delta t}{v_s\Delta t} \tag{8.45}$$

and so the opening angle of the shockwave is simply:

$$\theta = \sin^{-1}\left(\frac{c}{v_s}\right) \tag{8.46}$$

Problems

Q8.1: Two identical loudspeakers are connected to the same frequency generator such that they produce the same constant frequency of sound with the same phase. A person stands in front of the speakers and at an equal distance from both. When the first speaker is turned on the listener hears a sound with an intensity of 60 dB(SIL). What intensity do they hear when the second speaker is turned on?

Q8.2: A string with a mass of 100 g is placed under tension between two clamps which are place 0.5 m apart.
(a) What are the wavelengths of the first three modes of vibration of the string?
(b) The string is plucked in such a way that it vibrates with two nodes between the clamps. The frequency of the vibration is measured as 400 Hz. What is the tension in the string?

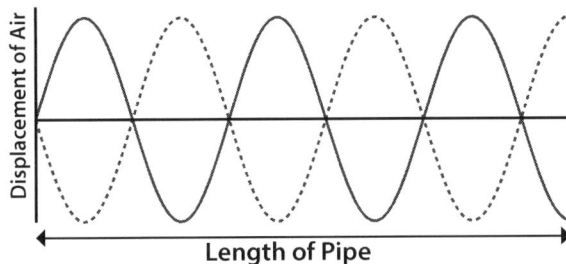

Fig. P8.1: *Displacement of air in a pipe for a particular standing wave mode.*

Q8.3: The air in a pipe which has one end closed and the other end open is made to vibrate by means of a loudspeaker. It is found that the lowest frequency which causes the pipe to resonate is f. The frequency is then increased until the standing wave shown in figure P8.1 is observed (the dashed line shows the displacement half a wave period later). What is the frequency of the speaker when this happens?

Q8.4: A tube with a cross-sectional area of 20 cm^2 has one end which is sealed and a second end which contains a moveable plunger which acts as an airtight seal. With the plunger set so that the tube is 1 m long the tube is filled

with a gas with a density of 5 kg m^{-3}. The gas in the tube is then made to vibrate by means of a loud speaker in the end of the tube and it is found that the tube resonates at both 150 Hz and again at 225 Hz. There are no resonances between these two frequencies but the loudspeaker cannot be driven much below 150 Hz so it is unknown whether there are any resonances with lower frequencies.
(a) Starting with the wavefunction $\Psi(x,t) = A\sin(kx)\sin(\omega t)$ show that the allowed wavelengths for acoustic waves in such a tube are $\lambda = 2L/n$ where L is the length of the tube and n is a positive integer.
(b) Calculate the speed of sound in the gas.
(c) For the 150 Hz resonance given above draw a diagram of the tube and label the approximate locations of *all* the nodes and anti-nodes.
(d) Given that the tube is sealed and assuming that the bulk modulus of the gas remains constant how much force is required to depress the plunger by 10 cm?

Q8.5: A long, thin piece of rubber with a mass of 1 g is 60 cm long and has a cross-section of 1 mm^2. Each end of this rubber band is then placed in a clamp and the clamps are moved so that the rubber band is under tension. Given that the Youngs modulus for rubber is 50 MN m^{-2} how far apart should the clamps be moved so that the rubber band has a fundamental mode with a frequency of 100 Hz?

Q8.6: Longitudinal waves travelling through a crystalline lattice obey a dispersion relation given by:
$$\omega = c\sin(ka/2)$$
where k is the wave number, a is the atomic spacing of the lattice and c is a constant.
(a) Derive the expression for the phase velocity, v_p, of the waves as a function of the wave number k and the constants a and c.
(b) Derive the expression for the group velocity, v_g, of the waves as a function of the wave number k and the constants a and c.
(c) What are the limiting values of both the

phase and group velocities for very long wave-length waves? [Hint: For small values of x, $\sin x = x - \frac{x^3}{3!} + \frac{x^5}{5!} \cdots$]

Q8.7: Submarine A is moving directly towards submarine B at a speed of $10 \, \text{m s}^{-1}$ relative to the water while submarine B is moving directly away from submarine A at a speed of $4 \, \text{m s}^{-1}$ relative to the water. Submarine A sends out an acoustic sonar pulse with a frequency of $2 \, \text{kHz}$ which travels with a speed of $1{,}480 \, \text{m s}^{-1}$ through the water and hits submarine B. What frequency does a detector on submarine B pick up when the sonar pulse hits it?

Q8.8: To avoid the issue of radar detectors a company decides to experiment with sonar, instead of radar, to measure a vehicle's speed. For the following situations calculate the change in frequency observed between the source and the wave reflected from a vehicle travelling at $20 \, \text{m s}^{-1}$ directly away from the source. Quote all answers to five significant figures. [Speed of sound in air is $340 \, \text{m s}^{-1}$, speed of light in vacuo is $3 \times 10^8 \, \text{m s}^{-1}$].

(a) A standard radar gun operating at $10 \, \text{GHz}$ both on a calm day with no wind and on a day when the wind is blowing at $10 \, \text{m s}^{-1}$ from the source towards the vehicle.

(b) A sonar gun generating ultrasound with a frequency of $25 \, \text{kHz}$ both on a calm day with no wind and on a day when the wind is blowing at $10 \, \text{m s}^{-1}$ from the source towards the vehicle.

(c) Why do these results suggest that a "sonar gun" might not be very practical for a police speed trap?

Q8.9: The IceCube detector is a cubic kilometre of ice approximately $2 \, \text{km}$ below the south pole. A subatomic particle called a muon passes through the detector travelling at approximately the speed of light in vacuum and as it does so it emits light. Since the muon is travelling faster than the speed of light in the ice it creates a shockwave of light called Cherenkov radiation. The opening angle of the shockwave produced is measured to be $44°$. What is the refractive index of the ice?

CHAPTER 9

Geometric Optics

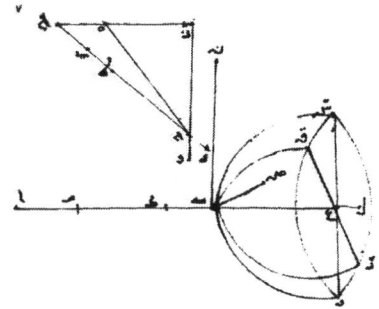

Fig. 9.1: Reproduction of a page from Ibn Sahl's manuscript where he uses the law of refraction to build a lens.
Ibn Sahl 984 AD, Wikimedia Commons.

Optics is one of the three fields, the other two being astronomy and mechanics, which were united together to form the subject of physics in ancient Greece by Aristotle who wrote the first book referring to the study of "Physics" in the 4th century BC. Optics itself dates back several centuries further.

The first study of geometric optics was performed by Euclid who used the concept of rays of light to explain perspective. However the Greeks were philosophers, not scientists, and their studies were based purely on logical reason, not experiment. It was not until the Middle Ages, just over 1,000 years later, that the study of optics was put on a firmer experimental basis by scientists in the Muslim world. In 984 AD Ibn Sahl (c. 940-1000 AD) described a law of refraction for light very similar to what we now call Snell's law and in the early 11th century Alhazen (Abu Ali AlHazan ibn Al-Hazan ibn Al-Haytham, 965-1040 AD) wrote the "Book of Optics" where he discussed reflection and refraction as well as attempting to explain vision and light. Sadly much of the latter is not correct but it was closer to the truth than the ideas it replaced! Crucially his work was based on observation and experiment and as a result, he is sometimes referred to as the father of optics.

Sadly Alhazen's work was largely ignored by the Arab world but around 1200 it was translated into Latin which made it accessible to scientists in Europe where it became the standard textbook on optics for over 400 years. By around 1286 Europeans invented the first wearable spectacles to correct poor eyesight and further experimentation led to the invention of the optical microscope (circa 1600) and the refracting telescope (1608)

Fig. 9.2: Portrait of Sir Isaac Newton (1642-1727) circa 1715 who was a major proponent of the corpuscular theory of light that described light as being made of tiny particles because it travelled in straight lines.
English School circa 1715-1720, Wikimedia Commons

which Galileo pointed at the sky a year later.

Despite these advances, the nature of light was still not known. In the 17th century British physicist Sir Isaac Newton (figure 9.2) described it as being transmitted by small particles which he called "corpuscles" while his contemporary, dutch physicist Christiaan Huyghens (figure 9.3) whom we previously met as at the inventor of the pendulum clock, proposed a competing wave theory of light. Interestingly, Newton attempted to prove Huyghens' wave theory but after failing to do so eventually gave up and declared that light had to consist of tiny particles and could not be a wave, a view he included in his book, "Opticks", published in 1704. The prestige of Newton meant that his corpuscular theory held sway until the 19th century when it was shown that light behaved as a wave.

We will start our discussion of light in this chapter with the early physics of light, a field called geometric optics, where light is modelled by rays which are drawn along the line of propagation of the light. We will delve a little into the wave nature of light when we use Huyghens' principle, based on his wave model for light, to explain the phenomenon of refraction.

9.1 Basic Principles

In geometric optics, we represent light by light rays which are perpendicular to the wave front of the light wave. These rays travel in straight lines until they encounter an object at which point they can be reflected, refracted or absorbed. These happen at the surface of an object and depend on the angle at which the light hits the surface. To quantify this behaviour we define the *angle of incidence* of the light to be the angle between the ray of light and a light perpendicular to the surface as shown in figure 9.4.

9.1.1 Reflection

Reflection is typically associated with smooth surfaces such as mirrors. However, any object which has a visible colour reflects light. For example, leaves are green because chlorophyll reflects green light and absorbs blue and red. The rough surface of a leaf at the scale of the wavelength of light means that the reflection is diffuse and does not form an image, as shown in figure 9.5, unlike a mirror which has a smooth surface and so will form an image.

Fig. 9.3: Portrait of Christiaan Huyghens (1629-1695) who presented his wave theory of light to the Paris Académie des Sciences in 1678.

Jacques Antoine Friquet de Vauroze, 1687-88.
(Rijksmuseum, Amsterdam)

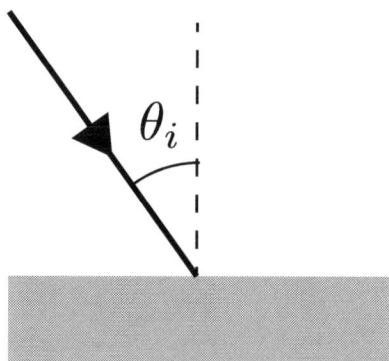

Fig. 9.4: Light ray striking a surface showing the angle of incidence, θ_i.

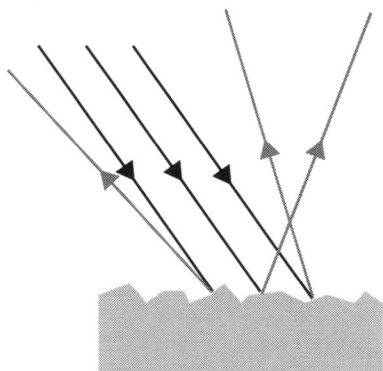

Fig. 9.5: Light ray striking the rough surface of a leaf showing show green light is reflected without generating an image.

To quantify reflection we define the *angle of reflection*, θ_r, as the angle between the normal to the surface and the reflected ray as shown in figure 9.6. The law of reflection then states that the angle of incidence is equal to the angle of reflection :

$$\text{Law of Reflection: } \theta_i = \theta_r \qquad (9.1)$$

This holds true for reflection at any surface. However in the case of diffuse reflection the changing angles of the rough surface result in the reflected light being scattered through many different angles because there any many different angles of incidence.

9.1.2 Refraction

Refraction occurs when a ray of light moves from one medium into another, for example from air to glass. This change in medium causes the ray to bend with the amount of bending, or *refraction*, being governed by Snell's law:

$$\text{Snell's Law: } \frac{\sin \theta_1}{\sin \theta_2} = \frac{n_2}{n_1} \qquad (9.2)$$

where θ_1 is the angle of incidence, θ_2 is the angle of refraction, n_1 is the *refractive index* for the first medium and n_2 is the *refractive index* for the second medium as shown in figure 9.7. The refractive index is a property of the material which we will learn more about in section 11.1. In general both refraction and reflection occurs at a surface which is why, when you look through a window, you can see a partial reflection superimposed on the view through the window. This phenomenon was used to great effect in Victorian theatre to make phantoms appear on stage through a system called "Pepper's Ghost". This technique is still used today in Disney theme parks and various museums around the globe.

It is important to note that if a ray strikes the surface perpendicular to it such that the angle of incidence is precisely $0°$ then the angle of refraction will be $0°$ as well for all materials. Refraction only occurs when there is a non-zero angle of incidence.

Example 9.1

A ray of light travelling through glass with a refractive index of 1.5 is incident on the surface of the glass at an angle of incidence of $30°$. Beyond the glass there is air with a refractive index of 1.0. What is the angle of refraction ?

Solution:

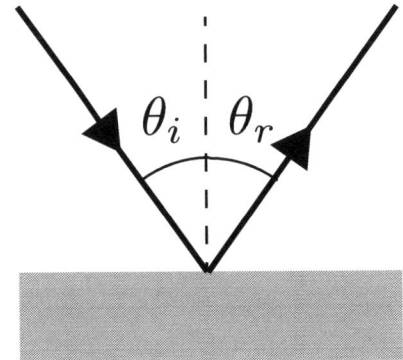

Fig. 9.6: Light ray striking a surface showing the angle of incidence, θ_i and the angle of reflection, θ_r. The law of reflection requires that $\theta_i = \theta_r$.

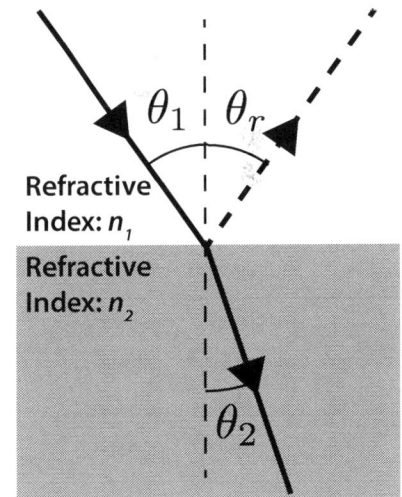

Fig. 9.7: Light ray being refracted at a surface between a medium with refractive index n_1 and another with refractive index n_2. Some of the light will be reflected as shown.

The light is starting in glass and so, using Snell's law as given in (9.2) we have $n_1 = 1.5$. The second medium it enters is air and so $n_2 = 1.0$. Putting the value for the angle of incidence into Snell's law and then rearranging we get:

$$\frac{\sin \theta_1}{\sin \theta_2} = \frac{1.0}{1.5} \implies \sin \theta_2 = 1.5 \sin 30°$$

and so the angle of refraction is just:

$$\theta_2 = 48.6°$$

Refractive Index: n_2

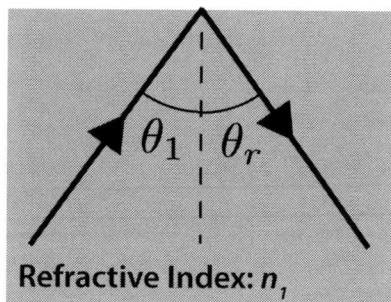

Refractive Index: n_1

Fig. 9.8: Total internal reflection of a light ray. For this to occur $n_2 < n_1$ and $\theta_1 > \theta_{\text{crit}}$. When it does occur by the law of reflection $\theta_r = \theta_1$.

9.1.3 Total Internal Reflection

For a ray propagating from a medium with a higher refractive index to one with a lower refractive index the angle of refraction is larger than the angle of incidence . So what happens if the angle of refraction is greater than 90°? In these cases the ray does not leave the medium at all and undergoes *total, internal reflection* as shown in figure 9.8.

To determine whether a ray will be refracted or internally reflected we can calculate the *critical angle* which is the angle at which a ray will have an angle of refraction, θ_2, of 90°. Putting this value into Snell's Law (9.2) and remembering that $\sin 90° = 1$ we get:

$$\sin \theta_{\text{crit}} = \frac{n_2}{n_1} \tag{9.3}$$

Note that (9.3) only has a solution if $n_2 < n_1$ which is the condition for the angle of refraction to be larger than the angle of incidence : you cannot get total internal reflection if the light ray is propagating into a medium with a higher refractive index than the one it is currently in.

9.2 Mirrors

The simplest way to form an image in optics is to use a mirror and the simplest of these is just the plane mirror. The earliest known manufactured mirrors are pieces of polished obsidian (volcanic glass) found in Turkey which date back to around 6,000 BC. The earliest mirrors made of metal consist of polished copper found in Mesopotamia and dating back to around 4,000 BC. The metal coated glass mirrors which we use today are believed to have been invented around the first century AD in what is now Lebanon with silver coated mirrors not being invented until 1835 by the German chemist Justus von Liebig.

Figure 9.9 shows how a plane mirror forms an image of an object placed in front of it. To determine where the image is formed we use a technique called *ray tracing*. Starting at the top of the object we draw a ray of light emanating from it and striking the surface of the mirror. Here we apply the law of reflection to determine where it goes after it is reflected. The reflected rays are is shown and, by tracing the path of these rays backwards, we see that they appear to emanate from a point behind the surface of the mirror. If we repeated this for each point on the object we would build up an image of the object behind the mirror and this is why you see an image of yourself behind the surface of the mirror when you stand in front of it.

The image produced in this way is called a *virtual image* because it doesn't really exist. The reflected rays only appear to come from an image behind the mirror but, as we can see in figure 9.9, this is an illusion because the rays all come from the surface of the mirror. This means that the image cannot be projected onto a screen and can only be seen when looking at the surface of the mirror.

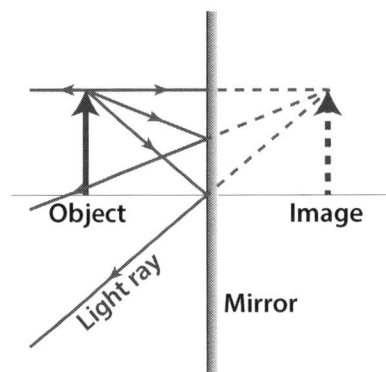

Fig. 9.9: Diagram of the light rays from an object striking a plane mirror and being reflected. Tracing the paths of the reflected rays backwards it is clear that they appear to emanate from a point behind the mirror which is where a virtual image will be produced.

Fig. 9.10: The concave face of the office building at 20, Fenchurch Street, London. Originally nicknamed the Walkie-Talkie because of its shape after the concave face focussed the sun's rays enough to melt plastic and burn carpets it got a new nickname: the Toaster.

9.2.1 Concave and Convex Mirrors

Concave mirrors are ones whose reflective surface is curved such that the edges of the mirror are nearer to the incident light rays than the centre of the mirror which means that the mirror can be used to focus light. An apocryphal story has it that Archimedes used a concave mirror to set fire to Roman ships which attacked Syracuse in 212 BC. Unfortunately, this is extremely unlikely to work with a single concave mirror and even tests using hundreds of smaller mirrors given to individuals to achieve the same result, as done in 2010 by the television show "Mythbusters", have failed to demonstrate it. However the physical principle is sound and will work with a sufficiently large surface area as shown by an office block at 20, Fenchurch Street, London, UK, shown in figure 9.10, which had a concave glass side so large that even the feeble British sun managed to melt plastic bodywork on parked cars and burn the floor mat of a shop until a protective awning was installed!

To understand how a concave mirror forms an image we first need to look at how such a mirror affects parallel rays of light. This is shown in figure 9.11 where several light rays travelling

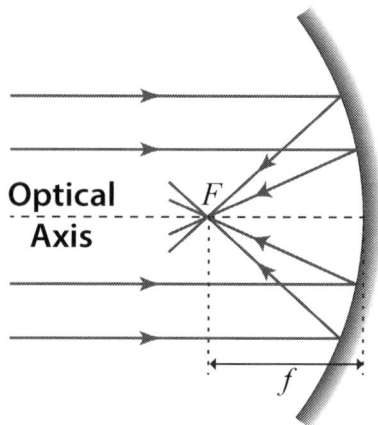

Fig. 9.11: A concave mirror showing how rays parallel to the optical axis are focussed to a point, F, called the focal point. The distance f between the focal point and the mirror along the optical axis is called the focal length.

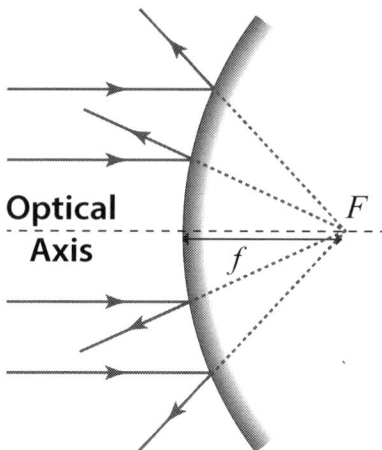

Fig. 9.12: A convex mirror showing how rays parallel to the optical axis are reflected so that they all appear to have come from a point, F, behind the mirror which is called the focal point. The distance f between the focal point and the mirror along the optical axis is called the focal length.

parallel to the axis of symmetry of the mirror, which is called the optical axis, are incident on the mirror. As shown these rays are reflected such that they all pass through a single point in front of the mirror. The point which they pass through is called the focal point which lies on the axis of symmetry of the mirror which is called the optical axis. The distance between the focal point and the mirror along the optical axis is called the focal length which is denoted as f in figure 9.11.

A convex mirror is one where the centre of the mirror is nearer to the incident light than the edges of the mirror. This has the opposite effect to a concave mirror in that rays parallel to the optical axis are reflected so that they all appear to emanate from a single point behind the mirror, as shown in figure 9.12. Like the concave mirror, this point is called the focal point the difference being that the rays only appear to come from a single point but actually do not.

9.2.2 Spherical Mirrors

The simplest type of concave or convex mirrors to consider are spherical mirrors. These are mirrors where the reflective surface is part of a sphere. To calculate the relationship between the focal length and the radius of curvature, which is the radius of the sphere, we need to consider some basic geometry for a single ray parallel to the axis of symmetry which reflects off a concave mirror as shown in figure 9.13. If we draw a line from the centre of the sphere to the point where the ray hits this line will be perpendicular to the surface of the mirror and so the angle between the incident ray and this radial line will be the angle of incidence, labelled as θ in figure 9.13. The law of reflection then tells us that angle of reflection equals the angle of incidence and so the angle between the reflected ray and this radial line is also θ as shown in figure 9.13.

Now since the incident ray is parallel to the optical axis the reflected ray will pass through the focal point, F, and the alternate angles theorem for a line intersecting two parallel lines tells us that the angle $A\widehat{C}B = \theta$ and the angle $A\widehat{F}B = 2\theta$ as shown on the diagram. Using the definition of an angle in radians and looking at the sector of a circle defined by ABC we have:

$$\theta = \frac{s}{r} \tag{9.4}$$

where s is the length of the arc between A and B. Now for a small value of θ the shape defined by the points ABF will also be approximately the same as a sector of a circle with F at the centre. For many concave mirrors this is not a bad approximation since the size of the mirror is usually a lot less than the radius of curvature which means $s << r$ which will make θ small. Using this approximation we can then write:

$$2\theta \approx \frac{s}{f} \qquad (9.5)$$

Combining equations (9.4) and (9.5) we get:

$$r \approx 2f \qquad (9.6)$$

which is the relationship between the radius of curvature and the focal length for a concave, spherical mirror.

The same situation for a convex, spherical mirror is shown in figure 9.14. Here the angles $\widehat{ACB}$ and $\widehat{AFB}$ are determined using the corresponding angles theorem for a line intersecting two parallel lines but have the same values as they did for the concave mirror. We can then look at the same sectors, ACB and AFB and make the same approximation to get the same result shown in (9.6). Hence if you wish to make a spherical mirror, either convex or concave, which has a focal length of f you need to use a sphere with a radius of twice this.

Fig. 9.13: A single ray travelling parallel to the optical axis is incident on a spherical, concave mirror. The angle of incidence is θ, the centre of the sphere is point C and the focal point is F. The sphere's radius, r, and the focal length, f, are also shown.

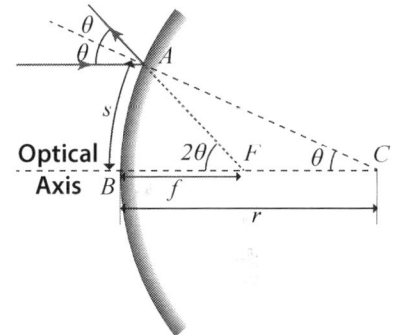

9.2.3 Ray Tracing

To study how a spherical mirror creates images we need to determine the paths of lights rays from an object which reflect off the mirror's surface to generate the image. This technique is called ray tracing and to find the location of the image produced we will trace the path of three special rays, called principal rays. These are rays which follow easily determined paths and where they meet or appear to meet, is where the image will be formed. It is important to remember that we only use the principal rays as a guide to determine the location of the image and that there is a ray for each point on the object and each angle from that point. These numerous rays create the entirety of the image. In addition, for spherical mirrors only, there is a fourth possible principal ray which comes from the centre of the sphere. However, as we shall discuss later, spherical mirrors are not the ideal shape for convex and concave mirrors and so in many cases, this fourth possible principal ray does not apply because the mirror will not be spherical.

For a spherical mirror the four principal rays are:

1. A ray parallel to the optical axis will be reflected to go through the focal point.

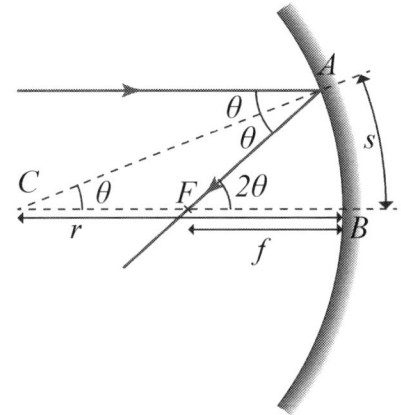

Fig. 9.14: A single ray travelling parallel to the optical axis is incident on a spherical, convex mirror. The angle of incidence is θ, the centre of the sphere is point C and the focal point is F. The sphere's radius, r, and the focal length, f, are also shown.

2. A ray which comes, or appears to come, from the focal point will be reflected to travel parallel to the optical axis.

3. A ray which hits the centre of the mirror where the optical axis touches the surface will be reflected such that the angle between the reflected ray and the optical axis is equal but opposite to the incident ray.

4. **[Spherical Mirrors Only]** A ray which comes, or appears to come, from the centre of the sphere will be reflected back along the same path.

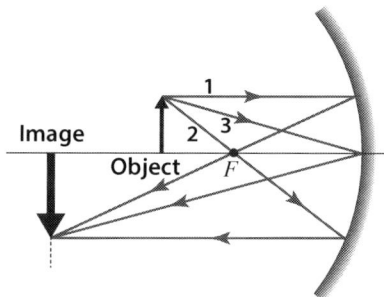

Fig. 9.15: A concave mirror showing that an object placed further away from the mirror than the focal point will generate a real, inverted and magnified image. The three principle rays are numbered using the same system given in the text.

All we now need to do is place an object in front of a mirror and then trace these rays from the tip of the object to see where they meet, or in the case of a virtual image where they appear to meet, and that will be the tip of the image. Figure 9.15 shows the principal rays for an object placed further from a concave mirror than the focal point. Each ray is labelled with the number from the list above. The ray tracing shows that the light rays actually meet at a single point so the image formed is a real image which could be projected onto a screen and, as shown, it will also be inverted.

A different type of image formation for a concave mirror is shown in figure 9.16 where the object is placed nearer to the mirror than the focal point. In this situation a virtual image is formed. This is different from a real image in that the rays of light only appear to come from the image but, in reality, do not. The image appears to be behind the mirror and can only be viewed when looking at the surface of the mirror. This type of image cannot be projected onto a screen. It is also worth noting that the although the principal ray from the focal point does not actually pass through that point it makes no difference to how it is reflected: reflection depends solely on the angle of incidence on the mirror, not the distance travelled by the ray.

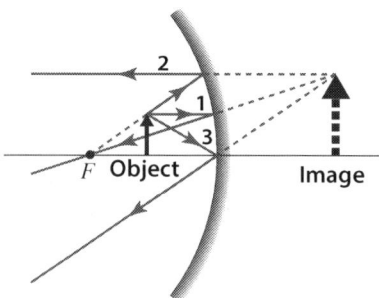

Fig. 9.16: A concave mirror showing that an object placed nearer to the mirror than the focal point will generate a virtual image which is not inverted. The three principal rays are labelled with the numbers given in the text.

Having considered an object on either side of the focal point the final situation to consider for a concave mirror is an object placed at the focal point. In this situation, the principal ray from the tip of the object through the focus point will be vertically down and so will not strike the mirror. While only two rays are needed to show where the image is formed in this case we will use the special principal ray which applies only to spherical mirrors and include the ray which passes through the centre of the sphere which is twice the focal length away from the centre of the mirror.

The ray diagram is for this situation is figure 9.17 which shows that the reflected rays are all parallel to each other. In this case, the image is said to be produced at infinity i.e. an infinite distance from the mirror. While this might seem a strange concept the reverse, taking an image at infinity and creating an image nearby is the principle behind all astronomical telescopes as well as photography of distant objects.

The final image generating scenario for curved mirrors is that of a convex mirror. Here the focal point is behind the mirror so there is only one situation to consider which is shown in figure 9.18. The image produced is always virtual and is always smaller than the object. This has the advantage that the mirror will compress a large field of view into a smaller area which is why wing mirrors of North American cars are slightly convex. The price paid for this increased field of view is that the image of an object is smaller and thus they appear further away compared to a flat mirror which is why these mirrors carry the warning "Objects in the mirror are closer than they appear". This is also why cars in Europe and elsewhere do not use convex wing mirrors on vehicles: the problem of the distance distortion effect is deemed to outweigh the benefits of a larger field of view.

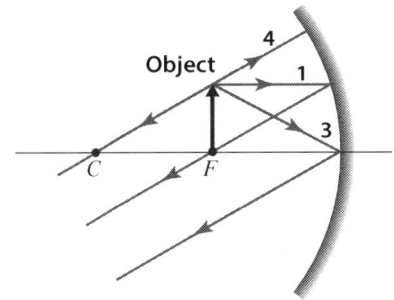

Fig. 9.17: A concave mirror showing that an object placed at the focal point will generate an image at infinity. The rays are all parallel and do not converge and so the image is effectively an infinite distance behind the mirror. This diagram includes the special, fourth principal ray which only applies to spherical mirrors and which passes through the centre of the sphere, labelled C.

9.2.4 The Mirror Equation

Now that we have the geometry of the lights rays determined we can derive an equation which links the positions of the object and the image to the focal length of the mirror. To do this we first need to define some quantities:

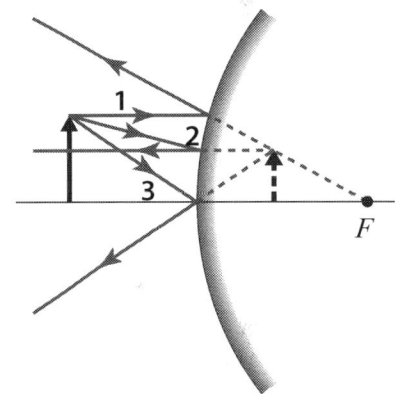

Fig. 9.18: A convex mirror showing that the image produced by an object is virtual and smaller than the object itself. The three principal rays are labelled with the numbers given in the text.

u the distance from the centre of the mirror to the object

v the distance from the centre of the mirror to the image

f the focal length of the mirror

O the height of the object

I the height of the image

Figure 9.19 shows the same ray diagram we had before but this time with all the relevant distances labelled using the scheme above. First lets consider the triangles formed by the third principle ray shown in figure 9.15 with the object and the image. This ray reflects off the centre of the mirror where the surface is vertical and perpendicular to the optical axis and so the angle above and below the optical axis will be equal by the law of reflection. If we take the tangent of this angle, labelled ϕ in

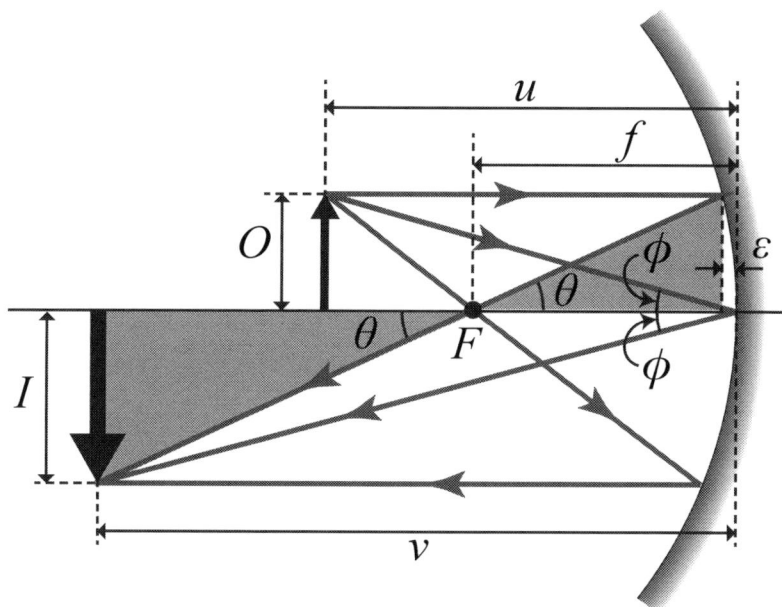

Fig. 9.19: A concave mirror with an object places further from the mirror than the focal point. The distances between the mirror and the object and the image are labelled as u and v respectively. The green triangles are used to derive the equation which relates the positions of the object an image to the focal length of the mirror.

figure 9.19, we get for both the triangles:

$$\tan\phi = \frac{O}{u} = \frac{I}{v} \implies \frac{O}{I} = \frac{u}{v} \tag{9.7}$$

Next we consider the two shaded, right-angled triangles shown in figure 9.19 which both have an angle θ in one corner. Using the definition for $\tan\theta$ and applying it to both of these triangles we get:

$$\tan\theta = \frac{O}{f - \epsilon} = \frac{I}{v - f} \implies \frac{O}{I} = \frac{f - \epsilon}{v - f} \tag{9.8}$$

However as we previously mentioned when deriving the relationship between the focal length and the radius of curvature the size of the mirror is typically a lot smaller than the radius of curvature and so $r \gg O$. In this situation the distance ϵ will be very small and so we can make the approximation that $\epsilon \approx 0$ and equation (9.8) becomes:

$$\frac{O}{I} \approx \frac{f}{v - f} \tag{9.9}$$

All that remains now is to combine equations (9.7) and (9.9) to eliminate the sizes of the object and image. This gives:

$$\frac{u}{v} = \frac{f}{v - f} \implies \frac{1}{u} + \frac{1}{v} = \frac{1}{f} \tag{9.10}$$

This is the mirror equation which relates the positions of the object and image to the focal length of the mirror. However to

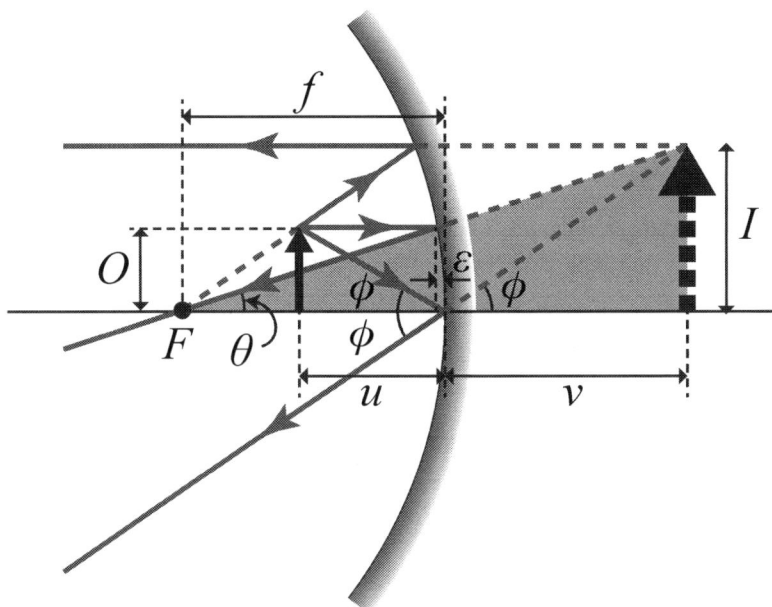

Fig. 9.20: A concave mirror with an object placed nearer to the mirror than the focal point. The distances between the mirror and the object and the image are labelled as u and v respectively. The green triangles are used to derive the equation which relates the positions of the object an image to the focal length of the mirror.

understand how it works in general we need to consider two more cases: an object closer than the focal point for a concave mirror (figure 9.16) and an object in front of a convex mirror (figure 9.18).

We will start by considering the virtual image produced by an object closer to a concave mirror than the focal length. Figure 9.20 shows the same ray diagram as before but now with angles and distances labelled using the same system as before. Again we will first consider the triangles formed by the third principle ray shown in figure 9.16 with the object and the image. This ray reflects off the centre of the mirror and creates two right-angled triangles with the same angle, ϕ, at the vertex which is located at the centre of the mirror. If we look at the tangent of the angle ϕ for the two triangles we get the equation:

$$\tan\phi = \frac{O}{u} = \frac{I}{v} \quad \Longrightarrow \quad \frac{O}{I} = \frac{u}{v} \qquad (9.11)$$

which is the same relationship we had before for a real image. Next, we consider the triangle shaded in green on the diagram. Using this triangle we can come up with two expressions for the tangent of the angle θ just as we did before:

$$\tan\theta = \frac{O}{f-\epsilon} = \frac{I}{f+v} \quad \Longrightarrow \quad \frac{O}{I} = \frac{f-\epsilon}{f+v} \qquad (9.12)$$

Once again we shall make the approximation that the radius of curvature of the mirror is a lot larger than the size of the object and so $\epsilon \approx 0$. Doing this we can now combine (9.11) and (9.12) to get:

$$\frac{u}{v} = \frac{f}{f+v} \implies \frac{1}{u} - \frac{1}{v} = \frac{1}{f} \qquad (9.13)$$

So we have almost the same equation again except that the v term is negative. The reason for this is that in this case the focal length and distance to the image add when considering the second triangle because the image is on the opposite side of the mirror to the object. This is a characteristic of all virtual images: real images must be formed on the same side as the object. Hence we can conclude that the same mirror equation works provided that we define v as being negative for virtual images.

The final situation to consider is the virtual image produced by a convex mirror. Figure 9.21 shows the same ray diagram as before but with the distances labelled using the same naming convention as we used in the two previous cases. Once again we will consider the triangle formed by the principle ray striking the centre of the mirror and the optical axis of the mirror. Taking the tangent of the angle ϕ we have the relationship:

$$\tan\phi = \frac{O}{u} = \frac{I}{v} \implies \frac{O}{I} = \frac{u}{v} \qquad (9.14)$$

which once again is the same relationship we had before. Repeating our method we now look at the green shaded triangle and consider the tangent of the angle θ which we can write down in two equivalent ways:

$$\tan\theta = \frac{O}{f-\epsilon} = \frac{I}{f-v} \implies \frac{O}{I} = \frac{f-\epsilon}{f-v} \qquad (9.15)$$

Again we make the approximation that the radius of curvature is a lot greater than the size of the object and $\epsilon \approx 0$. Doing this we then combine (9.14) and (9.15) to eliminate O/I and get:

$$\frac{u}{v} = \frac{f}{f-v} \implies \frac{1}{u} - \frac{1}{v} = -\frac{1}{f} \qquad (9.16)$$

Again this looks very similar to our original mirror equation and as before we see that v is negative for a virtual image. However in this case f is also negative and so we have the final rule which is that for convex mirrors we need to use a negative focal length. Hence we now have a general mirror equation which takes the form:

$$\frac{1}{u} + \frac{1}{v} = \frac{1}{f} \qquad (9.17)$$

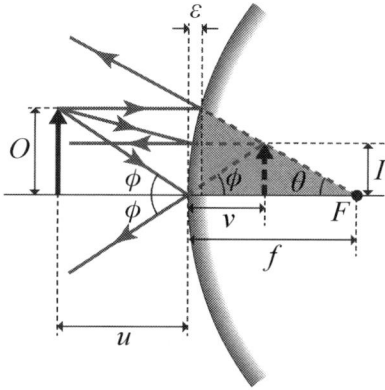

Fig. 9.21: A convex mirror generating a virtual image of an object placed in front of it. The distances between the mirror and the object and the image are labelled as u and v respectively. The green triangles are used to derive the equation which relates the positions of the object an image to the focal length of the mirror.

which applies to all spherical mirrors in the case where the radius of curvature is a lot greater than the size of the object provided that we use a negative value for v when dealing with virtual images and a negative value for f when using a convex mirror.

Using equation (9.7) which we showed held true in all three circumstances, we can define the magnification of the image as the ratio of the image size to the object size which gives:

$$\text{Magnification,} \ M = \frac{I}{O} = -\frac{v}{u} \qquad (9.18)$$

where u and v use the sign convention we established for the mirror equation. The height of the object is always taken as positive but, for an inverted, real image I will be negative. Hence for an inverted, real image, we have a negative magnification but for a virtual, non-inverted image we have a positive magnification.

9.3 Lenses

In the previous section, we have seen how reflection can be used to generate a modified image of an object. However, it is possible to generate exactly the same types of image using refraction instead of reflection. This is done using a lens which is a device with two refracting surfaces that is used to focus or defocus light rays.

The oldest known lens is the Nimrud Lens which dates back 2,700 years to ancient Assyria and which is currently located in the British Museum in London, UK. The simplest lenses consist of two spherical surfaces which are very close together so that the distance a light ray travels between the two of them is negligible. These are called *thin lenses* and are what we will discuss here. There are two basic types of thin lens: converging and diverging. Converging lenses have a thicker centre and narrow extremities whilst diverging lenses have thicker extremities and a narrow centre. As shown in figure 9.22 each basic type can come in several different forms.

The optical axis of a lens is the axis of rotational symmetry of the lens which passes through the centre of the lens such that it is perpendicular to both of the lens' surfaces. Somewhere along this axis, an equal distance on either side of the lens lie the focal points of the lens. For a converging lens, this is the point at which any ray parallel to the optical axis will be forced

Converging Lenses

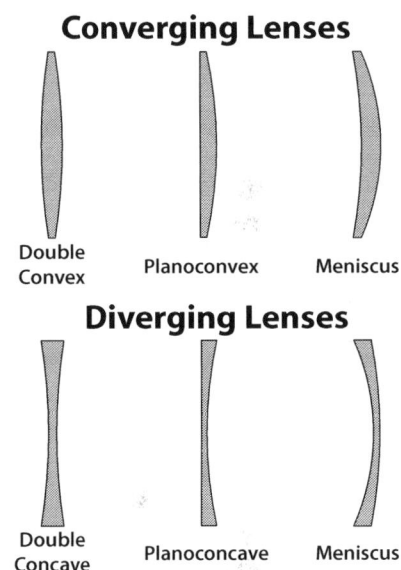

Double Convex   Planoconvex   Meniscus

Diverging Lenses

Double Concave   Planoconcave   Meniscus

Fig. 9.22: Different types of converging and diverging lenses.

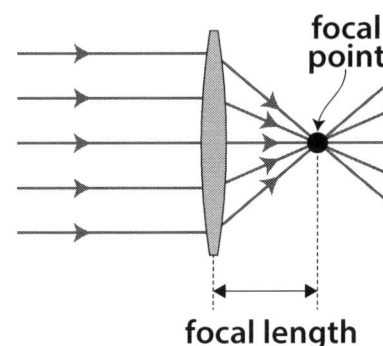

focal point

focal length

Fig. 9.23: Focal point of a converging lens.

Fig. 9.25: Empty lens diagram showing the focal points and optical axis. The vertical line represents the lens with the image of the lens there to indicate the lens type.

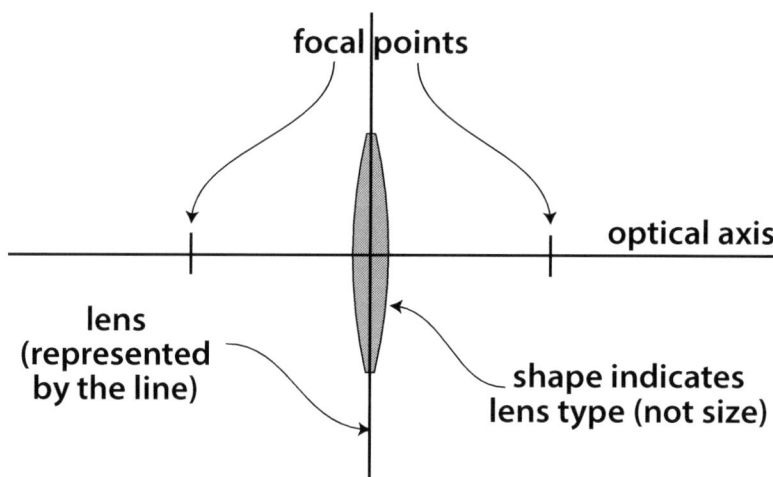

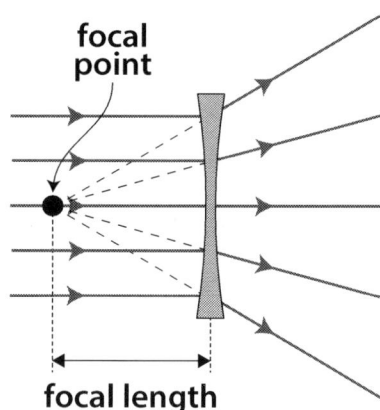

Fig. 9.24: Focal point of a diverging lens.

to pass through as shown in figure 9.23. For a diverging lens, the focal point is the point from which all rays parallel to the optical axis will appear to emanate as shown in figure 9.24. The distance between a focal point and the centre of the lens is called the focal length.

9.3.1 Ray Tracing

Just as we did with mirrors to study how a lens forms an image we first need to develop a systematic way to draw a lens and to trace the principal rays which form the image. To start with we draw a vertical line which represents the lens and we draw a small representation of the lens at the centre of this line. It is important to note that the line itself represents the thin lens, the small representation is only used to denote whether this is a converging or diverging lens. Next, a horizontal line is drawn through the centre of the vertical line which represents the optical axis of the lens. Finally, at an equal distance along the optical axis on either side of the lens, the focal point of the lens is marked as shown in figure 9.25.

Once such a diagram has been drawn an object can be placed on it and the three, principal light rays traced to determine where, if at all, an image will be created. The rules for drawing these rays are as follows:

1. Draw a ray parallel to the optical axis. This will be bent by the lens to pass through the focal point on the opposite side (converging lens) or so that it appears to come

from the focal point on the same side as the incident ray (diverging lens).

2. Draw a ray through the centre of the lens which will be undeflected by the lens because both surfaces of the lens are parallel at the centre.

3. Draw a ray which passes through the focal point on the same side (converging lens) or which would pass through the focal point on the remote side (diverging). The lens will deflect this ray so that it emerges parallel to the optical axis.

If the three principal rays coming from a single point of the object meet together at a single point elsewhere a real image will be formed at that point. Of course, there will be far more rays than just the three principal rays forming the image but these are the rays which follow easily drawn rules. To apply these rules let's consider the image produced by a converging lens for an object that is beyond the focal length of the lens. The three principle rays for this setup are shown in figure 9.26 with the numerical labels on the rays corresponding to the rules for the principal rays given above. The point where all the rays from the top of the object meet will be the top of the image. As is shown the lens produces a real image which is on the opposite side of the lens and which is inverted relative to the object.

Suppose that we now consider what happens when the object is placed closer to a converging lens than the focal point. The three principal rays, in this case, are shown in figure 9.27, with the third ray this time being projected as coming from the focal point, and the result is strikingly different. Instead of a real image, this has produced what is called a *virtual image*. The three principal rays from the top of the object do not converge at a point as before but instead diverge. However, if these rays are traced backwards in a straight line then they do converge at a point. Hence the three rays will appear to be coming from a point behind the lens and so this is where an image will appear. Unlike a real image, this image cannot be projected onto a screen. It is effectively an optical illusion created by the refraction of the lens and it can only be viewed by looking through the lens.

9.3.2 The Lens Equation

Having now established that a lens can generate either a real or a virtual image we want to know exactly where such an im-

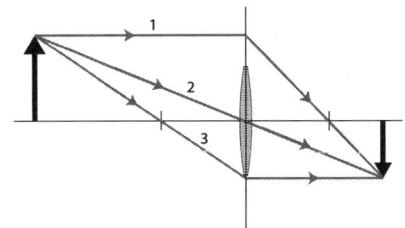

Fig. 9.26: Converging lens forming the image of an object which is placed at a distance greater than the focal length. The resulting image is real and inverted.

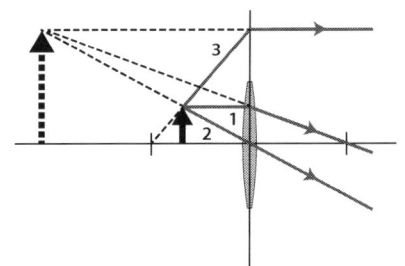

Fig. 9.27: A converging lens forming the image of an object which is placed at a distance less than the focal length. Note that the third principal ray is projected as coming from the focal point but does not actually do so. The resulting image is virtual but not inverted.

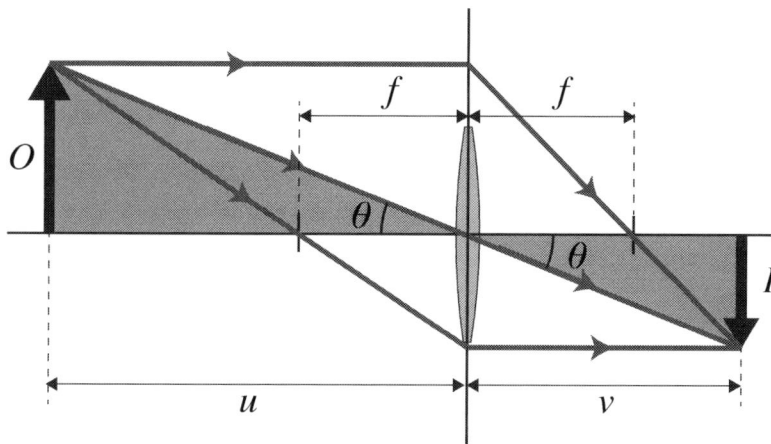

Fig. 9.28: Converging lens forming the image of an object with the distance to the object showing a pair of similar, right-angled triangles used in the derivation of the lens equation.

age will be produced, what type the image will be as well as how large the image will be compared to the original object. To derive the mathematical relationships we first need to define several quantities as shown in figure 9.28 and we will use exactly the same scheme that we used when discussing mirrors:

u the distance from the centre of the lens to the object

v the distance from the centre of the lens to the image

f the focal length of the lens

O the height of the object

I the height of the image

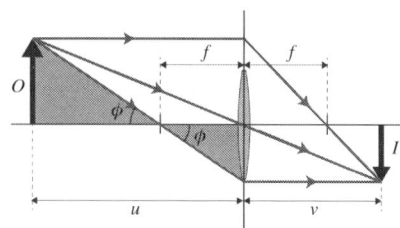

Fig. 9.29: Converging lens forming the image of an object with the distance to the object showing a second pair of similar, right-angled triangles used in the derivation of the lens equation.

To derive the relationship between these quantities lets consider the two, right-angled triangles shown in figure 9.28. Either by using similar triangles or by considering $\tan \theta$ we have the relationship:

$$\frac{I}{v} = \frac{O}{u} (= \tan \theta) \implies \frac{I}{O} = \frac{v}{u} \qquad (9.19)$$

Next we consider the two, right-angled triangles shown in figure 9.29. Again either by using similar triangles or by considering $\tan \phi$ we have the relationship:

$$\frac{I}{f} = \frac{O}{u - f} (= \tan \phi) \implies \frac{I}{O} = \frac{f}{u - f} \qquad (9.20)$$

Combining equations (9.19) and (9.20) to eliminate I and O we have:

$$\frac{v}{u} = \frac{f}{u - f} \implies uv - vf = uf \implies vf + uf = uv \qquad (9.21)$$

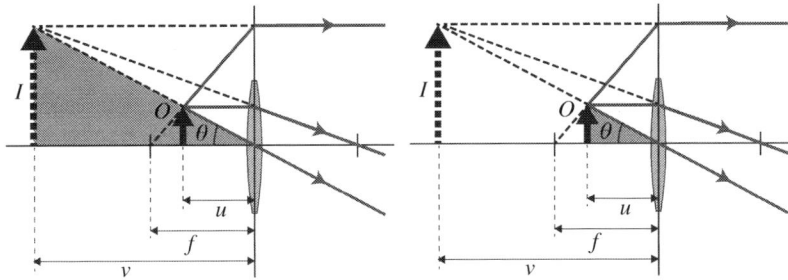

Fig. 9.30: The first pair of similar triangles which are used to derive the lens equation for a virtual image produced by a converging lens. These correspond to the equivalent triangles, which each have a vertex at the centre of the lens, for the case of a real image which are shown in figure 9.28.

All that remains is to divide (9.21) throughout by uvf and we get:

$$\frac{1}{u} + \frac{1}{v} = \frac{1}{f} \qquad (9.22)$$

which is called the lens equation. While we have shown the derivation here using a converging lens creating a real image the same equation applies to all thin lenses where a real or virtual image of the object is created provided that a suitable sign convention is used because the same pairs of triangles always exist.

To determine the sign convention lets consider a virtual image produced by a converging lens. The first pair of similar triangles, which each have a vertex at the centre of the lens, appear as shown in figure 9.30. Again the hypotenuse of these triangles is formed by the light ray from the object which passes through the centre of the lens. However, in this case, the triangles are both on the same side of the lens. Just as before this same pair of triangles gives us the same relationship as before which is given in equation (9.19). If we look at the second set of similar triangles, shown in figure 9.31, which have a hypotenuse formed by the ray from the object that appears to come from the focal point, then the equation is similar, but not quite the same as before:

$$\frac{I}{f} = \frac{O}{f-u}(= \tan\phi) \implies \frac{I}{O} = \frac{f}{f-u} \qquad (9.23)$$

which is not the same as (9.20). So substituting in for I/O from (9.19) we now have:

$$\frac{v}{u} = \frac{f}{f-u} \implies vf - uv = uf \implies \frac{1}{f} = \frac{1}{u} - \frac{1}{v} \qquad (9.24)$$

where we again divided through by uvf. This shows that we have something very similar to our original lens equation but with the sign of the $1/v$ term flipped which is due to the image being a virtual one and so being on the same side of the lens

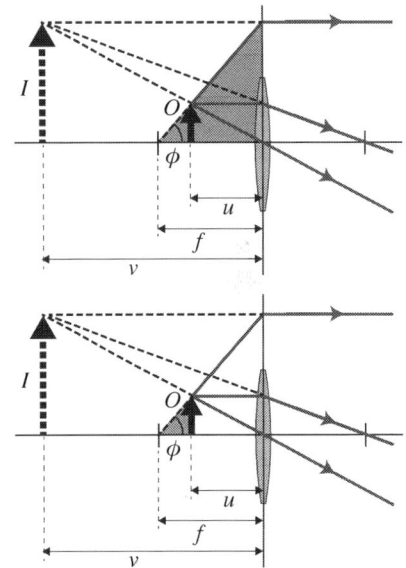

Fig. 9.31: The diagrams above give the second pair of similar triangles which are used to derive the lens equation for a virtual image produced by a converging lens. These correspond to the equivalent triangles, which each have a vertex at a focal point, for the case of a real image which are shown in figure 9.29.

Fig. 9.32: Production of a smaller, virtual image of an object by a diverging lens. This figure shows the first pair of similar triangles used to derive the lens equation.

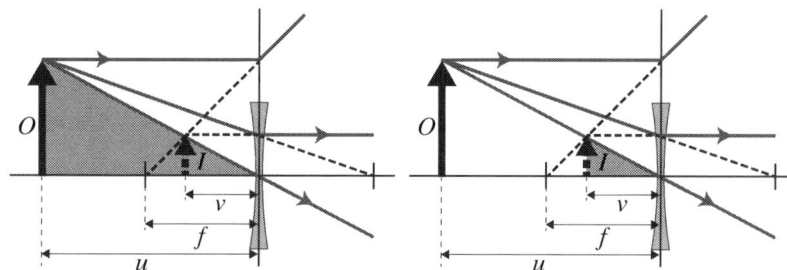

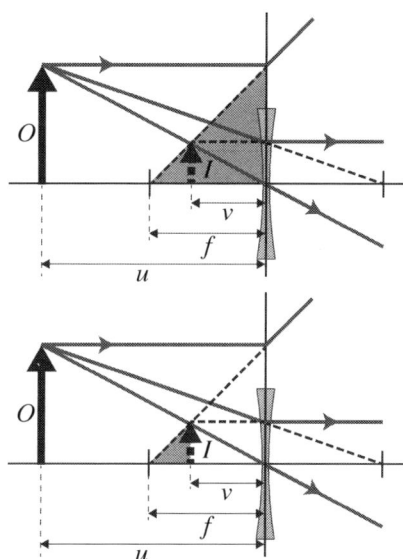

Fig. 9.33: Production of a smaller, virtual image of an object by a diverging lens. This figure shows the second pair of similar triangles used to derive the lens equation.

as the object. Hence the sign rule, in this case, is clear: v is to be taken as negative for virtual images.

The final case to consider is that of a diverging lens. In this case, for an object placed beyond the focal length of the lens, the result will be a virtual image which is smaller than the object and nearer to the lens. Again we have the same pairs of similar triangles and the ray diagram with the first pair highlighted, each of which has a vertex at the centre of the lens, is shown in figure 9.32. These triangles will give us equation (9.19) as before.

The second pair of triangles, which have vertices at the focal point, are shown in figure 9.33. Looking at the right hand diagram in figure 9.32 these triangles will give us the relationship:

$$\frac{I}{f-v} = \frac{O}{f} \implies \frac{I}{O} = \frac{f-v}{f} \tag{9.25}$$

Now substituting in for I/O from (9.19) we get:

$$\frac{v}{u} = \frac{f-v}{f} \implies vf = uf - uv \implies -\frac{1}{f} = \frac{1}{u} - \frac{1}{v} \tag{9.26}$$

Hence we again end up with something similar to the original lens equation. Since the image produced is virtual and on the same side of the lens as the object following our previous convention we would expect the $1/v$ term to be negative. However we also have a negative $1/f$ term which gives rise to our final sign convention: f is negative for a diverging lens.

Hence in summary the lens equation is:

$$\frac{1}{u} + \frac{1}{v} = \frac{1}{f} \tag{9.27}$$

where v is negative for a virtual image and f is negative for a diverging lens.

Just we did with mirrors we can also use equation (9.19), which we showed held true in all three circumstances, to define the magnification of the image. This is defined using the ratio of the image size to the object size and is given as:

$$\text{Magnification, } M = \frac{I}{O} = -\frac{v}{u} \qquad (9.28)$$

where the sign convention we used for the lens equation applies for u and v. The height of the object is always taken as positive but, for an inverted, real image I will be negative, as seen in figure 9.28. Hence for an inverted, real image, we have a negative magnification but for a virtual, non-inverted image we have a positive magnification.

Problems

Q9.1: A ray of light travelling through air ($n = 1.0$) strikes the surface of a transparent block at an angle of incidence of 40°. If the angle of refraction of the ray is 25° what is the speed of light in the block? [Speed of light in vacuo is $3 \times 10^8 \, \text{m s}^{-1}$]

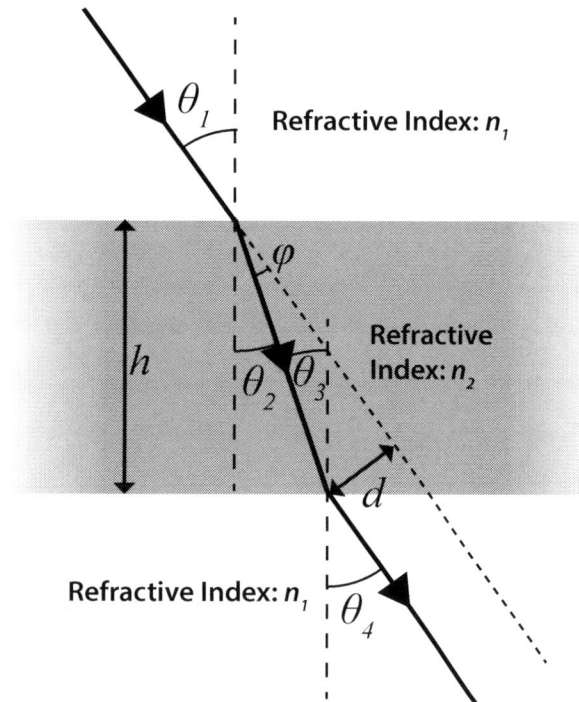

Fig. P9.1: *Diagram of a ray of light striking the face of a sheet of transparent material.*

Q9.2: A ray of light travelling through air ($n = 1.0$) strikes a sheet of transparent material at an angle of incidence of 35° and emerges again on the other side back into air as shown in figure P9.1. The sheet of material has a constant thickness of 2 cm.

(a) Prove that it emerges on the other side of the sheet with a direction parallel to the direction it had before it was incident on the sheet.

(b) If the refractive index of the sheet is 1.6 by what distance is the light ray displaced?

Q9.3: A shaving mirror has a spherical, concave surface with a radius of curvature of 45 cm. A man using it sees an upright image of his face

which has a linear magnification factor of 3.

(a) What type of image is he seeing and on which side of the mirror is it located?

(b) How far from the centre of the mirror is his face?

Q9.4: A small wheeled robot is moving along the optical axis towards a spherical mirror which has a radius of curvature, r. The robot contains a single, bright white LED which acts as a point source of light. If the distance to the centre of the mirror is u and the velocity of the robot is $\dot{u}$ what is the velocity of the image formed by the LED?

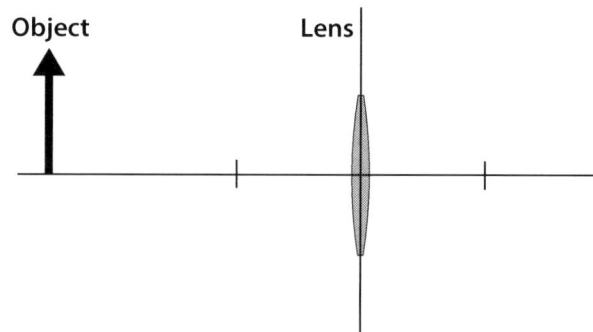

Fig. P9.2: *An object placed in front of a converging lens further away from the focal point.*

Q9.5: An object placed in front of a converging lens such that the object is further from the lens than the focal length of the lens as shown in figure P9.2.

(a) Draw a diagram showing the three principal rays from the top of the object. Draw and label clearly any image which is formed by these rays and state whether it is real or virtual.

(b) A converging lens has a focal length of 20 cm and an object with a height of 2 cm is placed a distance of 25 cm from the lens. For this case calculate the distance from the lens to the image and the height of the image.

Q9.6: A slide in a projector is placed in front of a lens and an image 75 times larger than the slide is projected onto a screen which is 5 m from the opposite side of the lens to the slide.

(a) How far from the lens is the slide?

(b) What is the focal length of the lens?

(c) Is this a converging or diverging lens and is the image formed real or virtual?

Q9.7: Sherlock Holmes is using a magnifying glass that consists of a single converging lens to examine a fingerprint which has a 1 cm diameter. Carefully positioning the magnifying glass and looking through it, towards the fingerprint on the other side he observes an image of the fingerprint which is 5 cm in diameter, has the same orientation as the fingerprint and is located 10 cm from the lens on the same side as the fingerprint.

(a) How far is the lens from the fingerprint?

(b) What is the focal length of the lens?

(c) Is the image real or virtual?

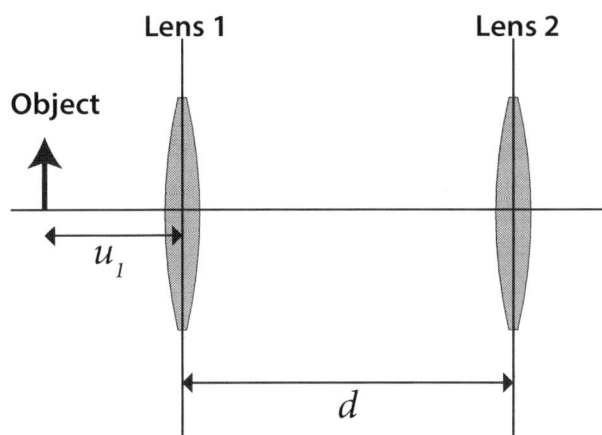

Fig. P9.3: *Two converging lenses with a common optical axis.*

Q9.8: Two converging lenses with focal lengths $f_1 = 17$ cm and $f_2 = 5$ cm are arranged as shown in figure P9.3 such that the distance between them $d = 35$ cm. If an object is placed at a distance $u_1 = 34$ cm to the left of the first lens, as shown, where is the final image formed?

CHAPTER 10

Optical Instruments

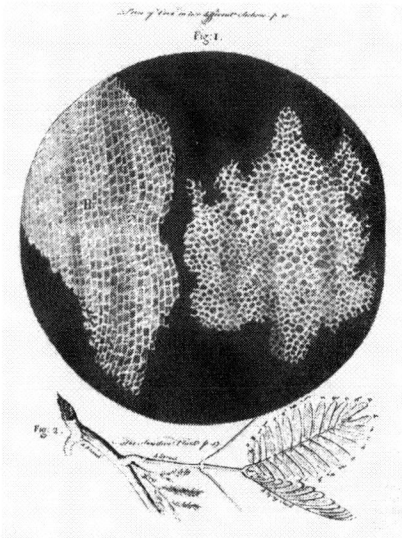

Fig. 10.1: Cork cells (top) as observed by Robert Hooke and published in his book, Micrographia, in 1665. The book contained various other detailed drawings of insects and plants all captured using his microscope. It was the first major publication of the Royal Society and became a scientific best-seller.

Micrographia, Robert Hooke, 1665, Wikimedia Commons

By the 17th century, our understanding of the principles of geometrical optics was well developed and the scene was set for the invention of optical instruments which completely revolutionized our understanding of nature. The first telescope using lenses was invented in the Netherlands in 1608 but within a year Galileo Galilei (1564-1642) built his own and turned it to the heavens.

Using his telescope in January 1610 Galileo was the first to observe the four largest moons of Jupiter, now called the Galilean moons in his honour. His observation of them orbiting around Jupiter lead him to conclude that, like these moons, the Earth must orbit around the sun and thus he changed our understanding of our place in the universe. However the telescope's discoveries were by no means over and just over 300 years later, in 1923, Erwin Hubble used a 2.5 m diameter, reflecting telescope to show that the universe did not consist of just the Milky-way (as was thought at the time) but instead contained billions of galaxies like the Milky-way making it far, far larger than anyone had ever conceived. This led to his future work on redshift mentioned in section 8.6.

At the other end of the size spectrum, physicists were busy using microscopes to explore the universe at far smaller scales. In 1665, two years before the Great Fire of London, our old friend Robert Hooke (1635-1703) built a compound microscope and used it to examine a piece of cork (see figure 10.1). He discovered that it was made up of a honeycomb of tiny structures which he named "cells". Only 9 years later in 1674 the dutch scientist, Antony Leeuwenhoek, using a much-improved micro-

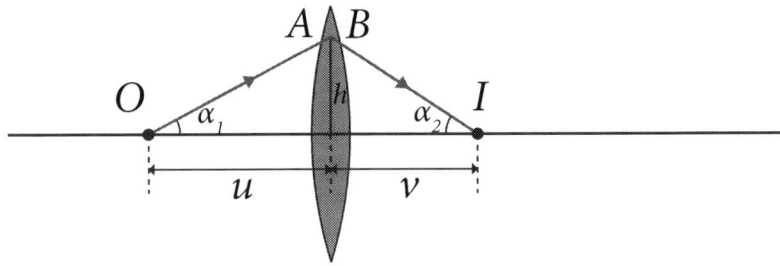

Fig. 10.2: Diagram of a converging lens focussing a ray from a point object, O a distance u from the lens to create a point image, I, a distance v from the lens. The point A is where the ray of light is incident on the lens and the point B is where the ray leaves the lens.

scope of his own design was the first person to observe the tiny, single-celled organisms we call bacteria and thus started the field of microbiology.

We will begin our discussion of optical instruments with how lenses are constructed. Next, we shall cover the simple magnifying glass and then go on to discuss microscopes and telescopes and conclude with a discussion of some of the common optical flaws which limit the performance of these instruments.

10.1 The Lensmaker's Equation

Lenses are an import component in many optical instruments and although we have discussed their properties we have not yet covered how to make one with the properties we need. To do this let's consider a very simple situation where we have a point object, O, on the optical axis of a converging lens and a distance u from the lens. This object produces a point image, I, a distance of v away from the lens along the optical axis as shown in figure 10.2.

The path followed by the ray of light is determined by the shape of the lens. To relate the path to the lens shape we need to consider the refraction both when the ray enters the lens at the point A and when it leaves at the point B. Starting with the refraction at point A we can draw another diagram, shown in figure 10.3 where instead of the entire lens we only consider the front half of the lens. To obtain the angles of incidence and refraction we need a line which is perpendicular to the surface. Since the lens surface is a spherical surface this line will be a radius of the sphere which will pass through the centre of the which is marked at the point C_1 on the figure.

Applying the law of refraction at the point A we obtain the equa-

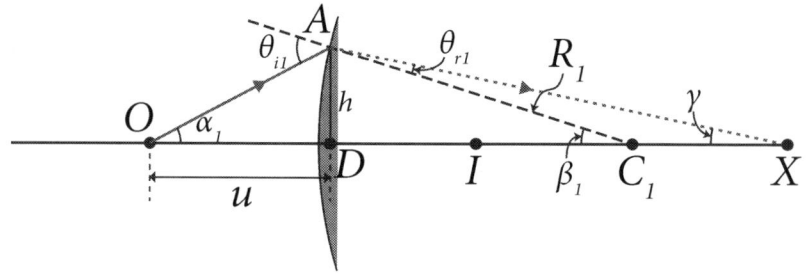

Fig. 10.3: The front half of a converging lens showing the refraction taking place at the point A where the ray enters the lens. The centre of the radius of curvature of the face is C_1 and θ_{i1} and θ_{r1} are the angles of incidence and refraction respectively at the point A.

tion:

$$\sin \theta_{i1} = n \sin \theta_{r1} \tag{10.1}$$

where n is the refractive index of the lens' material and we assume that the refractive index of air is approximately 1. For thin lenses the lens surfaces must be relatively flat and so the radius of curvature, r_1, must be large. This means that the angles of incidence and refraction will both be small and so we can make the small angle approximation where $\sin \theta \approx \theta$. Putting this into (10.1) we now get:

$$\theta_{i1} = n\theta_{r1} \tag{10.2}$$

Next we need to use a little geometry. The angle θ_{i1} is the exterior angle of the triangle AOC_1 and so it must be equal to the sum of the two opposite interior angles which gives:

$$\theta_{i1} = \alpha_1 + \beta_1 \tag{10.3}$$

and we can do the same with the triangle AC_1X where this time the exterior angle is β_1 which gives the relationship:

$$\beta_1 = \theta_{r1} + \gamma \implies \theta_{r1} = \beta_1 - \gamma \tag{10.4}$$

Now we can use the values for the angles of incidence and refraction in equations (10.3) and (10.4) and substitute them into equation (10.2) to get:

$$\alpha_1 + \beta_1 = n(\beta_1 - \gamma) \tag{10.5}$$

Now at this point we need to consider what happens to our refracted ray. Clearly it will not make it all the way from the point A to X because it will encounter the second surface of the lens and will be refracted a second time which is shown in figure 10.4. In this figure the ray is refracted at the point B and deflected so that it arrives at a point I on the optical axis and the point C_2 is the centre of the radius of curvature of the second face of the lens. Once again we will apply the law of refraction to get:

$$n \sin \theta_{i2} = \sin \theta_{r2} \implies n\theta_{i2} \approx \theta_{r2} \tag{10.6}$$

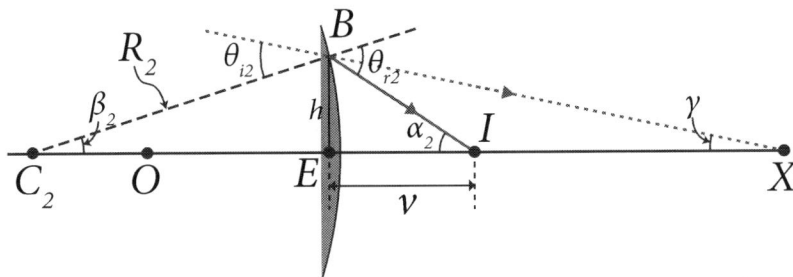

Fig. 10.4: The back half of a converging lens showing the refraction taking place at the point B where the ray leaves the lens. The centre of the radius of curvature of the face is C_2 and θ_{i2} and θ_{r2} are the angles of incidence and refraction respectively at the point B.

where now the refractive index of the glass, n, appears with the angle of incidence since the ray is moving from glass to air, which again we will assume has $n = 1$. Again we will use the exterior angles of triangles to give us expressions for the angles of incidence and refraction. First using triangle BC_2I we get:

$$\theta_{r2} = \alpha_2 + \beta_2 \qquad (10.7)$$

and from the triangle BC_2X we get:

$$\theta_{i2} = \beta_2 + \gamma \qquad (10.8)$$

Now we take the expressions for the angles of incidence and refraction from (10.7) and (10.8) and substitute them into (10.6) to get:

$$\alpha_2 + \beta_2 = n(\beta_2 + \gamma) \qquad (10.9)$$

At this point we can finally get rid of the angle γ by combining both (10.5) and (10.9) to get:

$$\alpha_1 + \alpha_2 = (n - 1)(\beta_1 + \beta_2) \qquad (10.10)$$

Now we need to make some further approximations. First we will assume that the lens is thin which means that there is negligible distance between the points A and B and so we will claim that both points are a distance h from the optical axis. Next we refer to figure 10.3 and if we define the radius of curvature of the first surface of the lens as R_1 then the line $\overline{AC_1}$ has a length equal to R_1 since it is a radius of the circle. Hence using triangle ADC_1 we can write the sine of the angle β_1 as:

$$\sin \beta_1 = \frac{h}{R_1} \implies \beta_1 \approx \frac{h}{R_1} \qquad (10.11)$$

where we take the small angle approximation for β which has to be valid if the lens is thin because the radius of the lens must be a lot less than the radius of curvature. Using the same figure 10.3 we can also derive an approximation for the angle α_1 from the triangle OAD:

$$\tan \alpha_1 = \frac{h}{u} \implies \alpha_1 \approx \frac{h}{u} \qquad (10.12)$$

where we use the small angle approximation for tangent that $\tan(\delta\theta) \approx \delta\theta$. Similarly looking at the figure 10.4 and using triangle BEC_2 we get an expression for β_2:

$$\sin \beta_2 = \frac{h}{R_2} \implies \beta_2 \approx \frac{h}{R_2} \qquad (10.13)$$

and using triangle IBE we get an expression for α_2:

$$\tan \alpha_2 = \frac{h}{v} \implies \alpha_2 \approx \frac{h}{v} \qquad (10.14)$$

Finally all that remains is to put (10.11), (10.12), (10.13) and (10.14) into the expression we originally derived (10.10). Since every term include h this cancels out and we are left with:

$$\frac{1}{u} + \frac{1}{v} = (n-1)\left(\frac{1}{R_1} + \frac{1}{R_2}\right) \qquad (10.15)$$

However we can simplify the left hand side of the equation using the lens equation (9.27) which finally gives us a relationship between the radius of curvature of the surfaces of the lens, R_1 and R_2, and the focal length of the lens, f:

$$\frac{1}{f} = (n-1)\left(\frac{1}{R_1} + \frac{1}{R_2}\right) \qquad (10.16)$$

This is known as the lensmaker's equation because it is the equation used to make a lens of a given focal length. The sign convention used is that the radii are positive for a convex face and negative for a concave face. This results in a sign convention for the focal length which is consistent with the lens equation: positive for a converging lens and negative for a diverging lens.

It is important to remember that this equation is only valid for thin lenses i.e. lenses whether the focal length is a lot longer than the radius of the lens and for lenses placed in a medium where the refractive index is approximately 1.0, such as air. The sign convention used here is that the radii of curvature are positive for a convex lens surface and negative for a concave surface.

The Big Picture

It is possible to derive a more general form of the lensmaker's equation which applies to thick lenses. In these cases a different sign convention is used for the radii of curvature. For the front of the lens the convention remains as it was before: $R_1 > 0$ for a convex face and $R_1 < 0$ for a concave face. However for the back face of the lens the convention is reversed: $R_2 < 0$ for a convex face and $R_2 > 0$ for a concave face. With this convention the more general form of the lensmaker's equation is:

$$\frac{1}{f} = (n-1)\left[\frac{1}{R_1} - \frac{1}{R_2} + \frac{(n-1)d}{nR_1R_2}\right] \qquad (10.17)$$

where d is the thickness of the lens measured at the optical axis.

10.2 Magnifying Glass

The magnifying glass is the simplest of all optical instruments and consists of nothing more than a single, converging lens. To obtain a magnified image the lens is placed so that it is closer to the object than the focal length. This will cause the lens to generate a virtual image as we have seen before in figure 9.27 and reproduced here in figure 10.5.

Looking at this figure we can see that the virtual image is larger than the object and so we conclude that the lens produced a magnified image. However we have to be careful because although the image is magnified it is also further from the observer's eye and if we look at the angular size of the image i.e. the angle at the observer's eye between the top and bottom of the image we can see that it is exactly the same as the angular size of the object. This means that the image has exactly the same perceived size as the object...so how does a magnifying glass magnify?

To understand why magnifying glasses work we need to consider the same observer looking at the object without a magnifying glass, as shown on the left in figure 10.6. In this case, the distance object can be no closer to the observer's eye than the near point which is the closest position at which the eye can focus a sharp image on the observer's retina. This is typically around 25 cm for young adults but can be less for short-sighted people and gets longer with age.

Now compare the two diagrams in figure 10.6. With the magnifying glass the virtual image is produced at a far greater distance from the eye than the object and so the eye can focus on the image. Hence the magnification comes not from the fact that the image is larger than the object but because with the lens the eye can now focus on the object when it is a lot closer to the eye.

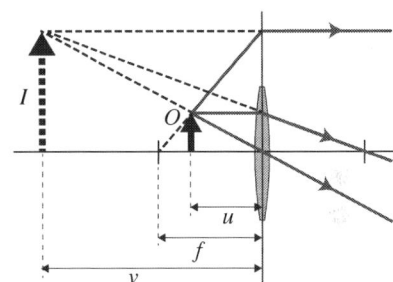

Fig. 10.5: Ray-traced diagram of a magnifying glass forming a virtual image of an object placed closer than the focal length of the lens. The three principle rays are shown.

Fig. 10.6: (Left): Observer looking at an object without any lens. The object cannot be closer to the observer than the near point, d, and the angular size of the object, θ, is shown.
(Right): Observer using a magnifying glass showing how both the object and image have the same angular size, θ', which is larger than with the naked eye because the object is now far closer to the observer who focusses on the distant image.

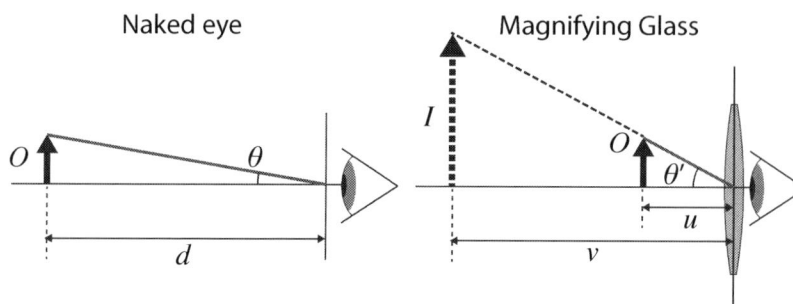

To determine the magnification factor we need to consider the angular magnification factor between the naked eye and the eye with a magnifying glass. This is defined as the ratio of the angular sizes of the image in each case. Since the rays from the bottom of the object pass along the optical axis of the lens and so are not deflected at all the angular sizes are simply the angles θ and θ' shown in figure 10.6. Hence we have an effective magnification factor, M, given by:

$$M = \frac{\theta'}{\theta} \tag{10.18}$$

To evaluate this we need to determine both angles. The simplest case to consider is that of the naked eye and so looking at the left diagram in figure 10.6 we have for the maximum angular size of the object:

$$\tan \theta = \frac{O}{d} \tag{10.19}$$

where d is the near point of the observer. Now for small objects the angle θ will be tiny and for small angles we can make the approximation that $\tan \theta \approx \theta$ The geometrical argument for this is very similar to the one we made for small values of sine when discussing small amplitude oscillations of a simple pendulum in section 5.5.1 (page 77). Using this approximation (10.19) becomes:

$$\theta \approx \frac{O}{d} \tag{10.20}$$

Now we need to consider the magnifying glass case. From the right of figure 10.6 we have:

$$\tan \theta' = \frac{O}{u} \tag{10.21}$$

Now if we look at the lens equation (9.27) then we have the relationship:

$$\frac{1}{u} + \frac{1}{v} = \frac{1}{f} \implies \frac{1}{u} = \frac{1}{f} - \frac{1}{v} \tag{10.22}$$

However since the image produced is a virtual image we use a negative distance for v and so the equation becomes:

$$\frac{1}{u} = \frac{1}{f} + \frac{1}{|v|} \qquad (10.23)$$

For the largest possible angular magnification we want the largest possible angular size for the image and if we look at equation (10.21) we can see this is when u is as small as possible which means we want $1/u$ to be as large as possible. If we now look at (10.23) we can see that this will happen when $|v|$ is as small as possible.

Now $|v|$ is the distance from the lens to the image and if our observer's eye is directly behind the lens then the closest image it can focus on is, by definition, an image at the near point. So if we define the near point of the observer to be a distance, d, then the maximum angular size of the object will occur when:

$$\frac{1}{u} = \frac{1}{f} + \frac{1}{d} = \frac{f+d}{fd} \qquad (10.24)$$

and hence the maximum angular size of the image as seen by the observer will be:

$$\tan \theta'_{max} = \frac{O(f+d)}{fd} \approx \theta'_{max} \qquad (10.25)$$

where once again we are making the approximation that the angular size of the image will be small.

Conversely we can also look at the minimum angular size for the image. This will occur when $|v|$ has it's maximum possible value. Since the eye can focus on objects at infinity i.e. so far away that the rays from the image are effectively parallel there is no upper limit to how far away the image and be. In this limit $1/|v|$ will go to zero which will happen when the object is placed one focal length away from the lens as shown in figure 10.7. In this case equation (10.23) will become:

$$\frac{1}{u} = \frac{1}{f} \qquad (10.26)$$

and so the minimum angular size of the image will be:

$$\tan \theta'_{min} = \frac{O}{f} \approx \theta'_{min} \qquad (10.27)$$

where again we assume that the angular size will be small. Now if we put all this together we can calculate the range of magnification factors which the observer will obtain using the magnifying glass. The minimum factor will be:

$$M_{min} = \frac{\theta'_{min}}{\theta} = \frac{O/f}{O/d} = \frac{d}{f} \qquad (10.28)$$

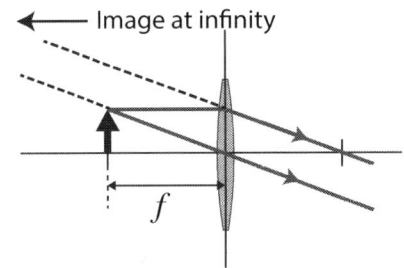

Fig. 10.7: Ray-traced diagram of a magnifying glass forming a virtual image of an object at infinity when the object is placed exactly at one focal length from the lens. This corresponds to the minimum magnification factor.

and the maximum magnification factor will be:

$$M_{\text{max}} = \frac{\theta'_{\text{max}}}{\theta} = \frac{O(f+d)}{fd} \times \frac{d}{O} = 1 + \frac{d}{f} \qquad (10.29)$$

Hence to get the highest magnification power from a magnifying glass the best technique is to hold the lens close to your eye and move both as close to the object as you can manage and still focus. However often it is more convenient to put the lens close to the object and have the eye a larger distance from the lens. In this case it is usual to have the object one focal length from the lens so that the image is produced at infinity so the focus is not sensitive to the position of the eye and the lowest magnification power is obtained.

10.3 Compound Microscope

The compound microscope, shown in figure 10.8, is an improvement on the simple magnifying glass and consists of two, converging lenses. The first lens, called the objective, is placed close to the object so that it is just over one focal length away. This causes the lens to produce a real, magnified image of the object which is arranged such that it is inside the focal length of the second lens, called the eyepiece or ocular lens because it is placed next to the observer's eye. The eyepiece then produces a virtual image at sufficient distance from the lens that the eye can focus on it.

The magnification of a compound microscope comes from two sources. First the objective produces a real image which has a linear magnification i.e. the size of the image is physically larger than the object. This is then magnified further by the angular magnification of the eyepiece. When combining the magnification factors we multiply them and so the total magnification for the microscope will be:

$$M = m_o m_e \qquad (10.30)$$

where m_o is the linear magnification of the objective and m_e is the angular magnification of the eyepiece. We have already derived the linear magnification factor for the objective in equation (9.28). Using the distances to the object and image shown in figure 10.8 this gives:

$$m_o = -\frac{v_o}{u_o} \qquad (10.31)$$

Now to get the largest magnification possible the object is placed close to the focal length, f_o, of the objective and hence we can

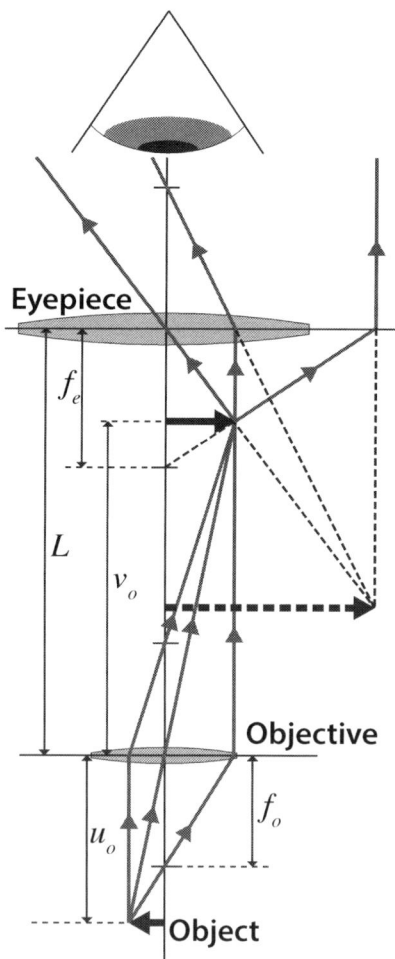

Fig. 10.8: Ray-traced diagram of a compound microscope.

make the approximation that $u_o \approx f_o$. Similarly to get the maximum magnification from the eyepiece we want the image produced by the objective to be as close as possible to the eyepiece and so we can make another approximation that $v_o \approx L$ where L is the distance between objective and the eyepiece i.e. the length of the microscope tube. Putting these into (10.31) we get a maximum magnification factor for the objective of:

$$m_o \approx -\frac{L}{f_o} \qquad (10.32)$$

For the eyepiece we have already calculated the maximum magnification factor and it is given in (10.29) so all we need to do put both values into (10.30) to get the maximum magnification for the microscope:

$$M = -\frac{L}{f_o}\left(1 + \frac{d}{f_e}\right) \qquad (10.33)$$

This means that the most powerful microscopes have the shortest focal lengths for both lenses. It is also worth noting that the magnification factor is negative which means that the image seen will be inverted compared to the object as shown in figure 10.8. The maximum possible magnification is limited by factor other than just the focal length of the lenses and one of the most important limits is the lighting. When an image is magnified the intensity of light it emits is reduced by the magnification factor and so microscopes are usually designed to provide very intense lighting for the sample under study.

10.4 Refracting Telescope

The first telescopes constructed were designed using lenses and since these work using the refraction of light this type of telescope is referred to as a refracting telescope. Like a compound microscope a refracting telescope consists of two lenses: the objective and the eyepiece. However, unlike a microscope, they are arranged to collect and magnify the parallel rays of light from a distant object and not one close to the lens. To achieve the two lenses are places so that their two focal points coincide as shown in figure 10.9.

The arrangement of the lenses means that the image produced will also be at infinity i.e. the rays of light will be parallel. However looking at figure 10.9 we can see that the angle between the light rays and the optical axis has been increased. The result is that, like the magnifying glass, we need to consider the

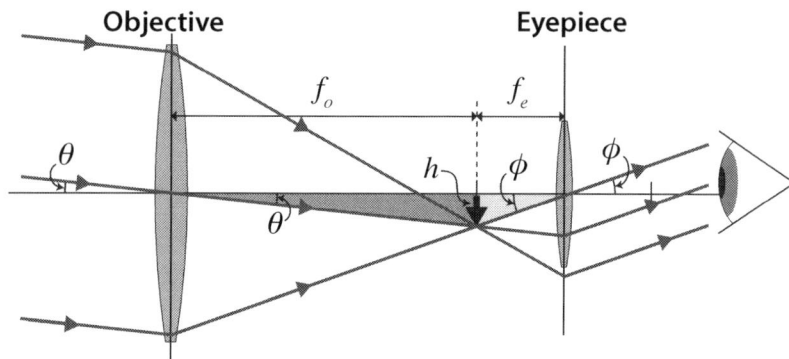

Fig. 10.9: Ray diagram for a refracting telescope where the objective and eyepiece are arranged so that their focal points coincide. This arrange works to provide an angular magnification for parallel rays of light from a distant object.

angular magnification of the telescope which will be defined as:

$$M = \frac{\phi}{\theta} \qquad (10.34)$$

To understand how this works consider two stars which are close to each other as shown in figure 10.10. If the telescope is aligned so that the light from the first star travels along the optical axis of the two lenses then it will remain undeflected. Light from the second star will arrive at the telescope a small angle to the optical axis but it will leave the eyepiece at a far larger angle. Hence to the observer looking through the eyepiece the separation between the two stars will have increased and so the image is magnified compared to just looking at the night sky. However as figure 10.9 shows the image is also inverted: the top incoming ray at the objective is the bottom ray leaving the eyepiece and vice versa.

To calculate the magnification factor we need to look at the two shaded triangles shown in figure 10.9. Looking both of these

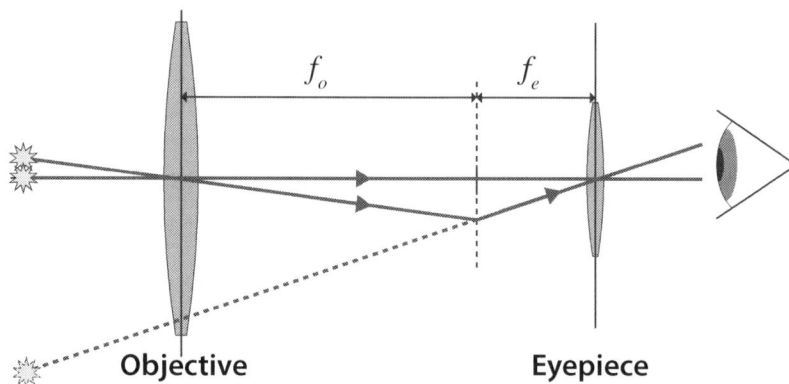

Fig. 10.10: Two stars viewed through a telescope showing the effect of the angular magnification which increases the apparent separation of the stars as shown.

we can determine the tangents of the two labelled angles:

$$\tan \theta = \frac{h}{f_o} \text{ and } \tan \phi = \frac{h}{f_e} \qquad (10.35)$$

Now since the focal length of the objective lens, f_o, is generally always a lot longer than the radius of the lens itself the angles θ and ϕ will be small and so we can use the small angle approximation that $\tan \theta \approx \theta$. Hence we can use (10.35) to provide values for θ and ϕ in (10.34) which gives the magnification factor for the telescope:

$$M = -\frac{f_o}{f_e} \qquad (10.36)$$

where the minus sign is added to indicate that the image is inverted. Hence the magnification is determined by the ratio of the two focal lengths of the lenses. Typically in most telescopes the eyepiece is removable and can be replaced by ones with different focal lengths to provide different magnification power. However, for an astronomical telescope, the magnification power is actually one of the least important factors in determining the performance and to discover what is typically the most important factor we need to discuss a different design of telescope.

10.5 Reflecting Telescope

A reflecting telescope is very similar in principle to a refracting telescope except that the objective is replaced by a concave mirror which acts as the primary optical element which collects and focusses the light from a distant object. As we have already seen in section 9.2.1 a concave mirror is very similar to a converging lens in that it has a focal point which rays of light parallel to the optical axis will pass through after being reflected. As before the eyepiece is placed so that the focal point of the lens coincides with the focal point of the mirror and the total magnification power of the telescope is:

$$M = -\frac{f_m}{f_e} \qquad (10.37)$$

where f_m is the focal length of the mirror.

However, for a reflecting telescope, the design is more complex because the focal point of the mirror is in front of the mirror and in the path of the incoming light. There are various ways to access this focal point each of which leads to a different design of reflecting telescope. One of the simplest designs and

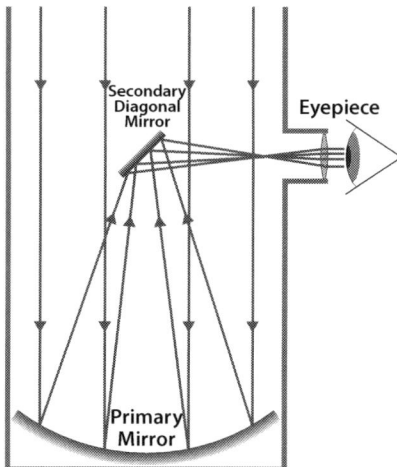

Fig. 10.11: Newtonian reflector telescope which uses a primary, convex mirror to collect and focus the light to a converging eyepiece lens. The secondary, diagonal mirror deflects the focal point out of the main tube and to the side.

one which is still in use today is the Newtonian telescope which uses a plane mirror positioned above the centre of the mirror to reflect the focal point to the side of the tube where the eyepiece is placed as shown in figure 10.11. This design was invented in 1668 by Sir Isaac Newton (1642-1727) and is a very simple design to build.

The main advantage of a reflecting telescope over a refracting telescope is that it is cheaper to construct a mirror of a given diameter than it is to construct a lens of the same dimensions and the diameter of the primary optical element defines the diameter of the aperture of the telescope. While the size of the aperture does not affect the magnification power of the telescope it does impact the amount of light which the telescope will collect from a particular source. The larger the aperture the larger the area which is collecting light as shown in figure 10.11 where the light incident on the large primary mirror is emitted from the far smaller eyepiece lens. This increases the intensity of the light and is very important for viewing faint objects which makes the aperture one of the primary performance indicators for telescopes and far more important than the magnification power. Current state-of-the-art telescopes have apertures on the order of 10 m in diameter but the next generation of machines due to start operating around 2022 will have apertures of 30 m or higher.

10.6 Aberrations

Until now we have assumed that the mirrors and lenses we have studied have all been made of idea materials and even then we have made certain approximations to keep the maths simple. However, when dealing with precision optical instruments these assumptions and approximations lead to a noticeable distortion and blurring of the image. Several of the most serious of these effects are described here as well as some of the methods employed to correct them.

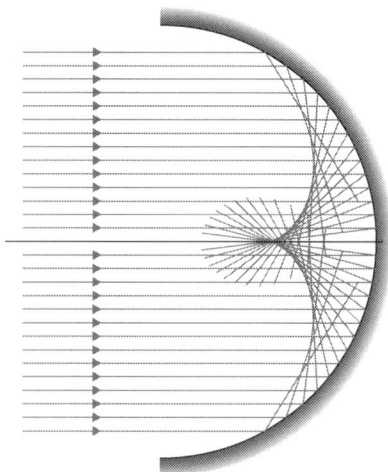

Fig. 10.12: Diagram showing the effects of spherical aberration in a spherical mirror with a radius comparable to the radius of curvature. Parallel rays impacting far from the optical axis are not reflected through a single focal point but parallel rays near the centre of the mirror are showing that the approximation used earlier is valid.

10.6.1 Spherical Aberration

When studying the formation of an image by a spherical mirror in section 9.2 we made an assumption that the radius of the mirror was significantly smaller than the radius of curvature of the mirror. Most notably we made this assumption when deriving equation (9.6) which showed that the focal length of the mirror was equal to half the radius of curvature.

This approximation is typically fine where a mirror is being used by itself e.g. in a North American car's wing mirror. However, in a precision optical instrument, such as a telescope, this produces an unacceptable blurring of the image which is called spherical aberration.

Figure 10.12 shows the actual pattern produced by light rays reflecting off the surface of a spherical mirror which has a significant curvature. Rays near the centre of the mirror are focussed to a small region in front of mirror which is why the approximation works for mirrors with a radius much less than the radius of curvature. To generate a perfect focal point for parallel rays the shape of mirror required is parabolic, as shown in figure 10.13. In this case, a ray parallel to the optical axis of the mirror will always be reflected so that it passes through the focal point regardless of the radius of the mirror.

Large mirrors in optical telescopes have to have a surface which is ground to a very high precision, of order 0.1 µm, which is far easier to achieve with a spherical mirror than a parabolic one. This then requires a correction which for modern amateur telescopes is often achieved through a Schmidt corrector plate which is a machined glass plate placed in front of the mirror that bends incoming parallel light rays so that the spherical mirror will focus them to a point, as shown in figure 10.14. In these telescopes it is often combined with a convex secondary mirror which mimics the effects of a very long focal length to create what is called a Schmidt-Cassegrain reflector as shown in figure 10.15.

Spherical aberration also occurs in lenses as well as a result of the approximations we made in deriving the lensmaker's equation, (10.16), where we assumed that the radius of the lens was a lot less than the radius of curvature of the surface so that the angles were small and the small angle approximations

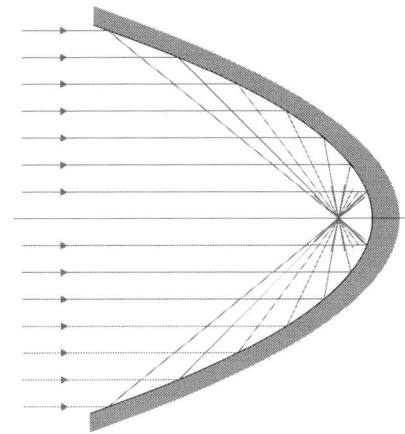

Fig. 10.13: Rays parallel to the optical axis incident on a parabolic mirror are all reflected through a single focal point as shown here.

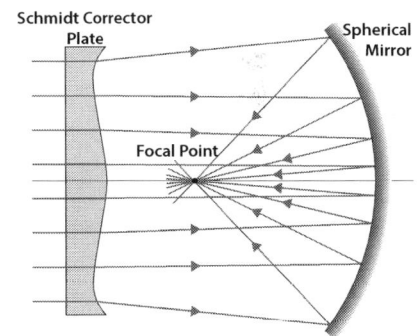

Fig. 10.14: One way to correct spherical aberration is to use a carefully machined sheet of glass called a Schmidt corrector plate. The profile of the plate has been greatly enhanced in this diagram for illustration and actual plates typically look just like a flat sheet of glass.

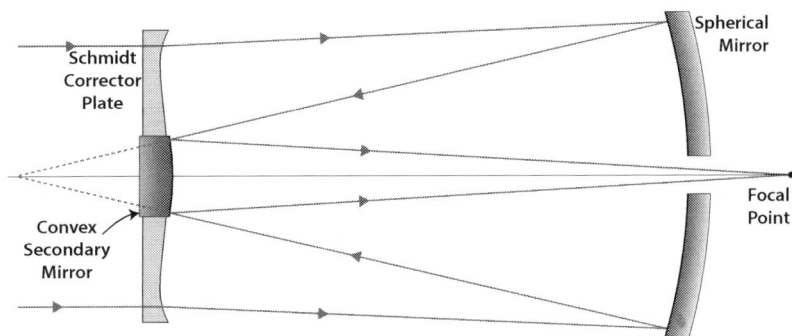

Fig. 10.15: A Schmidt-Cassegrain reflector which incorporates a Schmidt-corrector plate to reduce spherical aberration as well as a convex, secondary mirror to simulate a longer focal length. Unlike a Newtonian reflector the focal point is directly behind the primary mirror which has a hole in the centre to allow the light to pass through.

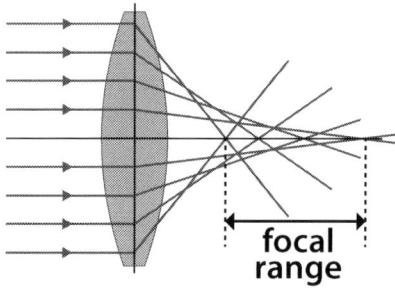

Fig. 10.16: Diagram showing the effects of spherical aberration in a lens. The focal point varies over the surface of the lens so that parallel rays far from the optical axis have a different focal point from those close to the axis.

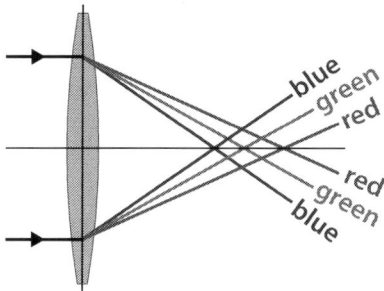

Fig. 10.17: Diagram showing the effects of chromatic aberration where each colour of light has a different focal length in a lens. The effect is caused by dispersion in the glass of the lens which means each wavelength has a different refractive index.

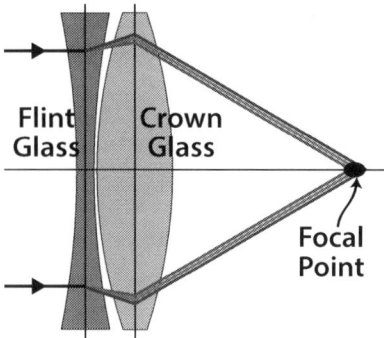

Fig. 10.18: An achromatic doublet showing how chromatic aberration can be corrected using two lenses, one with a high dispersion and the other with a low dispersion.

were valid. These approximations mean that the focal point for rays striking far from the centre of the lens is not the same as the focal point for those striking near the centre as shown in figure 10.16. To correct for this an aspherical lens is required which has a complex, non-spherical shape.

10.6.2 Chromatic Aberration

A different type of aberration arises because a lens is made of materials which do not behave in an ideal way. When deriving the lensmaker's equation we assumed that the refractive index of the lens material was a single value for all wavelengths of light. However, this is not the case and the refractive index changes as a function of the wavelength of light. This effect is called dispersion and is discussed in more detail in section 11.2.

For a lens, this effect means that each wavelength of light has a different focal length which results in slightly different images for each colour of light in the image which produces a colour-based blurring called chromatic aberration which is shown in figure 10.17. One means to correct for this is to use an achromatic doublet which is made from two lenses placed together. For a net converging lens, this consists of one diverging lens made from a material with a large dispersion i.e. a large difference in refractive index over the range of optical wavelengths and a second converging lens made from a material with a small dispersion, as shown in figure 10.18. This arrangement means the two achromatic effects of the different lenses roughly cancel to produce an apochromatic lens which has approximately the same focal length for all wavelengths of light.

Problems

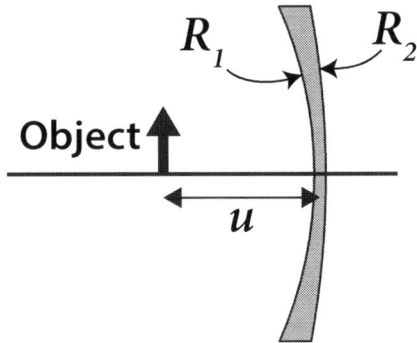

Fig. P10.1: *Lens which has spherical surfaces with the radii of curvature shown.*

Q10.1: An object is placed in front of a lens as shown in figure P10.1. The values of the parameters shown are $R_1 = 1.2\,\text{m}$, $R_2 = 2\,\text{m}$ and the distance $u = 50\,\text{cm}$. The refractive index of the lens' glass is 1.6. State whether the image produced is real or virtual, which side of the lens it is on and how far from the lens it will be.

Q10.2: A child, with a near point of 20 cm uses a magnifying glass with a focal length of 5 cm to look at an ant which is 1 cm long. He places the magnifying glass close to his eye and peers through it to see the ant.
(a) How much larger is the ant that it appears when the child look at it as close as possible with his naked eye assuming that he has made the image as large as possible?
(b) How close to the ant does the child need to place the magnifying glass to get this image?
(c) The child's friend comes along and wants to look at the ant as well. Both of them look at the ant with their eyes a long way from the magnifying glass. In this situation how much larger does the ant appear compared to looking at it closely with their naked eyes?

Q10.3: A compound microscope consists of an objective lens with a focal length of 1 cm and an eyepiece with a focal length of 5 cm attached to each end of a tube with a length of 25 cm. A scientist with a near point of 25 cm uses this microscope to view a sample on a slide.

(a) What approximate magnification factor will such a microscope achieve?
(b) What is the *exact* distance between the sample and the objective lens for the maximum magnification to occur?

Q10.4: An amateur astronomer has decided to make a simple telescope by grinding her own objective lens. She plans to use a standard eyepiece lens which has a 40 mm focal length and she will use a lens blank which has a refractive index of 1.6 to create the objective. The lenses will be mounted on either end of a 1 m long tube.
(a) What will the magnification factor of the telescope be?
(b) If she leaves one side of the lens blank flat what radius of curvature must she grind the other side to?

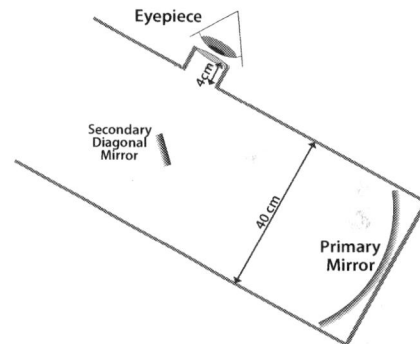

Fig. P10.2: *Newtonian reflector consisting of a primary, concave mirror and a single converging lens.*

Q10.5: A Newtonian reflector telescope shown in figure P10.2 consists of a primary, concave mirror with a diameter of 30 cm and an eyepiece lens with a diameter of 2 cm. The primary mirror has a radius of curvature of 1 m and is mounted in the bottom of a tube which has the same diameter as the mirror. The eyepiece has a focal length of 5 cm and is mounted 4 cm from the edge of the tube. The secondary, plane reflecting mirror is mounted in the centre of the tube and reflects the light through 90° out to the eyepiece.

(a) How far from the centre of the primary mirror should the centre of the secondary mirror be placed?

(b) The faintest objects in the night sky visible to the naked eye have a brightness equal to approximately 0.25% of the star Vega. What is the smallest intensity, expressed as a fraction of the brightness of Vega, which an eye can see with this telescope assuming that all optical components behave perfectly and the the blockage of the secondary mirror is negligible?

(c) Newton designed his reflector to avoid the problem of chromatic aberration which was a significant problem for refracting telescopes. Will this reflecting design suffer from any chromatic aberration at all? Explain.

Q10.6: A simple refractive telescope consists of lenses made from flint glass which has a refractive index which varies from 1.61 for red light to 1.66 for blue light. For red light the objective lens has a focal length of 40 cm and the eyepiece has a focal length of 5 cm.

(a) What magnification factor of the telescope for red light?

(b) What is the focal length of the objective lens for blue light?

(c) What magnification factor of the telescope for blue light if the telescope is adjusted to function properly for this wavelength?

Q10.7: An optical telescope which uses a spherical mirror to form the image is focussed on a distant object. When looking through the telescope it is found that the centre of the object is in almost perfect focus. However, at the edges of the image the object appears blurred while its colour remains the same.

(a) What type of aberration causes this behavour? Explain.

(b) Suggest two ways in which this aberration could be corrected.

CHAPTER 11

Light as a Wave

Fig. 11.1: Thomas Young (1773-1829) the British physicist who demonstrated the wave nature of light. He also described Hooke's law in terms of tensile stress and strain giving his name to a modulus of elasticity: "Young's Modulus".

Mid-nineteenth century engraving, published in "Six Lectures on Light" by John Tyndall, 1885.

On the 24th November 1803, in a talk to the Royal Society of London, a British physicist called Thomas Young (figure 11.1) performed a simple experiment which shattered the century-old view of Newton's that light was made of particles. He began this talk with his now famous quote:

> The experiments I am about to relate... may be repeated with great ease, whenever the sun shines, and without any other apparatus than is at hand to everyone.

He then described an experiment where he placed a thin strip of card, about 1 mm wide, in a beam of light from a narrow opening in a window and observed colour fringes on either side. He explained that these fringes are due to light waves interfering with each other and demonstrated the same effect using water waves in a ripple tank. He went further and explained the colours observed in thin films such as soap bubbles and oil on water as being caused by interference and even explained a phenomenon Newton himself observed, called Newton's rings, as being caused by the wave nature of light.

Despite this long list of experiments and observations one which is noticeably absent is showing the interference of light from two narrow slits. Today this experiment is called Young's double slit experiment because it is an excellent demonstration of the wave nature of light but rather ironically there is no evidence that Young ever performed it!

Interference is a phenomenon only exhibited by waves and Young's observation of interference in light showed that it was a wave and not a particle which completely revolutionized the understanding of light at the time. The true nature of light waves was revealed by another British physicist, James Clerk Maxwell (figure 11.2), who in 1862 noticed that the speed of propagation of an electromagnetic field was the same as the speed of light. He built on this simple observation and showed that simple electrical and magnetic experiments gave rise to equations that predicted the existence of waves of oscillating electric and magnetic fields that propagated at the speed of light. This unified theory of electricity and magnetism went through several incarnations before being published in 1873 in their modern form as four partial differential equations which today we call the Maxwell Equations.

Maxwell's work showed that light is part of the electromagnetic spectrum of waves. It occupies a very narrow range of frequencies between infrared and ultraviolet radiation which corresponds to wavelengths of 390 nm to 700 nm and frequencies of 430 THz to 790 THz. The range is defined by the sensitivity of our eyes and it lies at the peak of the sun's emission spectrum which is why our eyes evolved to be sensitive to this particular range of wavelengths.

The impact of Maxwell's work describing the nature of light had enormous repercussions for physics. With light explained as an electromagnetic wave the search began for the wave medium, something Maxwell referred to as the "luminiferous aether", whose existence was disproved by the famous Michelson-Morley experiment in 1887 and which ultimately lead to Einstein's development of Special Relativity. The Maxwell Equations also laid the foundation for the quantum mechanical explanation of light something which ultimately showed that both Newton and Huyghens were right: light is both a particle and a wave!

In this chapter, we will discuss the phenomena which arise from the wave nature of light starting with the work of Christiaan Huyghens who was one of the first physicists to suggest light was a wave.

Fig. 11.2: British physicist James Clerk Maxwell (1831-1879) who unified electricity and magnetism. He was also famous for developing the first durable colour photograph in 1861 and for a graphical method for analysis of trusses (Maxwell's diagram).
Campbell and Garnett, "The Life of James Clerk Maxwell", 1882.

11.1 Huyghens' Principle

Huyghens' principle states that we can consider each point on the front of a wave as a wave source. We can then construct the next wave front by summing together all the individual sources. This is shown in figure 11.3 for the simple propagation of a wave through a constant, uniform medium.

Where Huygen's principle is perhaps most interesting is in its description of refraction. Here the wavefronts from the incident ray enter the second material and, assuming it has a higher refractive index these waves slow down and start to travel at less than the speed of light in vacuum. The result of the waves slowing down is that the wavefront turns to become more perpendicular to the boundary. This is shown in figure 11.4.

Here Huyghens' principle shows the underlying cause of refraction, namely that the speed of light is not the same in all materials. It also provides the definition of refractive index which is:

$$\text{Refractive Index, } n = \frac{c}{v} \qquad (11.1)$$

where c is the speed of light in vacuo and v is the phase velocity of light in the medium. We can use this to derive Snell's law. Consider the diagram shown in figure 11.5. Each of the red lines represents a wave crest and so these are always one wavelength apart. The frequency of the wave remains the same, the colour of light is not changed by refraction, so because the speed of light is different in the two media the wavelength must also be different and so $\lambda_1 \neq \lambda_2$ which is why the wave is refracted. Looking at the two right-angled triangles on either side of the boundary we can write down the following relations

$$\sin \theta_1 = \frac{\lambda_1}{AB} \quad \text{and} \quad \sin \theta_2 = \frac{\lambda_2}{AB} \qquad (11.2)$$

Combining these to eliminate AB we get

$$\frac{\lambda_1}{\sin \theta_1} = \frac{\lambda_2}{\sin \theta_2} \implies \frac{v_1}{\sin \theta_1} = \frac{v_2}{\sin \theta_2} \qquad (11.3)$$

where v_x is the speed of light in the respective medium and we substitute for the wavelengths with $\lambda_x = v_x/f$ where the frequency f is the same for both media and so cancels from each side of the equation. Now can use the definition of refractive index given in (11.1) to convert the speeds into refractive indexes which gives

$$\frac{c}{n_1 \sin \theta_1} = \frac{c}{n_2 \sin \theta_2} \implies \frac{\sin \theta_1}{\sin \theta_2} = \frac{n_2}{n_1} \qquad (11.4)$$

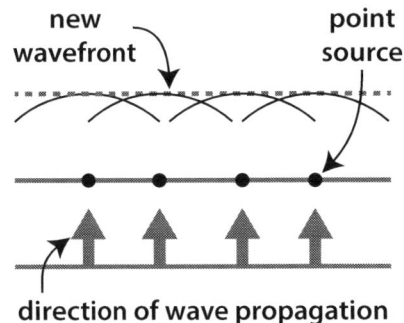

Fig. 11.3: Diagram showing how Huyghens' principle works for simple wave propagation. Each point on the wavefront acts as a point source which add to give the second wavefront.

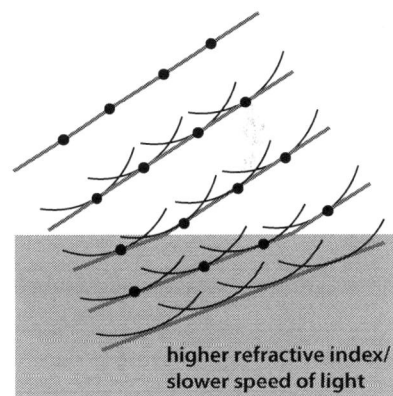

Fig. 11.4: Diagram showing how Huyghens' principle explains refraction. As the wavefront enters the medium where light travels move slowly the wavelets do not move as far causing the wave to turn towards the surface.

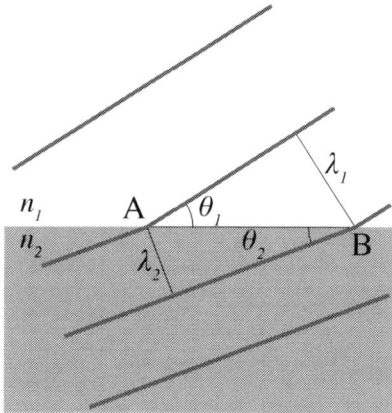

Fig. 11.5: Detailed look at wavefronts refracting at a boundary. Each red line represents a wave crest which are one wavelength apart. The refractive indexs of the media are n_1 and n_2 as shown.

which, since θ_1 is the angle of incidence and $\sin \theta_2$ is the angle of refraction, is Snell's law of refraction. Hence using Huyghens' principle we have proved that the law of refraction is a direct result of the change in the speed of light when moving from one material to another.

11.2 Dispersion

Up to now, we have assumed that the velocity of a light in a medium is the same for all wave frequencies. While the speed of light in a vacuum is absolute when light travels through a material the electromagnetic fields interact with that material and cause it to slow down. This interaction is generally dependent on the light's frequency and this results in a phenomenon called *dispersion* where the speed of light in a material varies with frequency or wavelength.

In section 11.1 we saw that the refractive index is a measure of the speed of light in that medium and so in a dispersive medium where the speed of light varies with wavelength the refractive index will also be a function of the wavelength of the light.

Now consider what happens when a beam of white light is refracted at the surface of a dispersive medium. From Snell's law of refraction we know that the angle of refraction depends on the refractive index but for a dispersive medium the refractive index varies with wavelength. Since white light consists of all visible wavelengths mixed together this means that each wavelength will have a different angle of refraction and so instead of a single beam refraction will produce a spray of colours as each wavelength is refracted through a different angle. The ray will be "dispersed" over a range of angles and hence the name of the phenomenon.

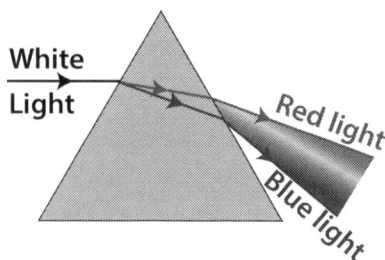

Fig. 11.6: Diagram of a triangular prism splitting white light into a spectrum of colours. Red light, which has the longest wavelength, is refracted the least and blue light is refracted the most.

To maximize the effect of dispersion a triangular prism is often used. By using a triangular prism the angle of incidence of the ray leaving the prism is not equal to the angle of refraction of the ray entering it, is it would be if the two surfaces were parallel, which prevents the dispersion caused when the light refracts upon entering the prism, from being reversed. This effect is shown in figure 11.6. Dispersion is also responsible for rainbows: the dispersion of sunlight hitting spherical water droplets separates the colours to give the rainbow and a double rainbow is caused when the light totally internally reflects inside the droplet before emerging.

To study this effect in more detail lets return to the phenomenon of beats which was discussed in section 8.4. There we considered the interference between two waves with slightly different frequencies at a single point. Returning to this lets consider the total wave and not confine ourselves to a single point. In this case we have two waves:

$$\Psi_1(x, t) = A\cos(k_1 x - \omega_1 t) \text{ and } \Psi_2(x, t) = A\cos(k_2 x - \omega_2 t)$$
$$(11.5)$$

where for the sake of simplicity we set the phase constant for each to zero because it has no effect on the result. Adding these together to get our interference pattern we have:

$$\Psi(x, t) = \Psi_1 + \Psi_2 = A\cos(k_1 x - \omega_1 t) + A\cos(k_2 x - \omega_2 t) \quad (11.6)$$

Now we need to use a trigonometric identity which we actually derived ourselves when dealing with beats in section 8.4. If we compare equation (8.15) with (8.17) then we showed that:

$$\cos \alpha + \cos \beta = 2\cos\left(\frac{\alpha + \beta}{2}\right)\cos\left(\frac{\alpha - \beta}{2}\right) \quad (11.7)$$

Hence we can rewrite (11.6) using (11.7) to get

$$\Psi(x, t) = 2\cos(\overline{k}x - \overline{\omega}t)\cos\left(\tfrac{1}{2}\Delta kx - \tfrac{1}{2}\Delta\omega t\right) \quad (11.8)$$

where $\overline{k}$ and $\overline{\omega}$ are the average wavenumber and angular frequency respectively and are defined as

$$\overline{k} = \tfrac{1}{2}(k_1 + k_2) \text{ and } \overline{\omega} = \tfrac{1}{2}(\omega_1 + \omega_2) \quad (11.9)$$

and Δk and $\Delta \omega$ are the difference in wavenumber and angular frequency respectively and are defined as

$$\Delta k = k_1 - k_2 \text{ and } \Delta\omega = \omega_1 - \omega_2 \quad (11.10)$$

Looking at (11.8) it is now clear that our solution when we add the two waves is a product of two wave functions. The cosine term gives the low frequency, long wavelength amplitude envelope for the high frequency, short wavelength oscillation. So the next question is what are the velocities of these two wave functions? In both cases the speed is just the angular frequency divided by the wavenumber which gives:

$$\text{Amplitude Velocity, } v_a = \frac{\Delta\omega}{\Delta k} = \frac{\omega_1 - \omega_2}{k_1 - k_2} \quad (11.11)$$

$$\text{Oscillation Velocity, } v_o = \frac{\overline{\omega}}{\overline{k}} = \frac{\omega_1 + \omega_2}{k_1 + k_2} \quad (11.12)$$

Now if we had an ideal medium where the refractive index was the same for all frequencies then the wave speed would be the same regardless of frequency and we would have the relationship

$$\frac{\omega_1}{k_1} = \frac{\omega_2}{k_2} = c \qquad (11.13)$$

If this were true then by making the substitutions $\omega_1 = ck_1$ and $\omega_2 = ck_2$ it is easy to show that (11.11) and (11.12) give the result that $v_a = v_o = c$. This would correspond to a non-dispersive medium and the amplitude and oscillation would move with the same speed.

Now consider the case where (11.13) is not true and the two waves both have difference phase velocities. In this case $v_a \neq v_o$ and the amplitude envelope moves at a different velocity compared to the oscillation velocity. Since the oscillation velocity is the velocity that a point of constant phase moves at this is, by definition, the *phase velocity* of the wave. The amplitude velocity is the velocity that the amplitude of a packet, or group, of waves moves at and so this is called the *group velocity* of the wave. Since we want to know these for a particular frequency of wave we take the limit as $\Delta k \to 0$ which gives the definitions as:

$$\text{Group Velocity, } v_g = \frac{d\omega}{dk} \qquad (11.14)$$

$$\text{Phase Velocity, } v_p = \frac{\omega}{k} \qquad (11.15)$$

For a single frequency the relationship $\omega = v_p k$ still holds true - one oscillation cycle of the wave must still take place in the time it takes the wave crest to move one wavelength - but the phase velocity now varies as a function of frequency or wave number.

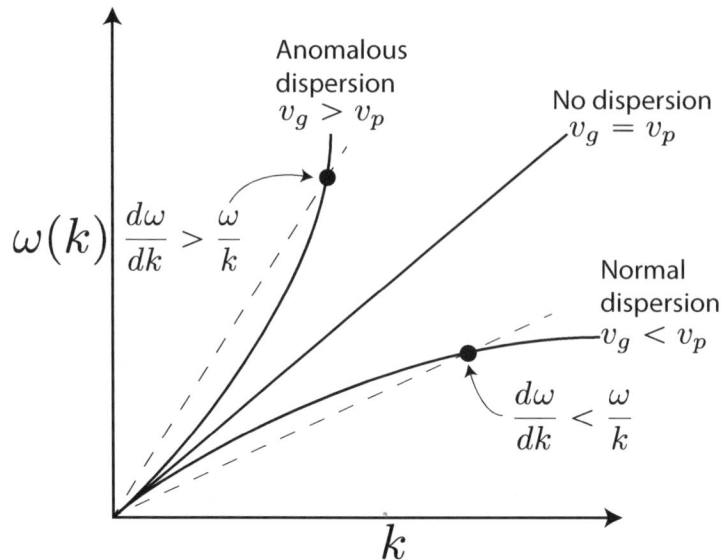

Fig. 11.7: Plot of ω vs. k showing the line for non-dispersive media as well as lines associated with both normal and anomalous dispersion.

Using this fact we can expand the differential form of the group velocity.

$$v_g = \frac{d\omega}{dk} = \frac{d(v_p k)}{dk} = v_p + \frac{dv_p}{dk}$$

$$\implies v_g = v_p - \lambda \frac{dv_p}{d\lambda} \tag{11.16}$$

where we use the definition of the wave number, $k = 2\pi/\lambda$ to transform the variables.

For most media $\frac{dv_p}{d\lambda}$ is positive i.e. the phase velocity increases with an increase in wavelength. Hence the group velocity is generally less than the phase velocity, $v_g < v_p$, and this is called *normal dispersion*. However this is not always the case and *anomalous dispersion* occurs when $v_g > v_p$. These two types of dispersion are easy to differentiate by plotting the wave number vs. frequency of waves for a given medium as shown in figure 11.7 which indicates the lines due to the two possible types of dispersion as well as no dispersion.

11.3 Polarization

Electromagnetic waves such as light are coupled oscillations of both an electric and a magnetic field, as shown in figure 11.8. As is evident these waves are transverse waves because both the electric and magnetic fields are perpendicular to the direction of travel of the wave. If we now concentrate on just the electric field then it is clear that there are two possible orientations of this field's oscillations. For a wave travelling along the z axis, the electric field can oscillate in either the x or y directions. Each of these possibilities is called a different *polarization* of the wave and for any transverse wave propagating in three dimensions there are two possible polarization states. For example, a transverse wave on a string can also vibrate in one of two possible orthogonal planes. For certain types of wave, both transverse and longitudinal waves are possible in which case the longitudinal wave can also be regarded as a third polarization state.

Figures 11.8 and 11.9 show the two possible linear polarization states for light so called because if viewed in the direction of propagation the electric field traces out a line. The direction of the polarization for an EM wave is, by convention, defined by the electric field's direction which is also shown in these figures. For light, the polarization vector is always perpendicular to the motion of the wave.

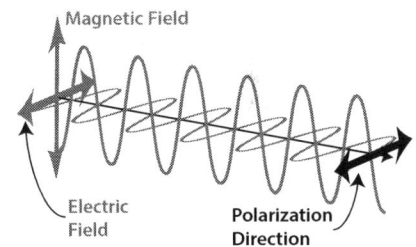

Fig. 11.8: The electric and magnetic fields which make up an electromagnetic wave such as light. By convention the direction of polarization is always taken to be that of the electric field.

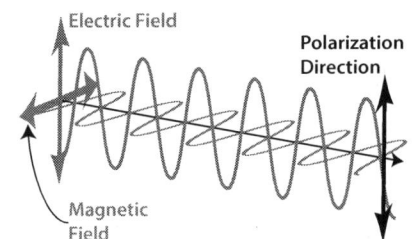

Fig. 11.9: The other possible linear polarization state for light (see fig. 11.8 for the other). The planes of the electric and magnetic fields are rotated about the axis of motion by 90°.

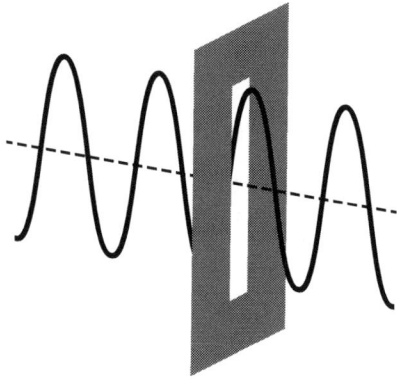

Fig. 11.10: Simple polarizer for transverse waves on a string: a board with a slot cut out of the centre. This allows the string to vibrate in one plane only and blocks the perpendicular polarization state.

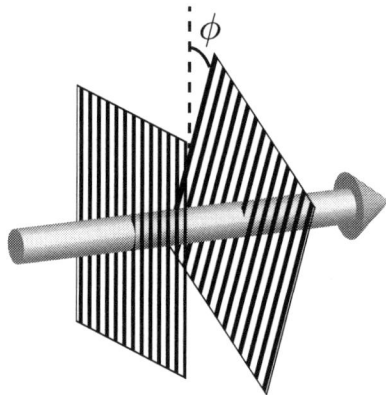

Fig. 11.11: Two polarizers with a beam of light passing through them. The second polarizers is rotated at an angle of ϕ with respect to the first one.

Most common sources of light waves are *unpolarized* which means that they produce are random mixture of all possible polarization states. The most common way to produce polarized light is to use a polarizer which is a filter that only allows one polarization of light to pass through it. For a wave on a string, the equivalent would be a sheet with a slot in it which will only allow the string to vibrate in a certain plane as shown in figure 11.10.

Having now produced linearly polarized light with one polarizer we can now ask what happens when this light falls on a second polarizer which has a polarization plane at an angle of ϕ to the original polarizer as seen in figure 11.11. In this case, we need to resolve the electric field into two components: one parallel to the second polarization plane and the other perpendicular to it and only the component parallel to the second polarizer's plane will pass. This means that the component of the electric field which will pass through the polarizer is:

$$E_{\parallel} = E \cos \phi \tag{11.17}$$

Hence the amplitude of the wave is reduced by a factor of $\cos \phi$. However what is measured is light intensity which, as we saw in section 7.8, is proportional to the square of the amplitude and so the reduction in intensity for light which has a polarization plane at an angle ϕ to that of the polarizer is:

$$\text{Intensity,} \ I = I_0 \cos^2 \phi \tag{11.18}$$

where I_0 is the intensity of the polarized light incident on the polarizer.

11.3.1 Brewster's Angle

When discussing refraction we mentioned that some of the light is not refracted but is actually reflected from the surface as shown in figure 9.7. This reflected light was studied by Sir David Brewster (1781-1868) who was a colourful British physicist being both an ordained minister in the Church of Scotland and inventor of the kaleidoscope child's toy!

What he discovered was that when the incident light was unpolarized the reflected light was generally partially polarized such that the electric field was perpendicular to the plane of reflection (this being the plane containing both the incident and reflected rays). In addition, for one particular angle called Brewster's angle or the polarization angle, the reflected light was completely polarized although the refracted ray remained only partially polarized.

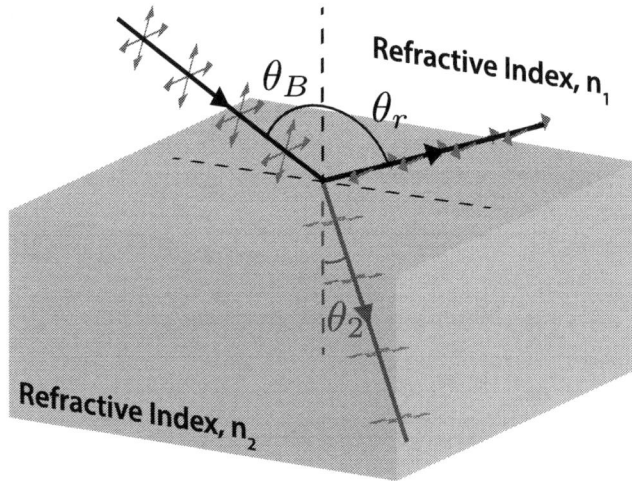

Fig. 11.12: Light reflecting and refracting at a surface. If the angle of incidence is equal to Brewster's angle the reflected beam is polarized perpendicularly to the plane of reflection and the refracted beam is partially polarized in the orthogonal direction.

Figure 11.12 shows this polarized, reflected ray which occurs when the angle between the reflected and refracted rays is precisely 90°. Hence for light incident at Brewster's angle we have the condition that:

$$\theta_r + \theta_2 = 90° \implies \theta_B + \theta_2 = 90° \tag{11.19}$$

as shown in figure 11.12 where θ_r is the angle of reflection and θ_2 is the angle of refraction . The second relation is derived using the law of reflection which requires the angle of reflection to equal the angle of incidence . In addition from Snell's law we have:

$$n_1 \sin \theta_B = n_2 \sin \theta_2 \tag{11.20}$$

Combining both (11.19) and (11.20) to obtain Brewster's angle we get:

$$n_1 \sin \theta_B = n_2 \sin(90° - \theta_B) = n_2 \cos \theta_B$$

$$\implies \theta_B = \tan^{-1}\left(\frac{n_2}{n_1}\right) \tag{11.21}$$

Since refractive index varies with wavelength differently for different materials this means that Brewster's angle is, in general, also a function of wavelength. For water ($n \approx 1.33$) and air($n \approx 1$) this angle is 53° for visible light. This means that sunlight reflected off water often has a high degree of polarization. Photographers over take advantage of this by using a polarized filter to remove light reflected from the water's surface so that objects underwater become visible.

11.4 Interference

As a wave light can interfere with itself. However for this to happen we need two sources of light which are coherent which is to say that they produce monochromatic light with a constant phase relationship. The overwhelming majority of light sources around us are neither monochromatic nor do they produce light of a constant phase with the result that interference phenomena are not readily visible in the everyday world.

11.4.1 Thin-film Interference

One of the simplest demonstrations of interference in light is the interference in a thin film of material, in particular, a soap bubble. If you look closely at the surface of a bubble you will often note that it shimmers with a variety of different colours when looked at from the correct angle. The film itself is transparent so these colours do not come from pigments in the film but rather from the interference of light reflecting from both the inner and outer surfaces of the bubble. A similar effect is visible in a thin film of oil on the surface of water.

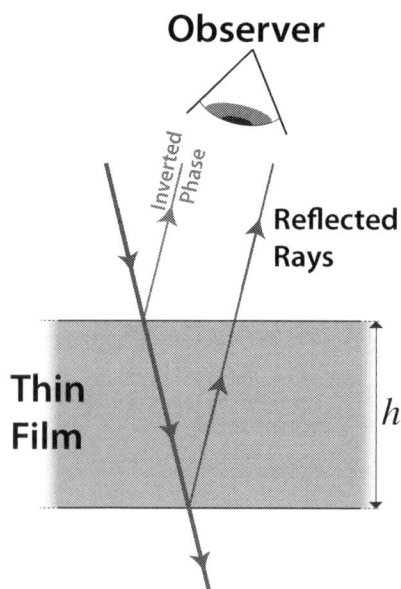

To understand what is happening we need to consider what happens when light reflects off the top and bottom surfaces of a thin film as shown in figure 11.13. Just as we saw with "Pepper's Ghost" the top surface of the film partially reflects some of the light and refracts the rest and the lower surface does the same. The result is that the original incoming ray is split into two reflected rays but one of these has a longer path length to get to the observer than the other.

However, a path length difference is not the only source of a phase difference between the two rays. Just as we saw when a wave pulse reflects from the end of a string light reflecting from the edge of a medium can have a phase change. For the string we saw that an open boundary resulted in no phase change for the reflected wave, see figure 8.3 but for a closed boundary there was a phase inversion, see figure 8.2. The same applies to light but, in this case reflecting off a boundary with a medium with a lower refractive index (faster) results in no phase change and reflecting off a boundary with a high refractive index (slower) causes a phase change of π.

For light rays perpendicular to the surface of the film the path length difference is just $2h$ where h is the thickness of the film but since the surrounding medium is usually air where the re-

Observer

Inverted Phase

Reflected Rays

Thin Film

h

Fig. 11.13: A ray of light incident on a thin film showing that how there is a reflected ray from both the top and bottom surface of the film.

fractive index is close to one the waves which reflect from the top surface have an addition phase change of π. Hence the wave functions for the two rays will be:

$$\psi_u(x,t) = A\cos(k'x + \pi - \omega t) \qquad \psi_l(x,t) = A\cos(k'x + 2hk - \omega t)$$
$$(11.22)$$

where ψ_u is the wave from the upper surface, ψ_l is the wave from the lower surface, k is the wavenumber in the film and k' is the wavenumber in the external medium. This means that the phase difference between the two rays is:

$$\Delta\phi = 2hk - \pi = \left(\frac{4h}{\lambda} - 1\right)\pi \qquad (11.23)$$

where λ is the wavelength in the film. Now if this phase difference is equal to $2n\pi$ where n is an integer then both reflected waves will have exactly the same phase and the two will add constructively to give bright a bright ray of reflected light. However if the phase difference is $(2n-1)\pi$ then the two rays will be exactly out of phase and so the two reflected rays will cancel each other and no light will be reflected.

Looking at the condition for cancellation we can put the required phase difference into (11.23) and get an expression for the wavelengths for which there will be no light reflected:

$$\left(\frac{4h}{\lambda} - 1\right)\pi = (2n-1)\pi \implies \frac{4h}{\lambda} = 2n \implies \lambda = \frac{2h}{n} \quad (11.24)$$

This is the reason why a thin film of liquid such as a soap bubble or oil on water appears multi-coloured. For any given thickness only certain wavelengths in white light will cancel out. The wavelengths which remain will then no longer appear white. In addition, the thickness of the film is not constant to an accuracy of the wavelength of visible light ($\sim$500 nm). This variation in thickness will mean that different wavelengths will cancel at different locations which will give a variety of colours. Note that here we have just considered the extreme, perfect cancellation condition. Wavelengths almost equal to the cancelled one will produce a phase difference between the reflected waves close to π and so will be almost, but not entirely, cancelled out too. Hence a range of wavelengths will have little to no reflection.

In general, we can split these thin films into one of two types: ones where the phase change of the reflected wave is the same at both boundaries and ones where the phase change of the reflected wave is different at each boundary. The case of the soap bubble where both boundaries are with air is clearly of the latter type: there is a phase inversion upon reflection at the first

	Constructive	Destructive
No/both phase change	$h = n\lambda/2$	$h = (2n - 1)\lambda/4$
One phase change	$h = (2n - 1)\lambda/4$	$h = n\lambda/2$

Table 11.1: Conditions for constructive and destructive interference for the light rays reflected from the two surfaces of a thin film. The films are categorized into ones where there is a phase inversion at either both or neither surface and ones where there is a phase inversion at only one surface. In all cases $n = 1, 2, 3, 4 \ldots$ and the wavelength is measured in the film.

boundary but there is none at the second. The conditions for destructive and constructive interference for both types of thin films is neatly summarized in table 11.1.

This type of thin film interference effect is used on cameras and other optical instruments where it is desirable to reduce reflections. To stop reflections a thin film of material is sprayed onto the lens. The refractive index of this film is between that of air and glass and so the film reflects light from both surfaces with a phase inversion because each successive layer has a lower speed of light. Looking at the conditions given in table 11.1 we can see that the conditions for cancellation of the reflected waves is $h = (2n - 1)\lambda/4$ and so for this thinnest film we have a thickness of $\lambda/4$. Obviously, this only provides perfect cancellation for a single wavelength but by picking a wavelength in the middle of the visible spectrum the reflection of most visible wavelengths can be suppressed. This is why, when you look at a good quality camera lens at an angle, you can often see a magenta tinted reflection. A more detailed calculation of this type of system is explored in example 11.1.

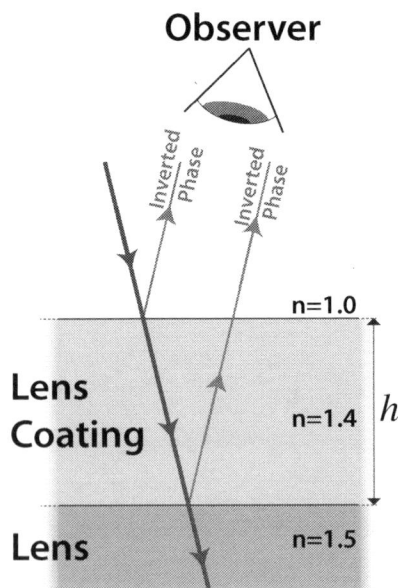

Observer

Inverted Phase   *Inverted Phase*

n=1.0

Lens Coating   n=1.4 h

Lens   n=1.5

Fig. 11.14: A ray of light incident on a thin film covering the surface of a glass lens. Both reflected rays are shown and each of them has a phase change of π due to the reflection.

Example 11.1

A telescope lens, made of glass with a refractive index of 1.5, is covered with a film of material 100 nm thick which has a refractive index of 1.4. If the refractive index of air is 1.0 what wavelength of visible light will not be reflected from the telescope lens? Given that the wavelengths for red, green and blue light in air are respectively, and approximately, 650 nm, 540 nm and 450 nm what colour will the lens appear when white light is reflected from its surface?

Solution:

Figure 11.14 shows a diagram of the lens coating with light reflecting from the surface of the lens. The ray reflected at the top surface is passing into a slower medium and so will have a phase change of π as will the ray reflecting from the interface between the coating and the glass since again *there is an increase in refractive index . For no light to be reflected we want to have perfect cancellation of the two reflected rays and so, looking at table 11.1 we require that:*

$$h = (2n - 1)\lambda/4$$

where h is the thickness of the coating. Rearranging this we get and expression for the wavelengths which will not be reflected of:

$$\lambda = \frac{4h}{2n - 1} = \frac{400}{2n - 1} \ nm$$

However this is the wavelength inside the coating. Since $c = f\lambda$ where f is constant and the speed of light c depends on the refractive index of the coating, $n_c = 1.4$ we have:

$$\lambda_{air} = n_c \lambda = \frac{560}{2n - 1} \ nm$$

At this point we need to determine the value of the integer n. The questions asks for the wavelength of visible light which is not reflected so the answer must lie in the range 400 nm to 670 nm. If we consider the first two values of n then for $n = 1$ we have $\lambda_{air} = 560$ nm and for $n = 2$, $\lambda_{air} = 187$ nm hence clearly the wavelength of visible light which is not reflected is 560 nm since 187 nm is well beyond the visible region of the spectrum and higher values of n will correspond to even shorter wavelengths.

Our answer for the first part of the question is very close to green light and so green light will not be reflected. As we move away from this value, either towards longer or shorter wavelengths, the cancellation will worsen. This means that the far blue/violet end of the spectrum and the far-red end of the spectrum will have the most reflection. Hence the reflected light, which will determine the colour of the lens, will a combination of red and blue which will appear purple or magenta in colour.

11.4.2 Thick Films

We now have an explanation for how light reflections can be partly, or entirely, cancelled when reflecting from a thin film. However, we do not observe this phenomenon with a thick film such as the glass in a window. There is no pattern of colours from a window as there is from a thin covering of oil or a soap bubble so why is this the case?

The reason for this apparent inconsistency is due to the typical, everyday sources of light which are not coherent. The phase of light emitted from any type of common light bulb, the sun or even a candle varies randomly with time: these are incoherent light sources. However, the phase, while varying randomly, does remain roughly constant for a short period of time. Thin films are typically on the order of micrometres (1×10^{-6} m) thick which is only 2-3 times the wavelength of the light. Over distances this small, the phase of the light will remain roughly constant and so the two reflected waves will have a constant phase relationship between them and so there will be interference.

Conversely, the pane of glass in a window is several millimetres thick which is roughly 10,000 times the wavelength of light. For distances this long the light is very definitely incoherent and so there are no interference effects observed.

Observer

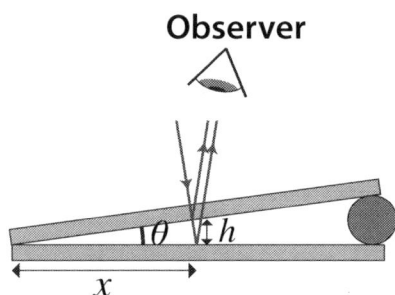

Fig. 11.15: A ray of light incident on a thin wedge showing that how there is a reflected ray from both the top and bottom surface of the wedge which will interfere. The thickness of the wedge depends on the distance from the thin end, x, and the opening angle of the wedge, θ.

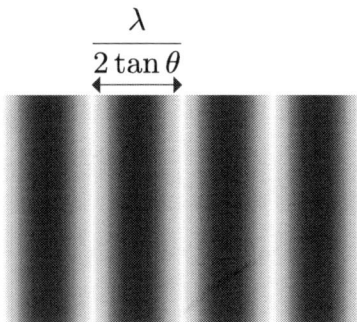

Fig. 11.16: The simulated pattern of fringes which will be observed from the wedge. The fringes are all equally spaced with the spacing being determined by the wavelength of the light as well as the opening angle of the wedge.

11.4.3 Thin Wedge

The thin films we have considered up to now all have a constant thickness but suppose we now consider what happens if we have a varying separation between the two reflected rays. Consider two sheets of glass which are touching at one end and have a very small separation at the other end to form a thin wedge as shown in figure 11.15. If monochromatic light is now incident on the system then light will be reflected from the surfaces forming the wedge. The top surface will be a glass-to-air boundary and so there will be no phase inversion however the bottom surface is air-to-glass and so there will be a phase inversion upon reflection because the refractive index of glass is higher than air. Hence, looking at table 11.1 we can see that the condition for there to be destructive interference is $h = n\lambda/2$.

Now the thickness, h, of the wedge varies with distance, x from the thin end. If the opening angle of the wedge is θ, as shown in figure 11.15, then the thickness at position x can be obtained using simple trigonometry and is:

$$h = x \tan \theta \tag{11.25}$$

Putting this into our condition for destructive interference we have:

$$\frac{n\lambda}{2} = x \tan \theta \implies x = \frac{n\lambda}{2 \tan \theta} \tag{11.26}$$

This is the condition for destructive interference which will generate a dark region in the interference pattern produced. However the condition for constructive interference will be satisfied at values of x where

$$(2n - 1)\lambda/4 = x \tan \theta \implies x = \frac{(2n - 1)\lambda}{4 \tan \theta} \qquad (11.27)$$

Hence for monochromatic light we will get a series of equally spaced bright and dark "stripes" where the light first cancels and then adds as shown in figure 11.16. These "stripes" which are a common occurrence from interference are called fringes. The spacing of the fringes depends not only on the wavelength of the light but also the opening angle of the wedge. To determine the width of a fringe we measure the distance from one point of cancellation to the next which, using equation (11.26) gives:

$$\text{Fringe Width, } \Delta x = \frac{\lambda}{2 \tan \theta} \qquad (11.28)$$

11.4.4 Newton's Rings

Newton's rings is an interference pattern produced when a spherical, convex lens is placed in contact with a flat pane of glass as shown in figure 11.17. The bottom surface of the lens and the top surface of the pane of glass act in a similar manner to the thin wedge discussed in the previous section.

To calculate the thickness of the wedge at a distance x from the point of contact of the lens and the glass consider figure 11.17 which shows a convex lens with a radius of curvature r lying on a plane sheet of glass. At a distance x from the point of contact of the two the thickness of the gap is t. To find the relationship between x and t we need to do a bit of geometry. Since OX is a diameter the angle OYX must be $90°$. Hence using Pythagorus on the triangles OXY, ONY and NXY we have, respectively:

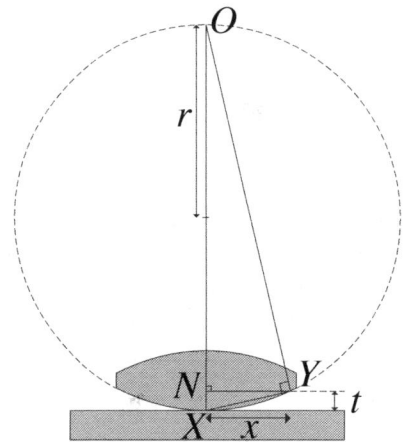

Fig. 11.17: A spherical lens resting on the surface of a pane of glass. The radius, r, shown is the radius of curvature of the lens and t is the thickness of the gap between the lens and the pane a distance x from the point of contact between the two.

$$XY^2 + OY^2 = OX^2 \implies XY^2 + OY^2 = 4r^2 \qquad (11.29)$$
$$ON^2 + NY^2 = OY^2 \implies ON^2 + x^2 = OY^2 \qquad (11.30)$$
$$NY^2 + NX^2 = XY^2 \implies x^2 + t^2 = XY^2 \qquad (11.31)$$

Now we combine equations (11.29) and (11.30) to eliminate OY:

$$4r^2 - XY^2 = ON^2 + x^2 \implies XY^2 = 4r^2 - ON^2 - x^2 \qquad (11.32)$$

and then we use equation (11.31) to eliminate XY to get:

$$4r^2 - ON^2 - x^2 = x^2 + t^2 \implies 4r^2 - ON^2 = 2x^2 + t^2 \qquad (11.33)$$

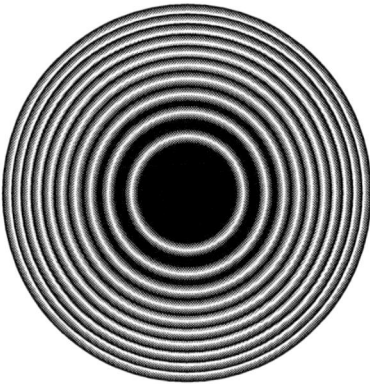

Fig. 11.18: Simulated picture of Newton's Rings formed by a convex lens resting on a pane of glass illuminated with monochromatic light using (11.37) to calculate the radii. Note that the central fringe is always dark due to the phase inversion of the reflected ray from the lower surface.

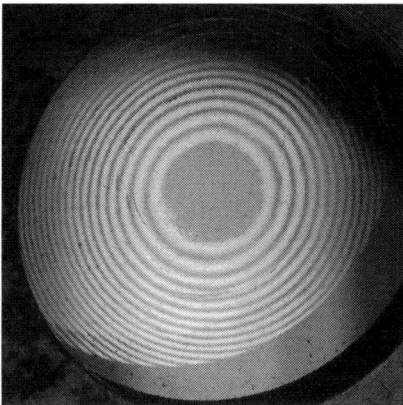

Fig. 11.19: Actual photograph of Newton's rings under a sodium light which is approximately monochromatic. Note the dark ring in the centre and the decreasing thickness of the rings with increasing radius as predicted, and simulated in figure 11.18.

Finally we look at ON since OX is a full diameter of the circle we can rewrite ON as $2r - t$. Putting this into (11.33) gives:

$$4r^2 - (2r - t)^2 = 2x^2 + t^2$$
$$4r^2 - 4r^2 + 4rt - t^2 = 2x^2 + t^2$$
$$2rt = x^2 + t^2 \qquad (11.34)$$

The final step is to make an approximation. The radius of curvature of the lens is very large, and always far larger than the physical size of the lens otherwise we would not have a thin lens. This means that $t \ll x$ and so in (11.34) the t^2 term will be far smaller than x^2 which gives the approximate result:

$$2rt \approx x^2 \implies t = \frac{x^2}{2r} \qquad (11.35)$$

Now that we know the thickness of the gap as a function of the distance from the point of contact we can determine at which distances there will be either constructive or destructive interference. Since the light reflected at the bottom surface will have a phase inversion but the light from the top surface will not we need to use the conditions from the 'one phase change' row of table 11.1.

Using this table we see that for destructive interference the condition is that $t = n\lambda/2$. Putting this into (11.35) gives:

$$\frac{n\lambda}{2} = \frac{x^2}{2r} \implies x = \sqrt{n\lambda r} \qquad (11.36)$$

which gives the horizontal distances from the contact point between the lens and the glass pane at which there will be a dark fringe. Since the lens has a rotational symmetry this dark fringe will actually form a ring and hence x is the radius of this circular, dark fringe. To determine the location of the bright fringes we need to return to table 11.1 and this time use the conditions for constructive interference: $t = (2n - 1)\lambda/4$ which give values of x where:

$$x = \sqrt{\left(n + \frac{1}{2}\right)\lambda r} \qquad (11.37)$$

At these radii there will be a bright, circular fringe.

The result is that there will be a series of concentric, circular fringes centred on the point of contact with the spacing between the fringes getting smaller as the radius increases as simulated in 11.18 and photographed in figure 11.19. This pattern is known as Newton's rings since it was studied extensively by Newton. The irony of this is that Newton tried for many years to prove that light was a wave, as suggested by Huyghens, before reluctantly concluding that light was actually transmitted by tiny particles rather than being waves. However, unbeknownst to him, this phenomenon was due to wave interference thus showing that light was actually behaving as a wave!

Newton's rings is also a very useful pattern to observe if you are making a lens. The pattern is very sensitive to any irregularities in the surface of the lens because the wavelength of light is short, $\sim 0.5\,\mu m$, and an irregularity on the scale of quarter of a wavelength is all it takes to invert the phase and turn a bright ring into a dark one. Also if the grinding of the lens is not perfectly circular the rings will deviate from their circular shape. As a result, the pattern is an excellent way to check the quality of the lens and it is used by lens makers to ensure that the lenses they make are accurate.

11.5 Interferometers

Interferometers are devices which use the interference of light to measure properties and extract information. They are at the heart of one of the most important failed experiments in the history of science and, more recently, have provided the ability to detect the collisions of black holes billions of light years away. Closer to home interferometers are used to image a cross-section of your eye at the opticians and in engineering they have a huge variety of uses from measuring small displacements to making gyroscopic sensors. Indeed the discussion in the previous section, 11.4.4, Newton's rings can be considered as a simple example of an interferometer which is used to determine the correct shape for the surface of a lens.

11.5.1 Michelson Interferometer

A very common design for an interferometer is the Michelson interferometer which was invented by Albert Michelson (1852-1931). It works by taking a single light source and reflects it off the surface of a half-silvered mirror angled at $45°$ to the incident beam, as shown in figure 11.20. This splits the beam into two: one refracts at both surfaces and passes straight through the mirror and the other reflects off the half-silvered surface and travels at $90°$ to the other beam. The two beams of light are then each reflected off two fully silvered mirrors such that they are again incident on the half-silvered mirror which this time recombines half of them (the other half being reflected/refracted

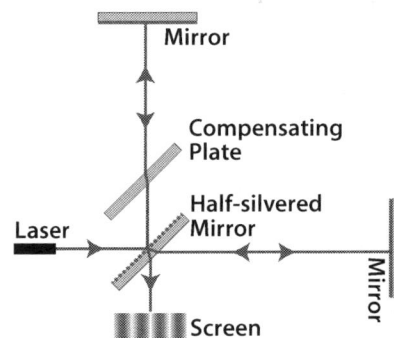

Fig. 11.20: Diagram of a Michelson interferometer which passes light from a source along two arms and recombines them to produce a series of fringes.

back to the source) into a single beam which is then sent to a screen.

Depending on which surface of the half-silvered mirror has the reflective coating one of the beams will be passed through a compensating plate. This is a pane of glass which is identical to glass used for the half-silvered mirror only it has no silvering at all. Its purpose is to compensate for the fact that one of the beams must pass through the glass of the half-silvered mirror twice more than the other beam in order to reach the half-silvered face of the glass to reflect. The different refractive index of the glass, compared to air, means that there is a different wavelength of light in the glass than in air which would cause a different phase shift if it were not corrected.

If both arms of the interferometer have identical lengths and the mirrors at the ends of the arms are perfectly perpendicular to the beams then the two beams will arrive in phase and will add together constructively. However the difference between constructive and destructive interference is a displacement of only half the wavelength of light, i.e. around 250 nm depending on the colour, so even a tiny misalignment will affect the pattern.

Typically this is the case and the mirrors at the ends are not perpendicular to the beam at this level of precision, as shown in figure 11.21. This results in the length of the beam varying linearly with the distance off-axis which produces a series of linear fringes as the two beams move into and out of phase with increasing distance from the axis in a direction parallel to the relative slope of the mirrors.

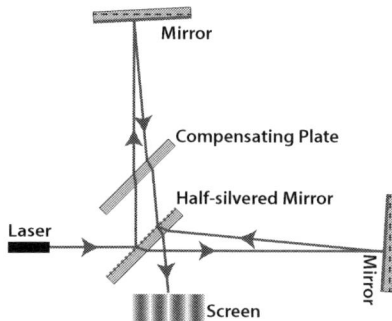

Fig. 11.21: Diagram of a Michelson interferometer showing how a tiny misalignment of the mirrors causes the fringe pattern.

11.5.2 Michelson-Morley Experiment

The Michelson-Morley experiment was conducted in 1887 and is perhaps the most famous of all failed experiments. In 1861 James Clark Maxwell had published his paper "On Physical Lines of Force" in which he had concluded that light was an electromagnetic wave. However, since all other waves needed a medium the question facing the physics community was what was the nature of this mysterious medium, called the *luminiferous aether*.

To detect the presence of the luminiferous aether Albert Michelson and Edward Morley devised a clever experiment using a Michelson interferometer. If light was transmitted by this medium then the interference pattern produced by the interferometer would be affected by the motion of the Earth through the aether. If we consider an interferometer with two arms of equal length with an aether wind with a velocity v parallel to one of the arms then the travel time for a light ray along the length of that arm, L, will be:

$$t_\parallel = \frac{L}{c-v} + \frac{L}{c+v} = \frac{2Lc}{c^2 - v^2} \qquad (11.38)$$

For the arm which is perpendicular to the aether wind part of the light's velocity is used to offset the velocity of the wind as shown in figure 11.22. This reduces the component of the velocity which is parallel to the arm and using Pythagoras on the right-angled triangle we get a speed of light in the lab of:

$$c' = \sqrt{c^2 - v^2} \qquad (11.39)$$

and hence the travel time for the ray perpendicular to the wind is:

$$t_\perp = \frac{2L}{\sqrt{c^2 - v^2}} \qquad (11.40)$$

because the speed of the light in the lab will be the same in both directions. The result is a propagation time difference between the arms of:

$$\Delta t = t_\parallel - t_\perp \approx \frac{Lv^2}{c^3} \qquad (11.41)$$

where we make the assumption that $v \ll c$ which for any reasonable model of the Earth's motion through the luminiferous aether is an extremely good assumption.

The interferometer arranged in this orientation is shown in the first image in figure 11.23. However what Michelson and Morley did was place the interferometer on a large, concrete slab which could be rotated so that the direction of the aether wind to the two arms could be changed. If the slab is rotated through $45°$ then the aether wind in this orientation is now at $45°$ to both rays of light, and the experiment in this configuration is shown in the second image in figure 11.23. In this configuration the speed of light in the lab will be c_a when moving against the aether and c_w when moving with the aether but this is the same for both beams of light as shown in the figure and so there is zero propagation delay between the beams i.e. $\Delta t = 0$.

This change in the propagation delay for the two beams depending on the orientation of the arms to the aether wind means that the presence of the wind can be detected by a change in the interference pattern between the beams. However, when they performed the experiment it failed to detect any shift in the pattern at all but this complete failure to detect the aether wind

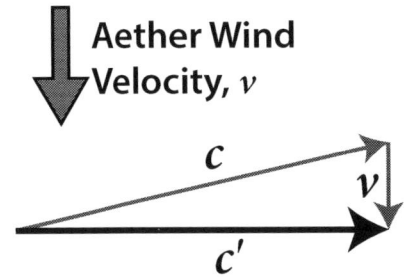

Fig. 11.22: Addition of velocity vectors required for a light ray travelling perpendicular to the aether wind. The velocity of light relative to the aether is c, the velocity of the aether is v and the velocity of light in the lab is c'.

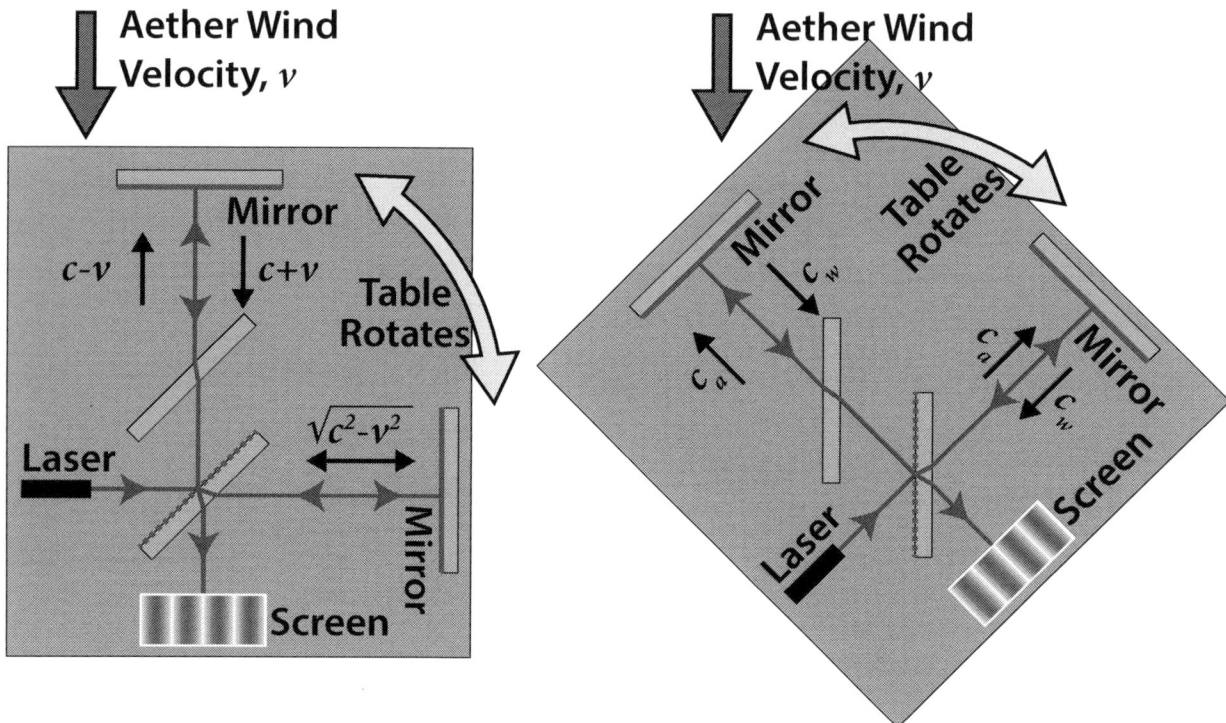

Fig. 11.23: *The Michelson-Morley experiment. In the first figure it is arranged so that one arm is parallel to the aether wind and the other is perpendicular which gives the maximum propagation difference between the two arms. In the second figure both arms are at 45° which gives a zero propagation difference.*

(something we now know does not exist) lead to another Albert, Albert Einstein, to postulate that the speed of light was a constant and never changed.

This idea, which is so simple to state, is one of the two core postulates of special relativity: the speed of light is the same in all inertial frames i.e. it does not matter how fast you may be travelling relative to a beam of light you will always measure its speed as being the same. This overturned 300 years of classical mechanics developed on Newton's laws of motion and caused a major paradigm shift in our understanding of space and time.

11.5.3 Gravitational Waves

The Michelson interferometer is also at the heart of an experiment called LIGO (Laser Interferometer Gravitational-wave Observatory) which, on the 11th February 2016 became the first experiment to ever detect gravitational waves. Gravitational waves are a prediction of general relativity and are a ripple in space-time where space is compressed in one direction and stretched in the other before the situation is reversed as shown in figure 11.24.

To detect these waves LIGO has two interferometers, both in the USA with one in Livingston, Louisiana and the other in Hanford, Washington State as shown in figure 11.25. When a gravitational wave passes through the interferometer the two arms, which are at 90° to each other, are compressed and stretched at difference times which causes the interference pattern to change. Each arm of the interferometer is 4 km long and light is passed multiple times up and down the arm to maximize the effective length and hence the sensitivity. This is needed because the wave only creates a change in length over 4 km of approximately 1 am=10^{-18} m which is less than a thousandth the diameter of a typical atomic nucleus or almost a billion times smaller than an atom.

To create gravitational waves requires an enormous amount of energy and the first signal detected, GW151226, came from the merger of two black holes approximately 1.4 billion light years away. As the black holes orbit each other they radiate gravitational wave energy causing the orbit to shrink and the frequency and amplitude of the waves to increase until they eventually coalesce. This generates a characteristic 'chirp' which is easier to detect above the background noise. Both of the LIGO interferometers observed the signal with the Hanford detector seeing it 1.1 ms later due to the propagation delay caused by the 3,002 km separation between the two interferometers.

11.5.4 Optical Coherence Tomography

Optical coherence tomography is a technique which uses a Michelson interferometer but instead of a coherent light source, such as a laser, it uses an incoherent source of light which means that the phase of the light is only constant for very short periods of time and, over longer periods of time, will randomly vary. Typically this type of light source is not very useful for interferometry because the random phase variations mean that there is only interference if both arms of the interferometer have exactly the same length but optical coherence tomography relies on this to great effect.

To image an object, such as the retina at the back of our eye, the eye is placed at the end of one arm of the interferometer to act as the reflective 'mirror'. Since the retina has several layers the distance which the light travels will depend on which layer of

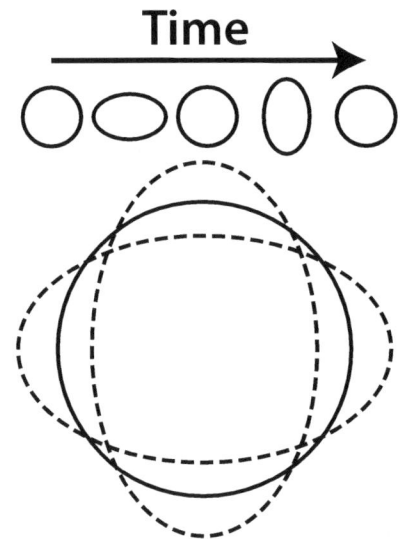

Fig. 11.24: The deformation of a circle due to a gravitational wave which alternately compresses and expands space in directions perpendicular to the propagation direction of the wave as shown in the figure above.

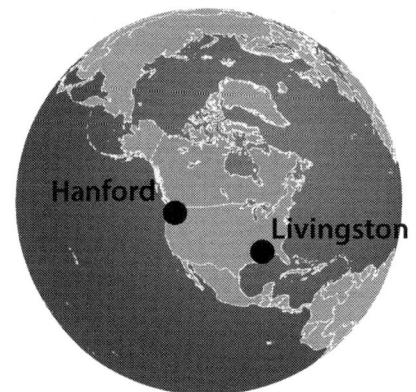

Fig. 11.25: Picture of the globe showing the locations of the two existing LIGO detectors. A third detector is planned for construction in India.

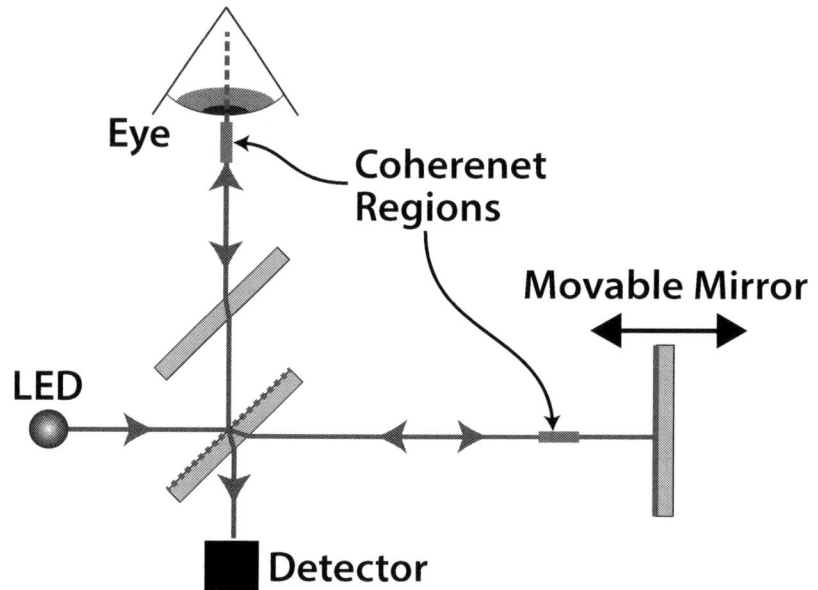

Fig. 11.26: Michelson interferometer with a movable mirror and an incoherent light source used to study an eye. For the two coherent regions of the beam to interfere they must reflect off something at equal distances. Hence by measuring the length of the mirror arm, you can measure the amount of light which is reflected at the same distance from inside the eye.

the retina it reflects from. The second arm has a regular mirror at the end to reflect the beam and this position of this mirror can be moved to change the length of the beam. However, because the phase of the light is always randomly changing there will only be an interference pattern produced when the length of the mirror beam is exactly the same as the length of the sample beam and so by changing the length of the mirror beam the light which reflects from different layers of the sample can be selected.

Using this technique it is possible to build up a three-dimensional picture of the sample without any invasive procedure being required. For the eye the light source used is near-infrared which avoids any damage to the retina and allows the retina to be studied directly for signs of any problems.

11.6 Diffraction

The phenomenon known as *diffraction* occurs when a wave encounters either a slit or other obstacle. It was first discovered in light waves by the Italian priest and physicist Francesco Grimaldi (1618-1663) although his observations were only published posthumously in 1665. Grimaldi observed that light passing through a small hole did not form a thin, pencil beam but instead spread out to form a cone and named this phenomenon diffraction.

What Grimaldi discovered was that waves passing through the slit spread out in accordance with Huyghens' principle. Each point on the wavefront acts as a source and so the wavefront will spread out to either side after passing through a narrow slit.

This approach using Huyghens' principle is precisely how we will calculate the diffraction patterns that we will get from different configurations of slits. In doing this we will confine ourselves to looking at the pattern at a large distance from the slits so that we may assume that the rays of light are parallel. Making this approximation gives what is called *Fraunhofer Diffraction* as opposed to *Fresnel Diffraction* which calculates the pattern a lot closer to the slits but which is far more complex mathematically to evaluate.

In making these calculations we will also assume that the slits are illuminated with a *coherent* source of light. This is a source which emits light with a single phase. Typical light sources, such as an incandescent or fluorescent lights, are not coherent since they produce light waves with many different phases. The most common coherent source of light is a laser. Before lasers coherent sources were created by putting a very narrow slit, with a width comparable to the wavelength of light, in front of an incoherent light source.

11.6.1 Single Slit

This simplest type of diffraction to consider is that from a single slit. Here we will assume that we have a single slit with a width w as shown in figure 11.27. Consider a length δy of the wave front which is a distance y from the centre of the slit. This will act as a point source of waves in the limit of small δy but compared to the light ray coming from the centre of the slit the waves from this source will have an additional path length of

$$\text{Path Length Difference, } \Delta r = y \sin \theta \qquad (11.42)$$

as shown in figure 11.27. To convert a path length difference into a phase difference we multiply by the wavenumber. Hence the phase difference of the waves from our source will be

$$\text{Phase Difference, } \Delta \phi = k y \sin \theta \qquad (11.43)$$

where k is the wave number for the waves incident on the slit. Now if we assume that each point on the wavefront has the same amplitude then the amplitude of the waves from this point source will be proportional to the size of the source. Hence if

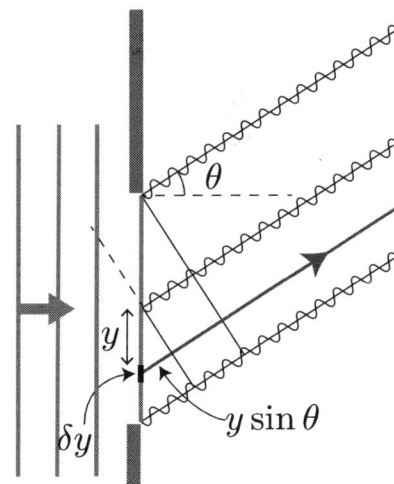

Fig. 11.27: Waves diffracting at a single, wide slit showing the path length difference as a function of y, the distance from the centre of the slit. Since the screen is considered to be at a large distance the outgoing waves are assumed to be approximately parallel.

the waves which are incident on the slit have the form $\cos(kr - \omega t)$ then the diffracted waves from our point source will be

$$\delta\psi(\theta) = \frac{A}{a}\delta y \cos(kr - \omega t + \Delta\phi) = \frac{A}{a}\cos(kr - \omega t + ky\sin\theta)\delta y$$

(11.44)

where r is the distance from the slit, A is the amplitude of the transmitted wave and a is the width of the slit so the first term in the above equation represents the amplitude per unit length of the wavefront. Now the amplitude of the wave observed on the screen at an angle θ will be the sum of all these point sources summed over the width of the slit in the limit that $\delta y \to 0$. The result is an integral

$$\psi(\theta) = \int_{-a/2}^{a/2} \frac{A}{a}\cos(kr - \omega t + ky\sin\theta)\,dy$$

(11.45)

where a is the width of the slit. To evaluate this we expand out the cosine term using our trig identity for $\cos(A + B)$ which gives:

$$\cos(kr - \omega t + ky\sin\theta) = \cos(kr - \omega t)\cos(ky\sin\theta)$$
$$- \sin(kr - \omega t)\sin(ky\sin\theta) \quad (11.46)$$

Now only the terms which include a y will be integrated and looking at the second of the terms this is a sine term which will give zero because of the symmetric limits of integration:

$$\int_{-a/2}^{+a/2} \sin(ky\sin\theta)\,dy = \left[-\frac{\cos(ky\sin\theta)}{k\sin\theta}\right]_{-a/2}^{+a/2} = 0 \quad (11.47)$$

Hence the only the first term from (11.46) matters and so we have:

$$\psi(\theta) = \frac{A}{a}\cos(kr - \omega t)\int_{-a/2}^{a/2}\cos(ky\sin\theta)dy$$
$$= \frac{A}{a}\cos(kr - \omega t)\left[\frac{\sin(ky\sin\theta)}{k\sin\theta}\right]_{-a/2}^{a/2}$$
$$= \frac{A\cos(kr - \omega t)}{ka\sin\theta}\left\{\sin(\tfrac{1}{2}ka\sin\theta) - \sin(-\tfrac{1}{2}ka\sin\theta)\right\}$$

(11.48)

Now we need to remember that because sine is an odd function we have the relationship that $\sin\theta = -\sin(-\theta)$ and so (11.48) simplifies to:

$$\psi(\theta) = \frac{A\cos(kr - \omega t)}{ka\sin\theta} \times 2\sin(\tfrac{1}{2}ka\sin\theta)$$
$$= A\cos(kr - \omega t)\frac{\sin(\tfrac{1}{2}ka\sin\theta)}{\tfrac{1}{2}ka\sin\theta}$$
$$= A\cos(kr - \omega t)\frac{\sin(\tfrac{\pi}{\lambda}a\sin\theta)}{\tfrac{\pi}{\lambda}a\sin\theta}$$

(11.49)

where in the last step we expand out the wave number in terms of the wavelength.

The result is a wave pattern whose amplitude varies as a function of $\sin \theta$ as shown in figure 11.28 and given by:

$$\psi(\theta) = A \frac{\sin\left(\frac{\pi}{\lambda} a \sin \theta\right)}{\frac{\pi}{\lambda} a \sin \theta} \cos(kr - \omega t) \qquad (11.50)$$

However when we project light from a diffraction slit onto a screen what is observed is the light intensity and not the amplitude. To find this we need to square the amplitude which gives:

$$\text{Intensity, } I = I_0 \frac{\sin^2\left(\frac{\pi}{\lambda} a \sin \theta\right)}{\frac{\pi^2}{\lambda^2} a^2 \sin^2 \theta} \qquad (11.51)$$

This gives an intensity pattern also shown in figure 11.28 which will be the pattern observed on a screen. There will be a zero intensity in the pattern every time that contents of the sine term is a non-zero integer multiple of π and so the condition for zero intensity is:

$$\frac{\pi}{\lambda} a \sin \theta = n\pi \implies \sin \theta = \frac{n\lambda}{a} \text{ where } n = 1, 2, 3 \ldots \qquad (11.52)$$

which will result in equally spaced dark bands whose separation is proportional to the wavelength and inversely proportional to the width of the slit. However the central bright band will be twice the width of the other bands since there is no minimum when $\sin \theta = 0$. To show this we need to consider the limit of $\sin x / x$ - called the sinc function - as $x \to 0$. To do this we need to start with the power series expansion of $\sin x$:

$$\sin x = x - \frac{x^3}{3!} + \frac{x^5}{5!} - \frac{x^7}{7!} + \cdots$$

$$\implies \frac{\sin x}{x} = 1 - \frac{x^2}{3!} + \frac{x^4}{5!} - \frac{x^6}{7!} + \cdots$$

$$\implies \lim_{x \to 0} \frac{\sin x}{x} = 1 \qquad (11.53)$$

Hence there is no minimum in the diffraction pattern when $\sin \theta = 0$ which gives a central, double width, bright band.

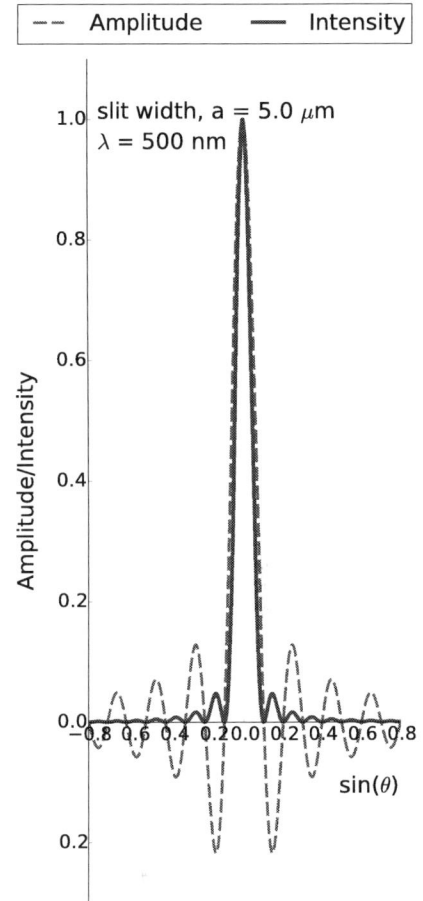

Fig. 11.28: The diffraction pattern from a single slit showing both the intensity and amplitude.

11.6.2 Double Slit

Having considered a single slit the next logical slit pattern to consider is a double slit. This is what became known as Young's double slit experiment despite the lack of any evidence that

Thomas Young ever performed it! However, it is such a clear demonstration of the wave nature of light that it is perhaps not too surprising that it ended up bearing his name.

To calculate the pattern produced let us first keep this simple by considering two narrow slits a distance d apart each of which will act as a point source of waves as shown in figure 11.29. This is actually an almost identical situation to one we have already considered back in section 8.3. In that section we considered the interference pattern from two point sources which were a distance d apart, as shown in figure 8.7. This is almost precisely what we have here! The only slight difference is that the waves from these point sources will only be emitted on one side of the source.

The interference pattern resulting from our double slits will be the sum of just two waves, one from each slit. However these will have a phase difference between them due to the different path lengths. If we call the phase difference between the two waves $\Delta\phi$ then one wave will have a phase of $+\Delta\phi/2$ and the other a phase of $-\Delta\phi/2$ which gives a combined wave of:

$$\psi(\theta) = A\cos(kr - \omega t + \Delta\phi/2) + A\cos(kr - \omega t - \Delta\phi/2) \quad (11.54)$$

This is just the sum of two cosine terms so to evaluate it we simply use the trig identity we already used with dispersion and given in (11.7). Using this identity greatly simplifies the expression to:

$$\psi(\theta) = 2A\cos(kr - \omega t)\cos\left(\frac{\Delta\phi}{2}\right) \quad (11.55)$$

We already calculated the phase difference in (8.9) or, alternatively, we can calculate it directly from figure 11.29, which gives:

$$\Delta\phi = k\Delta r = kd\sin\theta \quad (11.56)$$

Putting this into (11.55) gives:

$$\psi(\theta) = 2A\cos\left(\frac{\pi d}{\lambda}\sin\theta\right)\cos(kr - \omega t) \quad (11.57)$$

Again for determining the pattern observed on the screen what we are most interested in is the intensity of the light, not the amplitude. Since the intensity is proportional to the square of the amplitude this gives:

$$I = I_0\cos^2\left(\frac{\pi d}{\lambda}\sin\theta\right) \quad (11.58)$$

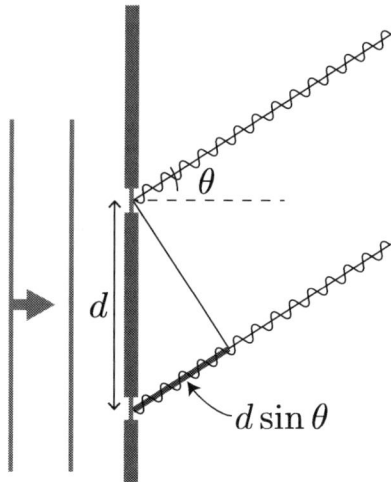

Fig. 11.29: Diffraction at two, narrow slits showing the path length difference as a function of the diffraction angle, θ.

This is a pattern of equal width bright and dark fringes as shown in figure 11.30. The condition for a bright fringe is that the cosine squared term must be one which will occur when:

$$\frac{\pi d}{\lambda}\sin\theta = n\pi \implies \sin\theta = \frac{n\lambda}{d} \text{ where } n = 0, \pm1, \pm2, \pm3\ldots$$

(11.59)

Note that unlike the single wide slit n here starts at zero but like the single slit it extends to negative angles i.e. the same pattern appears on both sides of the screen. This gives a bright fringe in the centre of the screen in front of the two slits. At this point the path length to both slits is equal and so there must be constructive interference.

This is the pattern for two narrow slits so what happens if we have two wide slits? In this case we need to replace the single, point source wave function that we used above with the wave function from a single slit that we derived in (11.50).

$$A\cos(kr - \omega t) \rightarrow A\frac{\sin\left(\frac{\pi}{\lambda}a\sin\theta\right)}{\frac{\pi}{\lambda}\sin\theta}\cos(kr - \omega t)$$

(11.60)

Putting this new wave amplitude into (11.57) gives:

$$\psi(\theta) = 2A\frac{\sin\left(\frac{\pi}{\lambda}a\sin\theta\right)}{\frac{\pi}{\lambda}a\sin\theta}\cos\left(\frac{\pi d}{\lambda}\sin\theta\right)\cos(kr - \omega t)$$

(11.61)

where a is the width of each slit and d is the separation between the centres of the two slits. To get the light intensity we simply square the amplitude to get:

$$I = I_0\frac{\sin^2\left(\frac{\pi}{\lambda}a\sin\theta\right)}{\frac{\pi^2}{\lambda^2}a^2\sin^2\theta}\cos^2\left(\frac{\pi d}{\lambda}\sin\theta\right)$$

(11.62)

The result is that we have the double narrow slit function multiplied by the single wide slit function. This means that while the fringes will maintain their even spacing the brightness of each fringe will vary and certain fringes may be missing if their maximum corresponds to a minimum in the single wide slit pattern. An example of such a combined pattern is shown in figure 11.31.

Example 11.2

Monochromatic light shines onto two, wide slits which have a spacing of 5 μm between their centres and the resulting diffraction pattern is viewed on a screen a long way from the slits. If each slit has a width of 1 μm how many bright fringes will there be in the centre of the pattern between the two locations where there is a missing fringe? What will happen to the pattern if the wavelength of the light is slowly increased?

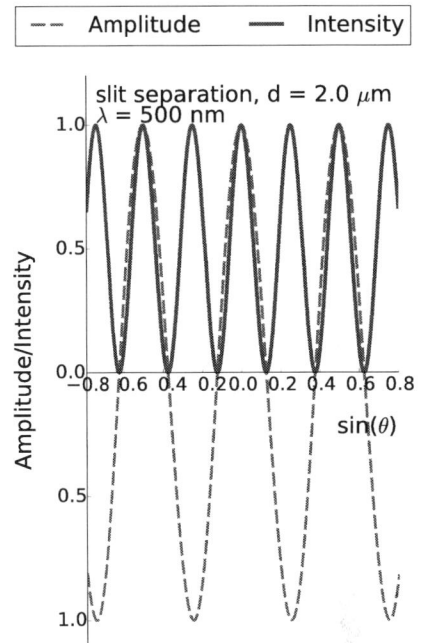

Fig. 11.30: The diffraction pattern from

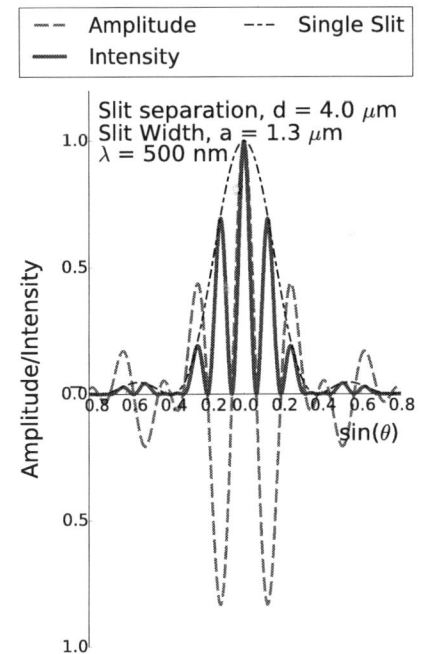

Fig. 11.31: The diffraction pattern from two, wide slits showing both the intensity and amplitude. The intensity for a single slit of the same width is also shown.

Solution:

The diffraction pattern for two, wide slits has two components as shown in equation (11.62). The fringe spacing is determined by the slit spacing and, starting with a bright fringe in the centre, the spacing between fringes is given in (11.59) as:

$$\sin\theta = \frac{n\lambda}{d} \implies \sin\theta = 0, \pm\frac{\lambda}{d}, \pm\frac{2\lambda}{d} \cdots$$

where d is the slit separation. However the amplitude of these fringes is multiplied by the single slit function which has minima, from (11.52), at:

$$\sin\theta = \frac{n\lambda}{a} \implies \sin\theta = \pm\frac{\lambda}{a}, \frac{2\lambda}{a}, \pm\frac{3\lambda}{a} \cdots$$

where a is the slit width. Now the slit width has to be less that the slit separation (otherwise we do not have two slits!) so we want to know how many of the narrow fringes will occur before the first minimum in the single slit pattern which will kill the intensity of that fringe and cause it to be missing from the pattern. Hence we construct the relationship below where n is the number of the narrow, double slit fringe which occurs at the same point as the first minimum in the single slit pattern:

$$\frac{n\lambda}{d} = \frac{\lambda}{a} \implies n = \frac{d}{a} = 5$$

This tells us that the amplitude of the $n = 5$ fringe will be zero and so it will not be seen. By symmetry this also applies to the $n = -5$ fringe. Hence the fringes which we will see correspond to $n = 0$ up to $n = \pm 4$ and so there will be 9 fringes in the centre of the patter before the first missing fringe on either side.

If the wavelength is increased then the width of both the narrow fringes from the double slits and the broader single slit amplitude function will both increase at the same rate: both are proportional to λ. Hence there will be the same number of fringes before one is missing but the spacing of those fringes will increase with the wavelength.

11.6.3 Diffraction Grating

The final pattern of slits we will consider is an infinite number of narrow slits with a spacing of d between them. Such a pattern is called a *diffraction grating* and is shown in figure 11.32. As can be seen in this figure each slit has a phase difference of:

$$\Delta\phi = kd\sin\theta \qquad (11.63)$$

Since this is the phase difference between adjacent slits the resulting wave amplitude at the angle θ will be a sum of an infinite number of waves each one with an integer multiple of this phase difference. The fact that there an infinite number of waves means is that unless this phase difference is an integer multiple of 2π we will always be able to find pairs of waves with opposite phases which cancel. However, if this phase difference is an integer multiple of 2π then all the waves will have exactly the same phase and there can be no cancellation.

The result is that for an infinite diffraction grating there is a bright line on the screen when:

$$kd\sin\theta = 2n\pi \implies \sin\theta = \frac{m\lambda}{d} \text{ where } m = 0, 1, 2, 3 \ldots$$
$$(11.64)$$

and zero brightness everywhere else as shown in figure 11.33. This makes the diffraction grating an excellent device to analyse the spectrum of a source because each wavelength will be diffracted through a different angle and produce a single line on the screen. Obviously an infinitely large diffraction grating is rather an impractical device to build but the effect of making the diffraction grating a finite size is to replace each perfect line with a the single slit, sinc function where the slit width is the width of the diffraction grating . For light with a wavelength of a few hundred nanometres a diffraction grating of a few centimetres width produces very narrow lines which have widths in the order of 10^{-5} in $\sin\theta$.

Rather than specify the separation between slits most diffraction gratings will specify the number of slits per metre, N. This simplifies the condition for a bright line to:

$$\sin\theta = mN\lambda \text{ where } m = 0, 1, 2, 3 \ldots \qquad (11.65)$$

Example 11.3

A red laser pointer shines light with a wavelength of 650 nm onto the surface of a CD. Bright red lines are observed on either side of the reflected ray at an angle of 23.3° to the reflected ray. What is the spacing of the tracks on the surface of the CD?

Solution:

What is happening here is that the light is being diffracted

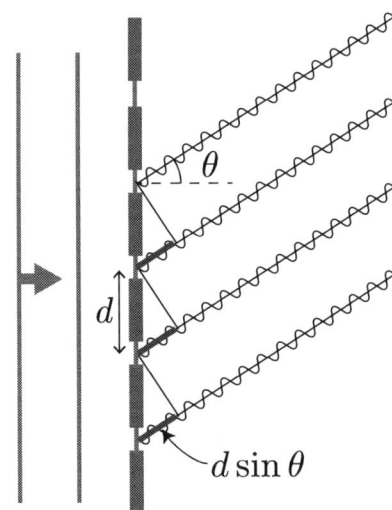

Fig. 11.32: A diffraction grating showing the path length difference between adjacent slits.

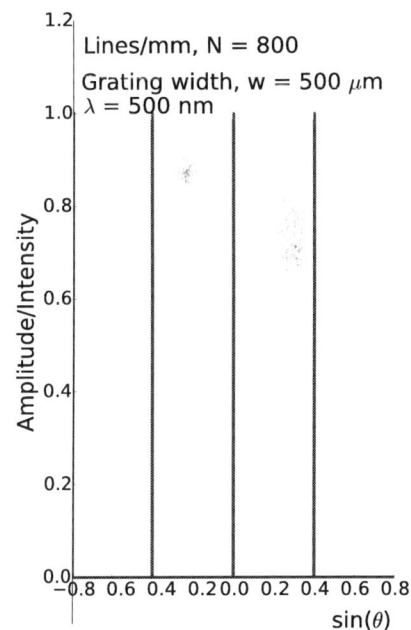

Fig. 11.33: Diffraction pattern from a diffraction grating which is 0.5 mm wide and has 800 lines/mm. The width of each line due to the finite size is too small to show.

by the tracks on the CD. The tracks form a diffraction grating with a spacing equal to the track spacing. We know how the spacing of the lines of a diffraction grating is related to the diffraction angle we can use this to calculate the spacing of the tracks. We can rearrange the diffraction grating equation to get the lines per metre in terms of the diffraction angle and the wavelength of the light. Since this is the first bright line we know that $m = 1$ so:

$$\sin\theta = mN\lambda \implies N = \frac{\sin\theta}{\lambda}$$

Putting in the numbers we get:

$$N = \frac{0.396}{650 \times 10^{-9}} = 608,532 \; m^{-1}$$

This is the number of lines per metre so to convert this into a line spacing we take the reciprocal:

$$Track\ Spacing = \frac{1}{N} = \frac{1}{608,532} = 1.64 \; \mu m$$

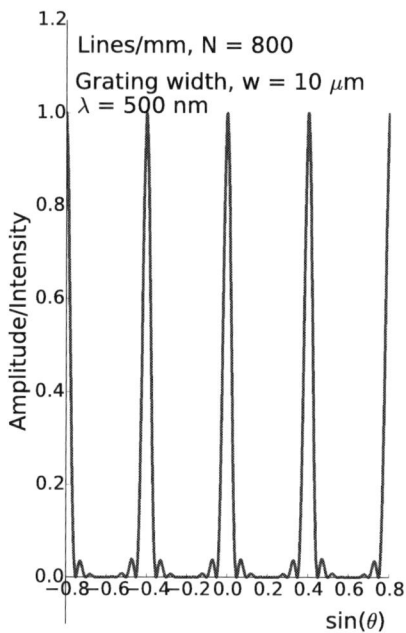

Fig. 11.34: Diffraction pattern from a diffraction grating which is 10 μm wide and has 800 lines/mm. The width of each line due to the finite size of the grating is clearly visible.

The type of grating we have discussed here is a transmission diffraction grating because light is transmitted through the diffraction grating . However reflection diffraction gratings also exist. These reflect light at certain positions and scatter it in between. A common example of this type of grating is an optical disc such as a CD, DVD or Blu-ray which is why such discs have a rainbow sheen to them when viewed under white light. The tracks of pits which store the digital data act as the grating and the different colours of white light diffract through different angles and so are split to give the entire spectrum.

The Big Picture

Mathematically the fringe pattern for Fraunhofer diffraction is something called a *fourier transform* of the slit function. A fourier transform of the function $f(x)$ is defined as:

$$F(\omega) = \int_{-\infty}^{\infty} f(x)e^{i\omega x}dx \qquad (11.66)$$

The slit function, $H(y)$, is simply equal to one where there is no obstruction and zero where there is an obstruction and the variable y is the position along the barrier with the slits.

For example our single, wide slit has the slit function:

$$H(y) = \begin{cases} = 1 \text{ if } |y| \leq a/2 \\ = 0 \text{ if } |y| > a/2 \end{cases}$$

Hence the limits of the integral given in (11.66) just become $\pm a/2$ as seen in (11.48).

A convolution of two functions is defined as:

$$(f * g)(t) = \int_{-\infty}^{\infty} f(\tau)g(t - \tau)d\tau$$

One of the properties of Fourier transforms is that the Fourier transform of the convolution of two functions equals the product of the Fourier transforms of each separate function and vice versa. If the two narrow slits are "convolved" with the single wide slit functions this gives a double wide slit function so the diffraction pattern of this is the product of the single wide slit pattern with the double narrow slit pattern.

Similarly, the slit function of a finite diffraction grating is the product of the single wide slit function with the slit function for the infinite diffraction grating . Hence the resulting diffraction pattern is the convolution of the infinite diffraction grating pattern with the single, wide slit pattern which is why each line becomes a sinc function.

Problems

Q11.1: A couple of hikers in the mountains look down upon a perfectly still lake. The first hiker, who is wearing sunglasses which use a neutral density filter, notes how beautiful the reflection of the sun looks. The second hiker, who is wearing polaroid sunglasses and standing next to her, replies that he can see absolutely no reflection at all. What is the angle of the sun above the horizon? [n(air)=1.0, n(water)=1.33]

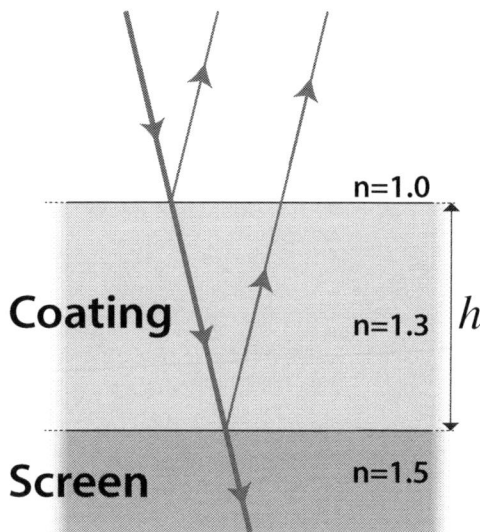

Fig. P11.1: *Layer of titanium dioxide added to the surface of a screen to act as an antiglare coating which reduces reflections.*

Q11.2: To reduce glare caused by reflections a thin coating with a refractive index, $n = 1.3$, is applied to the transparent surface of a computer screen ($n = 1.5$) as shown in figure P11.1. The thickness of the layer is designed to be the thinnest possible which will prevent reflection of green light with a wavelength of 530 nm when the device is operated in air.
(a) How thick is the layer of coating applied?
(b) When the screen is turned off and examined under white light. While reflections are greatly reduced they are still present and take on a purple colour. Why is this?
(c) Would the anti-glare coating work if the screen could be operated under water

[n(water)=1.33]? Explain your answer.

Q11.3: Observing a partial solar eclipse is tricky because it is dangerous to look directly at the sun even wearing normal sunglasses. One technique is to wear polaroid sunglasses and to take a second pair of polaroid sunglasses, hold them so that one eye sees through both pairs and to then rotate the held pair until sufficient light is blocked that the sun can be safely viewed through the one eye which looks through both pairs.
(a) If the light intensity has to be dropped by a factor of 10 from that observed without any sunglasses what is the minimum angle between the planes of the two polaroid filters allowed?
(b) If the light intensity drops by a factor of 250 from that observed without any sunglasses the eclipse cannot be observed at all so what is the maximum angle possible between the two polarization planes?

Q11.4: A soap bubble (n=1.4) is floating in air (n=1.0). The light reflected from the bubble is red with a wavelength that peaks at 650 nm in air. What is the minimum possible thickness for the wall of the bubble?

Q11.5: A piece of aluminium foil is placed under one end of a glass microscope slide which is 8 cm long. The slide is then placed onto a sheet of glass so that there is a small, air-filled wedge between the slide and the glass sheet. A red laser pointer with a wavelength of 650 nm is then shone onto the system and a series of horizontal fringes are observed as shown in figure P11.2.
(a) If two adjacent bright fringes are 1.6 mm apart what is the thickness of the aluminium foil?
(b) Will a bright or dark fringe be observed at the end where the slide is directly in contact with the glass sheet? Explain.

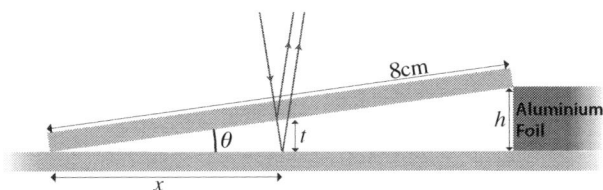

Fig. P11.2: *A vertically exaggerated diagram of a microscope slide resting on the edge of a piece of aluminium foil. When a red laser is shone onto the slide a series of fringes are produced.*

Q11.6: A very bright laser with a wavelength of 560 nm shines onto a slit with a width of 10 µm. How many dark fringes, where the cancellation is perfect, will be produced at a large distance from the slit? You should assume that you have a detector capable of observing the diffraction pattern at any angle and sensitive enough to observe any fringe regardless of contrast.

Q11.7: A laser with a wavelength of 650 nm shines onto a pair of slits which have variable but always equal widths and a fixed separation between their centres. Initially, the slits are almost closed and can be considered as thin slits. This produces a series of equally spaced, identical fringes on the screen which are each separated by an angle of 1.5 mrad.
(a) What is the separation between the centres of the slits?
(b) The width of the slits is increased until the third bright fringe from the centre of the pattern completely disappears while the first two fringes remain visible. What is the width of the slits at this point?
(c) Describe qualitatively what will happen as the width of the slits is increased beyond this point? Is there is limit to this and if so what happens in that limit?

Q11.8: A transmission diffraction grating consists of 20,000 lines in a 2 cm width and is used to measure the wavelengths in a spectrum of EM radiation.
(a) Explain briefly how the grating can be used to measure wavelength and why a diffraction grating, as opposed to a single slit, is particularly suited to this.
(b) What is the longest wavelength which can be measured?

Q11.9: A diffraction grating which is designed for extreme ultraviolet radiation also works for visible light and has red light with a wavelength of 650 nm shone onto it. The pattern produced shows only a central bright line at zero angle and there are no other lines observed. What conclusion can be drawn about the line spacing of the diffraction grating?

Q11.10: Hydrogen emits light at particular frequencies one of these, called the Lyman-α line, is in the ultraviolet part of the spectrum and has a wavelength of 121.6 nm. A ray of light from this emission line is passed through a diffraction grating such that it is incident perpendicular to the plane of the grating and the first maximum on the other side is observed at an angle of 10° to the original direction.
(a) How many lines per metre does the diffraction grating have?
(b) The diffraction grating is then fed light from the Lyman-α line emitted by a source in a distant galaxy. However this time the first maximum is observed at an angle of 60°. Given that there is no transverse movement what is the velocity of this distant galaxy relative to us as a fraction of the speed of light and it is moving towards or away from us?

CHAPTER 12

Quantum Mechanics

Fig. 12.1: A picture of Albert Einstein during a lecture in Vienna in 1921. This was the year he won his Nobel Prize for a discovery which lead to the development of quantum mechanics, a theory he never came to accept.

In the previous chapter, we started by discussing the two competing theories of light: light as a wave proposed by Huyghens and light as a particle proposed by Newton. Newton's theory of light as a particle was initially favoured mainly due to Newton's significant influence in the field but fell out of favour in 1803 when Thomas Young performed his double slit experiment which could only be explained by Huyghens' wave model of light. However almost exactly one hundred years later, at the start of the 20th century, the mystery surrounding the nature of light deepened considerably with the discovery of phenomena that required light to be a particle.

The solution of this mystery led to the discovery of a new field of physics that revolutionized our fundamental understanding of the universe and which made possible the modern electronic revolution. However, the implications of this new physics were so profound and strange that even Einstein himself could not bring himself to believe in it despite winning his Nobel Prize for one of the breakthroughs which lead to its discovery.

12.1 Atomic Spectra

By the mid 19th century light was firmly established as a wave. Diffraction of light had been known for more than 50 years and could only be explained by waves. Yet ironically it was the discovery of diffraction that would lead to the first indications that light was not behaving like any wave previously discovered.

The discovery of diffraction led to the use of the diffraction grating to analyse light spectra. These studies yielded surprising results: each element emitted a characteristic spectrum when it was heated. These spectra were each composed of a series of bright lines at fixed wavelengths with a pattern that was unique for a particular element. This was immediately seized on by chemists, including a Dr Robert Bunsen, as a means to analyse the composition of materials by heating them and measuring the spectrum of light emitted.

It was also found that elements would absorb these same particular wavelengths as well which lead to the discovery of a new element, Helium (from the greek Helios meaning sun), in the spectral absorption lines, called Fraunhofer lines, of the sun and figure 12.2 shows the emission lines for helium. However, useful as this discovery was, there was no understanding about why certain elements emitted only at certain frequencies nor was there any way to predict the frequencies.

The first breakthrough in understanding the lines came in 1885 from a Swiss school teacher, Johann Balmer (1825-98), who discovered that the wavelength of lines of the hydrogen spectrum obeyed a simple formula:

$$\lambda = C \frac{m^2}{m^2 - n^2} \qquad (12.1)$$

where C is 3.6456×10^{-7} m, $n = 2$ and $m = 3, 4, 5 \ldots$. This series was used to predict a new line corresponding to $m = 7$ which was promptly discovered.

The Balmer series, shown in figure 12.3, was in the visible-light part of the spectrum but further studies of hydrogen revealed more series of lines: the Lyman series in the ultraviolet and the Paschen, Brackett and Pfund series in the infrared. The result was that only three years later, in 1888, Swedish physicist Johannes Rydberg (1854-1919), found that the Balmer formula was a special case of the more general Rydberg formula:

$$\frac{1}{\lambda} = R_H \left(\frac{1}{n^2} - \frac{1}{m^2} \right) \qquad (12.2)$$

where R_H is the Rydberg constant for hydrogen and has a value of 1.097373×10^7 m^{-1}, $n = 1, 2, 3 \ldots$ and $m = n + 1, n + 2, \ldots$.

This formula is a mathematical description of the pattern of wavelengths in the hydrogen spectrum but it is not an explanation of why they have these values. A full and proper understanding of this would take a little longer to develop and lead to a revolution in our fundamental understanding of physics.

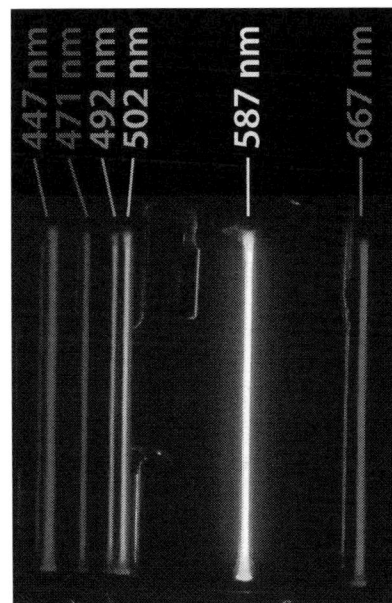

Fig. 12.2: Spectrum of helium taken by placing a diffraction grating in front of a camera lens.

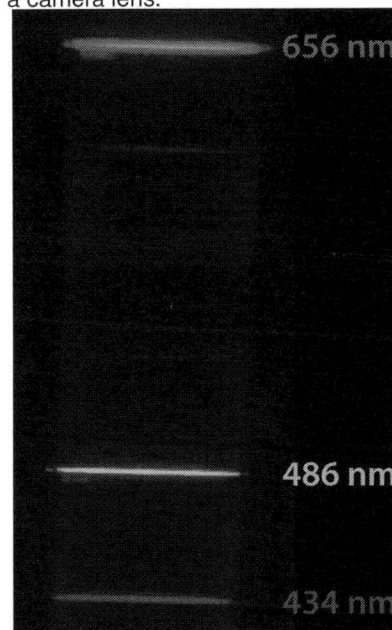

Fig. 12.3: Spectrum of hydrogen taken by placing a diffraction grating in front of a camera lens. The three lines shown belong to the Balmer series which lies in the visible spectrum. There is a fourth line in the extreme violet at 410 nm which was not captured by the camera.

12.2 The Blackbody Spectrum

Only shortly after the pattern in the hydrogen spectrum was discovered another problem with emission spectra was noticed in 1900. This problem arose when using classical physics to derive the spectrum of a blackbody which is an object that absorbs all wavelengths of radiation (hence "black") and emits radiation based solely on its temperature and not on the wavelengths of radiation it has absorbed.

It is rather surprising that a good example of a blackbody is the sun! The sun's plasma will absorb most wavelengths of EM radiation and the radiation emitted depends mainly on the temperature. The exception to this being the atomic absorption lines already noted in section 12.1.

An even better blackbody spectrum is that of the entire universe which is visible in the cosmic microwave background radiation, the so-called afterglow of the Big Bang. This radiation comes from all over the sky and only deviates from the perfect blackbody spectrum at one part in 100,000. It was formed 380,000 years after the Big Bang when the plasma which filled the entire universe at that point had cooled sufficiently to form hydrogen and helium gas.

The visible light photons released were then stretched by the expansion of the universe into the microwave region of the spectrum which we observe today and are shown in figure 12.4. The deviations from a perfect spectrum were essential to the formation of the universe we know since the were caused by density fluctuations in the plasma which, over time, collapsed to form the galaxies we see today. Without these, the universe would now be full of a low density, cold mixture of hydrogen and helium. The cause of these density fluctuations is one of the largest mysteries in physics today: dark matter.

12.2.1 The Ultraviolet Catastrophe

To understand the problem that a blackbody spectrum poses for classical physics we need to revisit standing waves. For a wave on a string, there is only one way to fit an allowed frequency onto the string: this is the case for all one-dimensional standing waves.

In two dimensions, for example, the allowed waves on a square sheet of material, it turns out that there is a separate integer

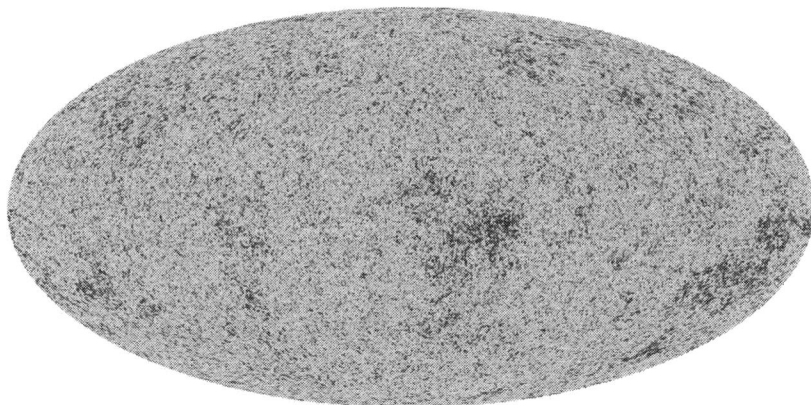

Fig. 12.4: The cosmic microwave background as measured by the Planck Satellite.
©2013 ESA and the Planck Collaboration.

for both the x and y directions. This means that there are two ways of putting the exact same wavelength onto the surface as shown in figure 12.5 and, for larger values of n_x and n_y, more states with almost the same wavelength.

However physical objects are three dimensional. If we model a blackbody as a cavity which traps all incident radiation then for any fixed wavelength there will be three ways of placing the wavelength in the box and, for large values of the wave's associated integers, many states with very similar frequencies. A careful analysis of these states by Lord Rayleigh and Sir James Jeans, for details, see appendix B, revealed that the number of allowed modes with a given frequency per unit volume of a cavity, $n_m(v)$ was (B.38):

$$n_m(v) = \frac{8\pi v^2}{c^3} \tag{12.3}$$

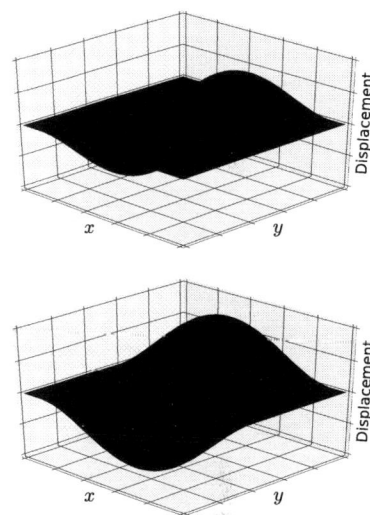

Fig. 12.5: A standing wave with the same wavelength can be arranged on a two dimensional sheet in two different ways. The top shows the wave arranged so that the x component of the wavelength is half that in y and the bottom plot shows the reverse.

where v is the frequency and c is the speed of light in vacuo. This means that the number of available modes for electromagnetic radiation dramatically increases with frequency.

The problem arises when you add in some thermodynamics. The equipartition theorem says that in thermal equilibrium each mode of storing energy in a system will, on average, have an equal amount of energy equal to the product of the Boltzmann constant, k, and the temperature, T i.e. kT. The emission spectrum will then depend on the total energy per frequency.

However, as (12.3) shows, the higher the frequency the more possible modes there are for that frequency and so the larger the total energy in the system at that frequency. Indeed this predicts that the total energy per unit volume in a cavity's standing waves as a function of frequency will be:

$$E(\nu) = n_m kT = \frac{8\pi \nu^2 kT}{c^3} \qquad (12.4)$$

Hence as $\nu \to \infty$ the energy per unit volume of the cavity goes to infinity too! Since the emission spectrum is proportional to the energy stored in each frequency this means that every object should be radiating in the extreme gamma ray end of the spectrum and more at even higher energies. Even worse if you ask for the total energy stored in the cavity then you have to integrate (12.4) over all possible frequencies. Since there is no upper limit on frequency this predicts that a cavity in thermodynamic equilibrium has an infinite amount of energy stored in it per unit volume!

At this point, we might conclude that there is a serious flaw somewhere in our derivation. However, the spectrum derived from these arguments, which was called the Rayleigh-Jeans spectrum, agrees well with the radio emission spectrum of objects and radio astronomers use this spectrum to describe radio sources today. Only when the higher frequencies/shorter wavelengths were examined is there a massive discrepancy between prediction and observation. Indeed the result is so strikingly wrong for higher frequencies that it was called the ultraviolet catastrophe and was a major problem for physics at the end of the 19th century!

Fig. 12.6: A picture of Max Planck in Berlin in 1933. His expression may hint at the political troubles of the time. In 1945 his son, Erwin Planck, was executed for taking part in the plot to assassinate Hitler.
Unknown author, Wikimedia commons, public domain.

12.2.2 Planck's Law

Fortunately the solution to the ultraviolet catastrophe was quickly found within the year by Max Planck, see figure 12.6, who proposed a radical idea: the energy of electromagnetic radiation was quantized. Rather than allowing the energy in a particular frequency to be any value, Planck said that the energy was quantized in units of:

$$E = h\nu \qquad (12.5)$$

where ν (the greek letter 'nu') is the frequency and h is the Planck constant:

$$h = 6.626 \times 10^{-34}\, \text{Js} \qquad (12.6)$$

This means that there is either zero energy in a mode or some integer multiple of this amount i.e. E=0, $h\nu$, $2h\nu$, $3h\nu \ldots$ etc. To understand how this solved the problem we need to know about the distribution of energy in each node. While the equipartition theorem says that the average energy per mode is kT not every

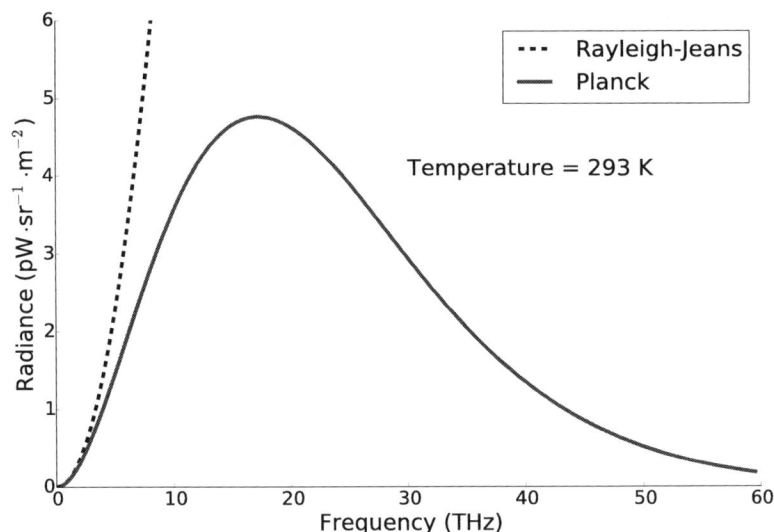

Fig. 12.7: The black body radiation spectrum for an object at room temperature, 293 K, as predicted by Planck's law. The dashed line shows the classical prediction using the Rayleigh-Jeans spectrum. This rapidly diverges from the observed Planck spectrum for higher frequencies.

mode will have exactly this much energy. Instead, the probability, $P(E)$, that a given mode has an energy E is given by the Boltzmann distribution:

$$P(E) \propto e^{-\frac{E}{kT}} \qquad (12.7)$$

Hence the chance for a mode to have more energy than the average kT is exponentially suppressed and the higher the energy excess the less likely it is to find a mode with that much energy.

Now consider what Planck claimed: the energy in the state either has to be an integer multiple of $h\nu$ or it has to be zero. In the low-frequency cases where $h\nu \ll kT$ there is no effect because the size of the energy steps is a lot less than the average energy. However, as the frequency increases the size of the energy steps gets a lot larger. The result is that for cases where $h\nu \gg kT$ the Boltzmann factor becomes vanishingly small: there is almost no probability that the mode will have an energy of $h\nu$ and so it will have zero energy. While the number of states grows with ν^2 the Boltzmann factor is an exponential suppression, $e^{-h\nu/kT}$, and easily wins. The result is that the total energy in higher frequency modes tends to zero as the frequency increases despite the fact that there are more available modes.

Deriving the spectrum now requires multiplying the number of modes by the probability that they will have more than the minimum energy quantum. The details of this calculation, up to the energy density as a function of frequency, are given in appendix B. The energy density as a function of frequency given in equation (B.49) can be used to derive the radiance of a small hole in the cavity with the result that the Planck radiation spectrum is given by:

$$\text{Radiance, } I(v, T) = \frac{2hv^3}{c^2} \frac{1}{e^{\frac{hv}{kT}} - 1} \qquad (12.8)$$

where the radiance is the intensity per unit frequency per unit solid angle and T is the absolute temperature of the object measured in kelvin. Figure 12.7 shows both the Planck spectrum and the Rayleigh-Jeans spectrum. The Planck spectrum agreed well with the solar spectrum as well as that of materials heated in the laboratory provided that the spectral lines were ignored.

The peak frequency of the Planck spectrum can be determined by using calculus to determine the maximum point of the emission spectrum. Details of this calculation are given in appendix B but the derivation results in an equation which can only be solved numerically with the result that the peak frequency is given by (B.52):

$$v_{max} = 5.879 \times 10^{10} \, \text{Hz K}^{-1} \cdot T \qquad (12.9)$$

This is known as Wien's displacement law and was actually derived before Planck's law in 1893 by Wilhelm Wien who used thermodynamics. This success of Planck's law to explain the observed spectrum and derive the already known quantitative laws associated with it was the first clear evidence that light was not a classical wave. The final nail in the coffin of the classical theory of light came from a phenomenon called the photoelectric effect.

12.3 The Photo-electric Effect

The photoelectric effect was first observed as early at 1887 when Heinrich Hertz noticed that electrodes illuminated with ultraviolet light emitted sparks more easily. Further study showed that ultraviolet light would cause the metal to emit electrons. However rather bizarrely the energy of the electrons depended on the type of metal used and the frequency of the light. If the light source was made more intense more electrons were emitted but at exactly the same energy as those produced by a fainter light source with the same frequency.

For a wave, this was a very strange result! As the intensity increases the amplitude and energy of the wave increase. This means that it has more energy to give to things it interacts with which should mean that the electrons gain more energy from a bright light source than a faint one. Instead, the energy remains the same but the number of electrons emitted increases with intensity.

It was not until 1905 when Albert Einstein had his *annus mirabilis* (year of wonders)[1] that this behaviour was explained.

Einstein explained that light was made up of small particles called photons (thankfully not corpuscles as Newton had originally called them!) that carried one quantum of energy, hf, just as in Planck's law. Each one of these particles interacted with a single electron in the metal giving it all the energy it possesses. Some of the energy is needed to extract the electron from the metal. This is called the work function and depends on the material. The remainder of the photon's energy is given to the electron as kinetic energy.

Since the energy per electron depends only on the energy per photon this will vary with frequency as shown in figure 12.8. The formula for the electron kinetic energy is simply:

$$E_e = hf - \phi \qquad (12.10)$$

where E_e is the electron's kinetic energy and ϕ is the material's work function. Increasing the intensity increases the number of photons and hence the number of electrons but not the energy available to each electron.

This was clear evidence that light was behaving as a particle and not a wave...so which was it? The picture that emerged was called wave-particle duality. Light was both a wave and a particle at the same time. One way to think about this is to regard each photon as a packet of waves as shown in figure 12.9. When lots of photons are added together the result is a wave and so light behaves as a wave composed of many individual photons that interact as particles.

> ## Example 12.1
>
> Sodium has a work function of 3.65×10^{-19} J. What is the longest wavelength of light which will cause sodium to emit electrons?
>
> ### Solution:
> *The energy of electrons emitted by the photoelectric effect is given by equation (12.10). When the photon has just enough*

[1] In this year Einstein published three papers explaining special relativity, the photoelectric effect and Brownian motion. In hindsight, each of these papers was clearly worthy of a Nobel Prize but only the paper on the photoelectric effect actually won it.

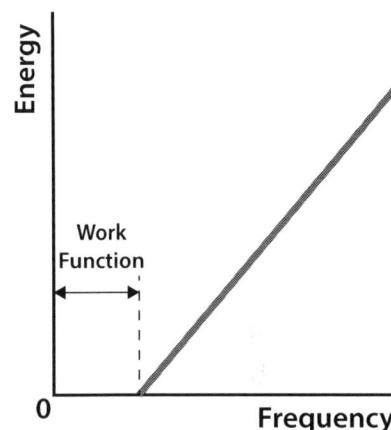

Fig. 12.8: Kinetic energy of an electron produced by the photoelectric effect as a function of the frequency of the incident photon. The minimum frequency at which an electron is emitted corresponds to the energy of the material's work function.

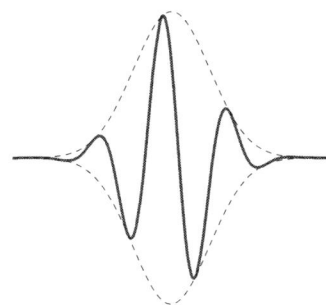

Fig. 12.9: A photon can be thought off as a small packet of waves, shown here with a gaussian amplitude envelope.

energy to kick out an electron but no more the electron energy, E, will be zero and so we have a relationship for the frequency of the photon:

$$hf - \phi = 0 \implies f = \frac{\phi}{h}$$

Now all we need to do is rearrange this to get an expression for the wavelength which we can then evaluate:

$$\lambda = \frac{hc}{\phi} = \frac{6.63 \times 10^{-34} \times 3 \times 10^8}{3.65 \times 10^{-19}} = 545 \ nm$$

12.3.1 The Electronvolt

When dealing with the energies in the photoelectric effect and in general with electrons in atoms the energies are all, because the electron has a tiny charge. For this reason, when dealing with these types of phenomena we define a new unit of energy called the electronvolt which has an abbreviation 'eV'. One electronvolt is the energy obtained by an electron when it moves through a potential difference of one volt. Since the electric potential is defined as the electrical potential energy per unit charge this corresponds to an energy difference of one volt times the electron's charge. Hence:

$$1 \, \text{eV} = 1.60217662 \times 10^{-19} \, \text{J} \tag{12.11}$$

which is equal in magnitude to the electron charge in coulombs.

This unit is commonly used throughout atomic, nuclear and particle physics. However, the extreme energies of particle physics mean that SI prefixes are used. For example, the energy of a proton in a beam of the Large Hadron Collider at CERN is measured in tera-electronvolts (TeV) and since energy and mass are interchangeable the masses of particles can also, in electronvolts provided an extra c term is added to ensure that the units are dimensionally correct. For example, the mass of a Higgs boson is $125 \, \text{GeV}/c^2$.

12.4 The Bohr Atom

Having now explained the continuous spectrum emitted by materials as due to the quantum nature of light it seems reasonable to suppose that this may also be capable of explaining the strange lines in spectra which were the first signs of the end for classical physics. The first explanation of the Rydberg spectrum came in 1913 when Niels Bohr proposed what became known at the Bohr model of the atom. This model formed was part of what is now called the old quantum theory and although it is now known not be correct it contains several key concepts which are still relevant to modern quantum theory.

Only a few years before in 1911 Rutherford had discovered that the atom contained a small, positive nucleus where almost all the mass of the atom concentrated. This was surrounded by negative electrons that filled the space in between the nuclei. Bohr proposed that the electrons orbited the nucleus in a classical solar system model where the centripetal force was the coulomb force. In such a model Newton's second law requires that:

$$\frac{(Ze)e}{4\pi\epsilon_0 r^2} = m_e \frac{v^2}{r} \implies \frac{Ze^2}{4\pi\epsilon_0 r} = m_e v^2 \qquad (12.12)$$

where Z is the number of protons in the nucleus, e is the electron charge, m_e is the electron mass, v is the orbital velocity of the electron and r is the radius of the electron's orbit. Since the angular momentum of the electron in the orbit is just:

$$L = m_e v r \qquad (12.13)$$

we can rewrite (12.12) in terms of the angular momentum which gives:

$$\frac{Ze^2}{4\pi\epsilon_0} = \frac{L^2}{m_e r} \qquad (12.14)$$

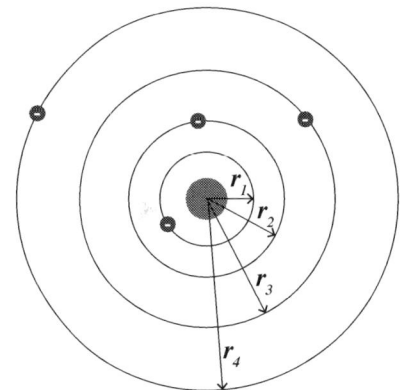

Fig. 12.10: Bohr atom showing how electrons can only orbit at certain orbital radii denoted by r_1 to r_4 on the diagram. Each orbit corresponds to a particular number of angular momentum quanta.

However this model has a problem because when you accelerate a charge it radiates energy. This means that these orbits will not be stable and the electrons will radiate electromagnetic waves and so will spiral into the centre. To prevent this Bohr proposed that the angular momentum of the electrons were quantized so that:

$$L = \frac{nh}{2\pi} = n\hbar \qquad (12.15)$$

where $n = 1, 2, 3 \ldots$ and $\hbar$, pronounced "h-bar", is defined in terms of the Planck constant as $\hbar = h/2\pi$. Putting this into (12.14) we get a formula for the orbital radii of electrons:

$$\frac{Ze^2}{4\pi\epsilon_0} = \frac{n^2\hbar^2}{m_e r_n} \implies r_n = \frac{4\pi n^2\hbar^2\epsilon_0}{Ze^2 m_e} \qquad (12.16)$$

Hence the consequence of quantization of the angular momentum is that electrons are only allowed to have certain orbital radii and so cannot spiral in radiating energy as shown in figure 12.10.

While Bohr was correct that the angular momentum of electrons in an atom is quantized we now know that this model of the electrons orbiting the nucleus like planets around a star is wrong. However what won Bohr the Nobel Prize for this model is what happens when we consider the energy of the electrons. Starting with (12.12) we can rearrange this to get the kinetic energy of the electron:

$$\frac{Ze^2}{4\pi\epsilon_0 r} = m_e v^2 \implies KE = \frac{Ze^2}{8\pi\epsilon_0 r} \tag{12.17}$$

Now the total energy of the electron, E, is just the sum of the potential and kinetic energies:

$$E = PE + KE = -\frac{Ze^2}{4\pi\epsilon_0 r} + \frac{Ze^2}{8\pi\epsilon_0 r} = -\frac{Ze^2}{8\pi\epsilon_0 r} \tag{12.18}$$

Now we can substitute into this our expression for the radius of the orbit from (12.16):

$$E_n = -\frac{Ze^2}{8\pi\epsilon_0}\frac{Ze^2 m_e}{4\pi n^2 \hbar^2 \epsilon_0} = -\frac{Z^2 e^4 m_e}{32\pi^2 \epsilon_0^2 n^2 \hbar^2} \tag{12.19}$$

This shows that the electron energy levels in the atom are quantized and suddenly we have an explanation for the lines we see in the hydrogen spectrum! When an electron moves from one orbit to another it will emit a photon of light which has an energy difference equal to the difference in energy between the orbits. Similarly the atom can only absorb certain frequencies of light which correspond to the energy difference between two orbits which explains absorption lines. Both of these processes are shown in figure 12.11.

For example if an electron moves from a state m to a state n the energy difference will be:

$$\Delta E = \frac{Z^2 e^4 m_e}{32\pi^2 \epsilon_0^2 \hbar^2}\left(\frac{1}{n^2} - \frac{1}{m^2}\right) \tag{12.20}$$

The energy of the photon produced when an electron moves between orbits equals this energy difference between the orbits. However, the photon's energy is related to its frequency and hence wavelength:

$$E_\gamma = hf = \frac{hc}{\lambda} \tag{12.21}$$

Combining this with the energy difference expression in (12.20) we get:

$$\frac{1}{\lambda} = \frac{Z^2 e^4 m_e}{64\pi^3 \epsilon_0^2 \hbar^3 c}\left(\frac{1}{n^2} - \frac{1}{m^2}\right) \tag{12.22}$$

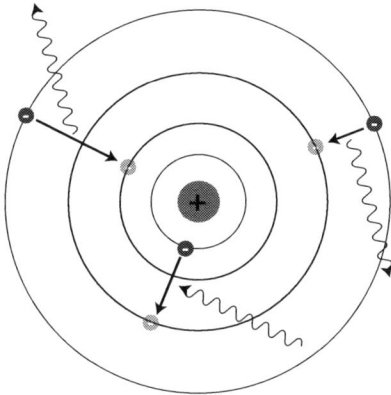

Fig. 12.11: Bohr atom showing how electrons moving between orbits emit, or absorb fixed wavelengths of radiation.

This result is the Rydberg formula only now we have an expression for the Rydberg constant which agrees with the measured value. This was the huge success of the Bohr model: it explained why there were spectral lines. While this is an improvement there were still many questions remaining: why is the angular momentum quantized? Why do the electrons not all collapse into the lowest energy state? Why does this not work well for elements other than hydrogen? To answer these questions a better understanding about the nature of an electron was required.

Example 12.2

What is the wavelength of light emitted when an electron in a Bohr hydrogen atom transitions from the $n = 4$ state to the $n = 2$ state?

Solution:
To calculate this we state with the formula for the energy levels in a Bohr atom given in (12.19) which we can simplify because we are dealing with hydrogen and so the atomic number, which is just the number of protons in the nucleus, $Z = 1$:

$$E_n = -\frac{Z^2 e^4 m_e}{32\pi^2 \epsilon_0^2 n^2 \hbar^2} = -\frac{e^4 m_e}{32\pi^2 \epsilon_0^2 n^2 \hbar^2} = -\frac{2.18 \times 10^{-18}}{n^2}$$

In this case the electron starts in the $n = 4$ state and goes to the $n = 2$ state so the energy of the emitted photon is the difference in energy between these two states:

$$\Delta E = 2.18 \times 10^{-18} \left(-\frac{1}{4^2} + \frac{1}{2^2} \right) = 4.09 \times 10^{-19}$$

To convert this into a wavelength we use Planck's relationship for the energy and frequency of a photon:

$$E = hf \implies \lambda = \frac{hc}{\Delta E} = \frac{6.63 \times 10^{-34} \times 3 \times 10^8}{4.09 \times 10^{-19}} = 486 \ nm$$

If you look at figure 12.3 this is the pale blue line which is actually observed in the visible light spectrum of hydrogen and is one of the lines of the Balmer series.

12.5 Matter as a Wave

Finding that light behaved as a particle was not the only surprise at the start of the 20th century. In 1924, in his PhD thesis, french physicist Louis de Broglie (1892-1987) postulated that electrons were the same as photons and so could behave as a wave with a wavelength of:

$$\text{de Broglie Wavelength, } \lambda = \frac{h}{p} \qquad (12.23)$$

where h is Planck's constant and p is the electron's momentum. This was an astounding claim and you can only imagine the sort of questions that his PhD examiners raised in his defence of it! One of the most obvious apparent problems with this is that if electrons are waves, why had nobody seen them diffract?

To account for this we need to consider the wavelength of a typical electron. The kinetic energy of an electron at room temperature is of the order of kT where k is Boltzmann's constant and T is the temperature in kelvin. The classical energy, mass momentum relationship works well at this energy and so we have that the momentum for an electron at room temperature is:

$$p = \sqrt{2mE} = \sqrt{2m_e kT} \implies \lambda = \frac{h}{\sqrt{2m_e kT}} \qquad (12.24)$$

Putting in the values for these constants we end up with a wavelength of 7.72×10^{-9} m. This was on a smaller scale than anyone had measured before and so it was not surprising that electrons behaved as particles when looking at distance scales about 1,000 times larger than the wavelength.

It was 3 years before de Broglie's postulate was proven to be correct. In 1927 two groups, George Thomson in the UK and Joseph Davisson in the US won the Nobel Prize for showing that electrons diffracted when passed through matter where the arrangement of atoms acted in a similar manner to a diffraction grating. This discovery confirmed that it was not just waves that behaved as particles, particles appeared to behave as waves too! It also provided a technique to study the arrangement of atoms in a material which later, in 1953, enabled Crick and Watson to determine the helical structure of DNA.

Example 12.3

A beam of electrons is incident on a crystal lattice which acts as a diffraction grating with a spacing of 10 nm between "slits". The first line of the resulting diffraction pattern is found at an angle of 2.5 mrad. What is the energy of electrons in the beam?

Solution:
First we use the diffraction grating equation (11.65) to calculate the wavelength of the electrons:

$$\sin\theta = mN\lambda \implies \lambda = \frac{\sin\theta}{N} = d\sin\theta$$

where d is the slit spacing of the grating and since we are looking at the first line of the pattern $m = 1$. This gives a wavelength of:

$$\lambda = 0.0025 \times 10 \times 10^{-9} = 2.5 \times 10^{-11} \ m$$

Now rearranging the de Broglie relationship given in (12.23) we can calculate the electron momentum:

$$p = \frac{h}{\lambda} = \frac{6.63 \times 10^{-34}}{2.5 \times 10^{-11}} = 2.65 \times 10^{-23} \ kg \ m/s$$

Next we use the classical form of kinetic energy and momentum to express the electron energy in terms of its momentum and then we just use the value of the momentum we calculated to determine the electron energy:

$$E = \frac{1}{2}m_e v^2 = \frac{p^2}{2m_e} = \frac{(2.65 \times 10^{-23})^2}{2 \times 9.11 \times 10^{-31}} = 3.86 \times 10^{-16} \ J$$

12.5.1 Momentum of a Photon

While the de Broglie wavelength equation had implications for the nature of matter it also had an interesting implication for the properties of a photon. If we consider the de Broglie wavelength as applying to a photon then we can rewrite it in terms of the photon's frequency, v:

$$\lambda = \frac{h}{p} \implies \frac{c}{v} = \frac{h}{p} \tag{12.25}$$

If we now rearrange this and remember that the energy of a photon, $E = hv$ as determined by Planck then we get:

$$hv = pc \implies E = pc \tag{12.26}$$

Now this relationship implies that the photon has a momentum, but we already knew this from Einstein's energy-mass-momentum relationship from Special Relativity:

$$E^2 = p^2c^2 + m^2c^4 \underset{m=0}{\implies} E = pc \tag{12.27}$$

So de Broglie's wavelength agrees with the result from Special Relativity that photons carry momentum.

12.6 The Schrödinger Equation

The Schrödinger Equation was developed in 1925 and gave a far fuller picture of the structure of an atom. Erwin Schrödinger derived it from the classical energy-mass-momentum relationship. If we can consider matter as a wave then we can write down a wavefunction to describe it which in one dimension will look something like:

$$\Psi(x, t) = Ae^{i(kx-\omega t)} \tag{12.28}$$

This is a complex wave, there is no implied real operator and the full complex solution is required. The square of the modulus of this wave function, $|\Psi(x, t)|^2$, gives the probability density i.e. the probability of finding the particle at a particular location at a particular time. In one dimension this is the probability per unit length as a function of position, x, and time, t.

Now if we differentiate the wavefunction with respect to x then we get:

$$\frac{\partial \Psi}{\partial x} = ikAe^{i(kx-\omega t)} = ik\Psi = \frac{2\pi i}{\lambda}\Psi = \frac{ip}{\hbar}\Psi \tag{12.29}$$

where we have used de Broglie to introduce p as the momentum of the particle and as before $\hbar$ is defined as $\hbar = h/2\pi$. If we repeat this then we get:

$$\frac{\partial^2 \Psi}{\partial x^2} = -k^2 Ae^{i(kx-\omega t)} = -\frac{p^2}{\hbar^2}\Psi \tag{12.30}$$

Now consider the classical relationship between kinetic energy, T, and momentum, p, which is:

$$T = \frac{1}{2}mv^2 = \frac{p^2}{2m} \tag{12.31}$$

and hence we can write down an expression which will give the kinetic energy times the particle wavefunction:

$$-\frac{\hbar^2}{2m}\frac{\partial^2 \Psi}{\partial x^2} = -\frac{\hbar^2}{2m} \times -\frac{p^2}{\hbar^2}\Psi = \frac{p^2}{2m}\Psi = T\Psi \tag{12.32}$$

This gives the kinetic energy but what about the total energy? To find this we need to return to our original wave and consider what happens when we differentiate with respect to time, t:

$$\frac{\partial \Psi}{\partial t} = -i\omega Ae^{i(kx-\omega t)} \implies i\hbar\frac{\partial \Psi}{\partial t} = \hbar\omega\Psi = E\Psi \tag{12.33}$$

where $E = \hbar\omega$ is the total energy. Now the total energy of a classical particle is the sum of its kinetic and potential energies due to conservation of energy and so we can write down an equation:

$$-\frac{\hbar^2}{2m}\frac{\partial^2 \Psi}{\partial x^2} + V(x,t)\Psi(x,t) = i\hbar\frac{\partial \Psi}{\partial t} \qquad (12.34)$$

where $V(x,t)$ is the potential energy of the particle. This is the time dependent Schrödinger Equation and since it is based on the classical relationship between energy, mass and momentum it can only be used to describe non-relativistic particles. The solutions to this equation depend on the form of the potential, $V(x,t)$, and will be probability waves. In other words it is no longer possible to calculate precisely where a particle will be at any given time, the best that can be done is to describe the probability distribution of where a particle might be.

To use this to understand the atom we need to consider the Schrödinger solutions for an electron wave trapped in a spherically symmetric coulomb potential. Since the nucleus is so much more massive than the electron we can assume that the coulomb potential of the nucleus is constant with respect to time, i.e. $V(x,t) \rightarrow V(x)$. This allows us to use the separation of variables technique, the same one which we used to solve the wave equation in appendix A, to derive what is called the time independent Schrödinger equation:

$$-\frac{\hbar^2}{2m}\frac{\mathrm{d}^2 \psi}{\mathrm{d}x^2} + V(x)\psi(x) = E\psi(x) \qquad (12.35)$$

where $\Psi(x,t) = \psi(x)T(t)$ and E is the particle's energy which is constant. However, this is in one dimension and for the case of an atom, we need to consider three dimensions using spherical polar coordinates due to the symmetry so the maths required is quite complex.

The result is a 3D standing wave which is trapped by the coulomb potential of the nucleus. Each possible wave solution is called an electron orbital and the shape of the wave depends on both the energy of the electron (Bohr's orbital radius number) and the orbital angular momentum. Since there are three dimensions there are three quantum numbers: n which gives the energy, l the total angular momentum and m the z-component of angular momentum. The standing wave is a 3D structure and describes the probability per unit volume of finding the electron. Due to the difficulty of displaying a fully 3D structure with varying values at each coordinate typically a surface corresponding to the regions with the highest probability densities is shown as seen in figure 12.12. However, it is important to realize that

n=2, l=2, m=1

n=6, l=4, m=2

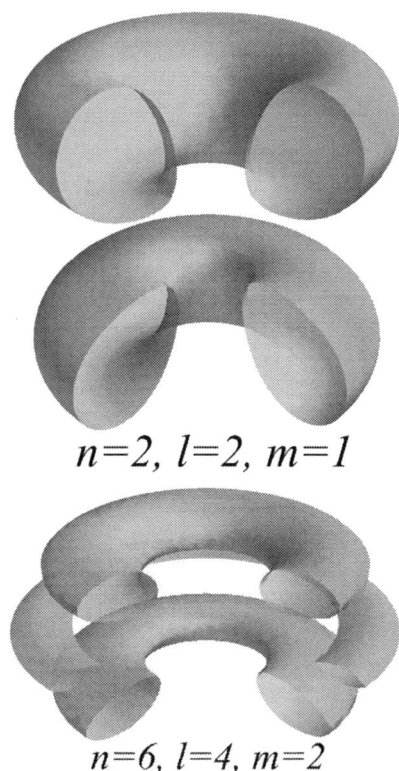

Fig. 12.12: Wavefunctions for electrons in a hydrogen atom. The pictures show the surface of maximum probability with the actual distribution being a cloud which fills a three dimensional volume around the nucleus. The three quantum numbers of each orbital are given below the image and a 90° segment is cut away to allow viewing of the cross-section. Also note that there is no relative scale: the n=6 orbital is larger in radius than the n=2 one shown.

the wave is not constrained to these regions and fills the whole space around the nucleus, even overlapping with the nucleus in some cases which is how nuclear decay via electron capture is possible.

12.7 Quantum Phenomena

This new understanding of matter as a probability wave produces some very strange, and counter-intuitive results which have far reaching consequences for our basic understanding of the universe. While Newtonian mechanics suggested a perfectly deterministic and predictable universe, quantum mechanics tells is that the universe is not deterministic: the identical starting state will not always lead to the same final state. This simple result has philosophical implications: a Newtonian universe would preclude the existence of free will whereas a quantum universe allows for, but does not guarantee, its existence. Restricting ourselves to science quantum mechanics has some startling predictions which run counter to our intuitive grasp of physics.

12.7.1 Single Particle Diffraction

Let's return to our two-slit diffraction experiment which we considered for light waves. When light passes through the slits we get interference and hence diffraction because we have waves passing through both slits which interfere with each other to generate a series of bright fringes.

We know that light consists of photons so let's drop the luminosity such that we know we are sending photons one at a time through the slits. With only a single photon in the experiment at a time intuition tells us that the photon passes through one slit or the other and that there is no photon passing through the other slit with which it can interfere. Hence we would expect two spots on the screen, one for each slit. However what we observe is that the single photons show an interference pattern.

If case this is not convincing enough you can also do this with electrons and get the same result: an interference pattern even when only a single electron is in the experiment at any instant in time. To understand what is happening you might attach a sensor to one of the slits which will let you know which slit the electron passed through. The result of this is that in this case you get what you expected the first time: two single spots, one for each slit and no interference at all. In fact, any observation which differentiates between the slits prevents interference even if it occurs by replacing the screen with a final detector which can measure the electron's direction as well as position.

There is no easily grasped, intuitive explanation for this phenomenon which provides satisfactory answers under all situations. The way it is generally described is using what is called the "Copenhagen Interpretation" which was developed by Niels Bohr and Werner Heisenberg between 1925 and 1927. This says that the electron's probability wave expands and passes through both slits so that the electron can interfere with itself. By observing it halfway through the experiment you collapse the probability wave to one or the other slit and this collapses the interfere pattern to two points. If you don't observe it until it hits the screen then you collapse the wavefunction at this point and get a diffraction pattern.

12.7.2 Schrödinger's Cat

Although a popular way to explain Quantum Mechanics the Copenhagen Interpretation has problems with the discontinuous nature of the wavefunction collapsing when observed and raises all sorts of questions about what counts as an observation. The most well-known argument against it was a Gedankenexperiment (thought experiment) proposed by Erwin Schrödinger which become known as Schrödinger's Cat.

The experiment is to place a cat in a box with a mechanism which will release some poison when it detects a radioactive decay. Once the box is sealed you cannot know which state the cat is in, alive or dead, and so the argument goes that Copenhagen Interpretation says that it is both alive and dead until the box is opened and the wavefunction collapses into either the alive or dead state.

This is very commonly misinterpreted as an actual prediction of quantum mechanics which is absolutely not the case! Clearly the cat is either alive or dead and not in some mysterious quantum state. Schrödinger intended this to show that the Copenhagen Interpretation was incorrect using everyday objects that people had some clear intuition about. However, the Copenhagen Interpretation does not require a physical observer to collapse the wavefunction: any interaction which requires the wavefunction to collapse will do hence the detector when it de-

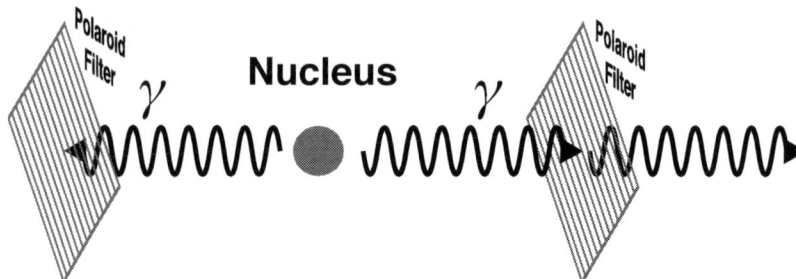

Fig. 12.13: Two photons are emitted in opposite directions by the decay of a nucleus. Each passes through a polaroid filter with a parallel alignment. Experimentally the probabilities of each photon to pas its filter are perfectly anti-correlated: one photon will always pass and one will always be absorbed as shown.

tects or fails to detect a radioactive decay makes the observation and so the cat is either dead or alive depending on the observation of the detector.

12.7.3 Quantum Entanglement

One of the most confusing and yet useful quantum phenomena is that of quantum entanglement. This can be demonstrated by considering an excited nucleus which decays by emitting two photons in opposite directions with opposite polarization states. What is interesting is that these photons are emitted in what is called an entangled state such that if two aligned polarizing filters are placed in the path of each photon one photon will always pass through and one will always be absorbed as shown in figure 12.13.

On the face of it this seems pretty obvious and unsurprising but when you look at the details it gets increasingly puzzling. Suppose that the plane of polarization of one of the photons is at 45° to the angle of its polarizer. The other photon will be polarized at 90° to the first photon but its polarizer will be parallel to the other polarizer so it too will hit its polarizer at 45°.

Now we know the probability of a photon to pass through a polarizer at 45° to its plane of polarization is 50% so each photon has a 50% chance to pass its polarizer. However, if we calculate these probabilities independently then one photon will be absorbed and one photon will pass only half the time, the other half of the times either both photons will pass or both will be absorbed. Experimentally this *never* happens and the chance of one photon to pass the filter is perfectly anti-correlated with whether the other photon passes. However, these are two photons moving in opposite directions at the speed of light so relativity prevents any information travelling between them so how is such a correlation possible? This is known as the Einstein-

Podolsky-Rosen paradox (or EPR paradox for short) which Einstein described as requiring "spooky action at a distance" to explain.

One possible explanation for this effect would be to assume that each photon already has a predetermined value for every possible measurement which can be made and that the photons have essentially pre-determined which will pass the filter and which will not. However, in a 1964 paper, John Bell derived an inequality, called Bell's inequality, for arbitrary angles between the two polarizers in the EPR experiment arriving at different predictions for both quantum mechanics and hidden variables. Experimental measurements followed which showed perfect agreement with quantum mechanics and ruling out hidden variables as a means to explain observations.

While this paradox appears to rely on communicating information between the photons faster than the speed of light it is important to note that even if this were the case the effect cannot be used to transmit external information faster than light and so despite appearing to it does not actually violate relativity. However, it can be used to securely transmit messages without any possibility of a man-in-the-middle attack. Whether a photon passes the filter or is absorbed can produce a binary string that forms a cryptographic key. Anyone who intercepts this transmission will collapse the quantum entangled state and so it is impossible for someone to intercept the key transmission and then re-transmit it without one or the other party knowing that this has happened.

Another use of quantum entanglement is in quantum computing. Here entangled quantum bits, qubits, can be used to run algorithms which would take a standard computer centuries to complete. One example of this is the factorization of large numbers. The inability of standard computers to solve this sort of problem efficiently is a key factor in current cryptographic algorithms such as those used to secure financial institutions. Hence the development of a viable quantum computer with a sufficiently large number of qubits will have major impact...and will probably make its inventor a lot of money!

Problems

Q12.1: The Lyman series in the hydrogen spectrum contains the shortest wavelength lines. Using only the Rydberg constant calculate the wavelengths of the three longest wavelength lines in the Lyman series. [R_H= 1.097×10^7 m^{-1}]

Q12.2: Visible light has wavelengths in the range of 400 nm to 700 nm. A particular metal sample emits electrons for all wavelengths of visible light, what limit(s) can be set on the value of its work function? [h=6.626×10^{-34} J s, c=3×10^8 m s^{-1}]

Q12.3: A Higgs boson has a mass equal to 133 times that of a proton. One is produced in the ATLAS experiment at CERN travelling with a velocity of 1.5×10^6 m s^{-1}. What is its wavelength? [m_p = 1.673×10^{-27} kg, h=6.626×10^{-34} J s]

Q12.4: In an old cathode ray tube based television set electrons would strike the screen with a typical energy of 3.2×10^{-15} J. [m_e= 9.109×10^{-31} kg, m_p = 1.673×10^{-27} kg, h=6.626×10^{-34} J s]

(a) Given that the electrons can be treated as non-relativistic what is the wavelength of such an electron?

(b) If a beam of protons had the same energy what would their wavelength be?

Q12.5: A beam of non-relativistic electrons with an energy of 1.6 fJ is incident on a crystal lattice. The planes of atoms in the crystal act as a diffraction grating and the electron beam is diffracted to produce a series of bright spots on a fluorescent screen. If the angle between the central maximum and the first spot is 8 mrad what is the spacing between the planes of atoms in the crystal?

Q12.6: A green laser pointer emits photons with a wavelength of 520 nm and has a power output of 1 mW. [h=6.626×10^{-34} J s, c=3×10^8 m s^{-1}]

(a) How many photons per second does the laser pointer emit?

(b) What is the momentum of one of the photons emitted?

(c) What force is exerted on the laser pointer due to the emission of photons?

Q12.7: Lithium has a nucleus with three protons. A doubly charged lithium ion, Li^{2+}, therefore only has a single electron and in many regards behaves in a similar fashion to a hydrogen atom. [ϵ_0=8.85×10^{-12} F m^{-1}, $\hbar$=1.05×10^{-34} J s, m_e=9.11×10^{-31} kg, e = 1.6×10^{-19} C]

(a) What is the lowest energy state for the electron in the lithium ion and how does this compare to the ground state energy in the hydrogen atom?

(b) What is the ionization energy for the lithium ion and how does this compare to the ionization energy for the hydrogen atom?

(c) What is the longest wavelength line in the emission spectrum of this lithium ion which corresponds to a transition of an electron to the ground state?

Q12.8: Consider the one-dimensional, time independent Schrödinger equation for a particle with energy E in a constant potential V that does not change with time or position:

$$-\frac{\hbar^2}{2m}\frac{\mathrm{d}^2\psi}{\mathrm{d}x^2} + V\psi(x) = E\psi(x)$$

(a) Show that in the case where $E > V$ this gives a solution corresponding to a wave.

(b) Show that in the case where $V > E$ this gives a solution corresponding to an exponential decay.

(c) A quantum particle with energy E in a region where $V = 0$ encounters a thin potential barrier where $V = 1.5E$. By considering your answers to the two parts above explain qualitatively what will happen to the particle and compare this to the case of a classical particle.

APPENDIX A

Wave Equations

Any wave can be written down as a function of position and time where the value of the function gives the displacement of the medium at the given space-time coordinates. If a one-dimensional wave has a phase velocity of $+c$ then after one second the function representing the wave will have moved a distance of c as shown in figure A.1. Hence the wave must be a function of $x - ct$ where x is the position and t is the time. Similarly for a wave with a velocity of $-c$ it must be described by a function of $x + ct$ for the same reasons.

This is the most general description possible for a wave. We have made no assumption about its shape nor its frequency only that it has a constant shape which moves through the medium. This minimal set of assumptions is all that we need to both derive and then solve the wave equation.

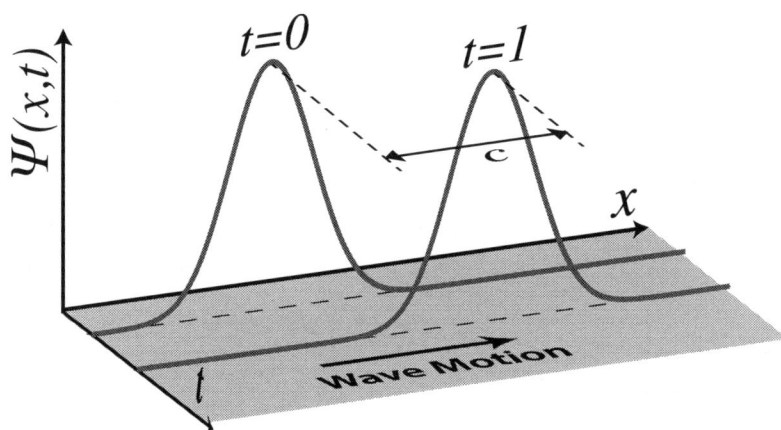

Fig. A.1: Propagation of a general wave through a medium. After a time of one second the wave has maintained its shape and move a distance of c. Hence the function describing the displacement of the wave, Ψ, must be a function of $x - ct$ for a wave moving in the positive x direction (as shown) or $x + ct$ if is moving in the negative x direction (opposite to the one shown).

A.1 General 1D Wave Equation

In one dimension a wave only has two possible directions of motion, either the positive or negative x directions, and hence the phase velocity can only be one of two values: $+c$ or $-c$. Hence the most general possible description for the displacement of a one-dimensional medium with waves that could be propagating in either, or both, directions is the sum of the two possible wave functions:

$$\phi = f(x - ct) + g(x + ct) \tag{A.1}$$

Now let's make a variable substitute such that $\xi = x - ct$ and $\eta = x + ct$. This gives:

$$\phi = f(\xi) + g(\eta) \tag{A.2}$$

Now lets take a partial derivative with respect to η. Since $f(\xi)$ is purely a function of ξ it is treated as a constant and will differentiate to zero so we have:

$$\frac{\partial \phi}{\partial \eta} = \frac{\partial g(\eta)}{\partial \eta} \tag{A.3}$$

Now take the partial derivate with respect to ξ. Since $g(\eta)$ is only a function of η so is its derivative and so it is treated as a constant and will differentiate to zero so we have:

$$\frac{\partial^2 \phi}{\partial \xi \partial \eta} = 0 \tag{A.4}$$

The final step is to get rid of our ξ and η which we can do using the partial derivative chain rule which states that:

$$\frac{\partial \phi}{\partial \xi} = \frac{\partial \phi}{\partial x}\frac{\partial x}{\partial \xi} + \frac{\partial \phi}{\partial t}\frac{\partial t}{\partial \xi} \quad \text{and} \quad \frac{\partial \phi}{\partial \eta} = \frac{\partial \phi}{\partial x}\frac{\partial x}{\partial \eta} + \frac{\partial \phi}{\partial t}\frac{\partial t}{\partial \eta} \tag{A.5}$$

with a similar equation for η. Now using our definitions of ξ and η we have the following expressions for x and t:

$$x = \frac{1}{2}(\eta + \xi) \quad \text{and} \quad t = \frac{1}{2c}(\eta - \xi) \tag{A.6}$$

Using these we can see that:

$$\frac{\partial x}{\partial \xi} = \frac{1}{2} \;,\; \frac{\partial t}{\partial \xi} = -\frac{1}{2c} \;,\; \frac{\partial x}{\partial \eta} = \frac{1}{2} \quad \text{and} \quad \frac{\partial t}{\partial \eta} = \frac{1}{2c} \tag{A.7}$$

So lets start first with the derivative with respect to η:

$$\frac{\partial \phi}{\partial \eta} = \frac{1}{2}\frac{\partial \phi}{\partial x} + \frac{1}{2c}\frac{\partial \phi}{\partial t} \tag{A.8}$$

Next we want to take the partial derivative with respect to ξ and then tidy up the algebra afterwards:

$$\frac{\partial^2 \phi}{\partial \xi \partial \eta} = \frac{1}{2} \frac{\partial}{\partial x} \left(\frac{1}{2} \frac{\partial \phi}{\partial x} + \frac{1}{2c} \frac{\partial \phi}{\partial t} \right) - \frac{1}{2c} \frac{\partial}{\partial t} \left(\frac{1}{2} \frac{\partial \phi}{\partial x} + \frac{1}{2c} \frac{\partial \phi}{\partial t} \right)$$

$$= \frac{1}{4} \frac{\partial^2 \phi}{\partial x^2} + \frac{1}{4c} \frac{\partial^2 \phi}{\partial t \partial x} - \frac{1}{4c} \frac{\partial^2 \phi}{\partial x \partial t} - \frac{1}{4c^2} \frac{\partial^2 \phi}{\partial t^2} \qquad (A.9)$$

Now we need to use another general result of partial derivatives which is that the result is independent of the order of differentiation, in other words:

$$\frac{\partial^2 \phi}{\partial x \partial t} = \frac{\partial^2 \phi}{\partial t \partial x} \qquad (A.10)$$

Hence (A.9) reduces to:

$$\frac{\partial^2 \phi}{\partial \xi \partial \eta} = \frac{1}{4} \frac{\partial^2 \phi}{\partial x^2} - \frac{1}{4c^2} \frac{\partial^2 \phi}{\partial t^2} \qquad (A.11)$$

Now looking at (A.4) we see that the left hand side of (A.11) is just zero and so:

$$\frac{\partial^2 \phi}{\partial x^2} = \frac{1}{c^2} \frac{\partial^2 \phi}{\partial t^2} \qquad (A.12)$$

This is the general wave equation for one dimensional waves. We have proved this using general properties of partial derivatives and have made no assumptions about the shape of the wave. Hence, although we are primarily concerned with sinusoidal waves, this is a completely general result that is valid for all waves. The equation was first derived in 1746 by Jean-Baptiste le Rond d'Alembert (see figure A.2) and within ten years it was extended to three dimensions by Euler.

A.2 Solving the Wave Equation

To solve the wave equation we will need a function of both x and t. To start let's make the assumption that we can write this as a product of a function of x and a function of t i.e. that we can separate the variables such that:

$$\Psi(x, t) = X(x)T(t) \qquad (A.13)$$

Fig. A.2: Jean-Baptiste le Rond d'Alembert (1717-1783) was the French mathematician and physicist who first discovered the one dimensional wave equation in 1746.
by Maurice Quentin de La Tour, 1753.
(Musée du Louvre, Paris), public domain.

This technique for solving a partial differential equation is called *separation of variables* and is by far the most common method for solving partial differential equations in physics. Making this assumption we can now calculate the second order partial derivatives:

$$\frac{\partial^2 \Psi}{\partial x^2} = \frac{\partial}{\partial x} \left(\frac{\partial X}{\partial x} T + X \frac{\partial T}{\partial x} \right) \qquad (A.14)$$

However $T(t)$ is just a function of t and so $\frac{\partial T}{\partial x} = 0$. Hence we have:

$$\frac{\partial^2 \Psi}{\partial x^2} = \frac{\partial}{\partial x}\left(\frac{\partial X}{\partial x}T\right) = \frac{\partial^2 X}{\partial x^2}T + \frac{\partial X}{\partial x}\frac{\partial T}{\partial x} = \frac{\partial^2 X}{\partial x^2}T \quad \text{(A.15)}$$

Now since $X(x)$ is purely a function of x then the partial derivative will equal the ordinary derivative and so we can write:

$$\frac{\partial^2 \Psi}{\partial x^2} = \frac{d^2 X}{dx^2}T \quad \text{(A.16)}$$

Similarly for the second order differential with respect to t we end up with:

$$\frac{\partial^2 \Psi}{\partial t^2} = X\frac{d^2 T}{dt^2} \quad \text{(A.17)}$$

Putting this into the wave equation given in (A.12) we get:

$$\frac{d^2 X}{dx^2}T = \frac{1}{c^2}X\frac{d^2 T}{dt^2} \implies \frac{d^2 X}{dx^2}\frac{1}{X} = \frac{d^2 T}{dt^2}\frac{1}{c^2 T} \quad \text{(A.18)}$$

Written in this way the lefthand side is purely a function of x and the righthand side is purely a function of t. Since there is no correlation between x and t - we can ask what the displacement is at any point in space or at any instant of time - the only way that (A.18) can possibly be correct is when each side is equal to a constant which we will call $-k^2$. Hence we now have two differential equations:

$$\frac{d^2 X}{dx^2} = -k^2 X(x) \quad \text{and} \quad \frac{d^2 T}{dt^2} = -k^2 c^2 T \quad \text{(A.19)}$$

To solve these we need a function which, when differentiated twice, gives the same function i.e. an exponential. By similarity to the simple harmonic oscillator equation of motion (5.77) we can write down the solutions as:

$$X(x) = A_x e^{ikx} + B_x e^{-ikx} \quad \text{and} \quad T(t) = A_t e^{i\omega t} + B_t e^{-i\omega t} \quad \text{(A.20)}$$

where A_x, A_t, B_x and B_t are complex constants and $\omega = ck$. Hence our full solution to the wave equation is:

$$\Psi(x,t) = \left(A_x e^{ikx} + B_x e^{-ikx}\right)\left(A_t e^{i\omega t} + B_t e^{-i\omega t}\right) \quad \text{(A.21)}$$

$$= A_x A_t e^{i(kx+\omega t)} + B_x B_t e^{-i(kx+\omega t)} \quad \text{(A.22)}$$

$$+ A_x B_t e^{i(kx-\omega t)} + B_x A_t e^{-i(kx-\omega t)} \quad \text{(A.23)}$$

Combining the constants we can write our solution as:

$$\Psi(x,t) = A_+ e^{i(kx+\omega t)} + B_+ e^{-i(kx+\omega t)} + A_- e^{i(kx-\omega t)} + B_- e^{-i(kx-\omega t)}$$
$$\text{(A.24)}$$

Since this represents the displacement of the medium as a function of position and time we need a real solution. The conditions for a real solution are identical to those of the single oscillator ($A_+ = \overline{B_+}$ and $A_- = \overline{B_-}$) and the result is the same:

$$\Psi(x, t) = A\cos(kx + \omega t + \phi_A) + B\cos(kx - \omega t + \phi_B) \qquad \text{(A.25)}$$

where A, B, ϕ_A and ϕ_B are all real constants.

Now we have the real solutions to the wave equation we need a physical interpretation of the constants. We started with c being the velocity of the wave. If we consider the solution at a fixed point in space, $x = 0$, then we have:

$$\Psi(x, t) = A\cos(\omega t + \phi_A) + B\cos(-\omega t + \phi_B) \qquad \text{(A.26)}$$

and so by comparison to the solution for the single harmonic oscillator ω is clearly the angular frequency of the wave. To interpret the constant k, which we introduced when separating the two equations, we return to our definition of ω:

$$\omega = ck \implies k = \frac{\omega}{c} = \frac{2\pi f}{c} \qquad \text{(A.27)}$$

where f is the wave frequency which is related to the wavelength by the equation, $c = f\lambda$ and so k becomes:

$$k = \frac{2\pi}{\lambda} \qquad \text{(A.28)}$$

which is just the wave number!

Now that we have a physical interpretation for all the constants we need interpret the two cosine functions we derived. If we take out a factor of k then we have:

$$\Psi(x, t) = A\cos(k[x + ct] + \phi_A) + B\cos(k[x - ct] + \phi_B) \qquad \text{(A.29)}$$

because $\omega/k = c$. Hence we have a function of $x - ct$ which represents a wave with a positive phase velocity and a function of $x + ct$ which represents a wave with a negative phase velocity:

$$\Psi_\rightarrow(x, t) = A\cos(kx - \omega t + \phi) \qquad \text{(A.30)}$$
$$\Psi_\leftarrow(x, t) = A\cos(kx + \omega t + \phi) \qquad \text{(A.31)}$$

It also also worth pointing out that we can also write these functions using the implied real operator in the same manner as the simple harmonic oscillator. This gives the more commonly seen notation:

$$\Psi_\rightarrow(x, t) = Ae^{i(kx - \omega t)} = Ae^{ikx}e^{-i\omega t} \qquad \text{(A.32)}$$
$$\Psi_\leftarrow(x, t) = Ae^{i(kx + \omega t)} = Ae^{ikx}e^{i\omega t} \qquad \text{(A.33)}$$

where A is the complex amplitude that contains both the real amplitude and the constant phase information.

Note that when solving the wave equation we assumed that the constant we obtained when splitting into the two equations for x and t was negative. We could have equally have assumed a positive constant and this would have given real exponentials. These solutions represent a special type of wave, called an *evanescent wave*, that does not propagate but which exponentially decays in a similar fashion to a damped harmonic oscillator.

A.3 The Acoustic Wave Equation

An acoustic wave is a longitudinal wave which travels through a solid or fluid. As with any mechanical wave it can be represented as a displacement of the medium from its equilibrium position. However acoustic waves in fluids can also be represented as a pressure wave.

To determine the relationship between these two representations we will derive the wave equation for each here, starting with the pressure wave.

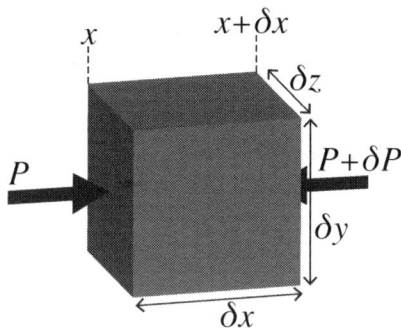

Fig. A.3: Small element of a fluid showing the pressure difference acting on it due to an acoustic pressure wave.

A.3.1 1D Pressure Wave

To keep things simple we will only consider motion in one direction, x, as shown in figure A.3. Both the pressure P and the x component of velocity, u, vary with both time and position in the fluid and so they are functions of both x and t i.e. $P(x, t)$ and $u(x, t)$.

The net force acting on the element in the x direction at any given time t will depend solely on the pressure difference across the element. The rate of change of pressure with respect to position, x, is $\partial P / \partial x$ where we have to use a partial derivate because we are keeping the time, t, constant. Hence the pressure difference on the element at time t is:

$$\delta P = \frac{\partial P}{\partial x} \delta x \tag{A.34}$$

To convert this into the net force in the x direction we just multiply it by the area of the element in the direction perpendicular to x:

$$\delta F_x = -\frac{\partial P}{\partial x} \delta x \delta y \delta z \tag{A.35}$$

where the minus sign is required because if the rate of change of pressure with respect to x, $\frac{\partial P}{\partial x}$, is positive it means that the pressure at $x + \delta x$ will be greater than the pressure at x and hence the net force is in the negative x direction.

Now that we know the force acting on the fluid element in the x direction we can use Newton's second law to get the equation of motion for this element. The x component of the fluid velocity changes with both position and time so the acceleration of the element at a given position will be $\partial u/\partial t$ because we are differentiating the velocity with respect to time while keeping the position, x, constant. The mass of the fluid element will just be the density of the fluid, ρ, multiplied by the volume of the element and so, putting these into Newton's second law, we get:

$$F_x = ma_x$$

$$-\frac{\partial P}{\partial x}\delta x \delta y \delta z = \rho \delta x \delta y \delta z \frac{\partial u}{\partial t}$$

$$\frac{\partial P}{\partial x} = -\rho \left.\frac{\partial u}{\partial t}\right|_x \qquad (A.36)$$

Now we must consider the compression of the fluid as the wave passes through it. Figure A.4 shows the same fluid element but now we are considering the difference in the fluid velocity between the two sides of the element. If there is a velocity difference between the sides then the element will either be compressed or expanded. Since this results in a change in volume we need to consider the bulk modulus, B of the fluid which is defined by:

$$\text{Bulk Modulus, } B = -\frac{\Delta P}{\Delta V/V_0} \qquad (A.37)$$

For our element the change in pressure and volume will be small so we will use the representation δP and δV respectively. The difference in the fluid velocity from one side of the element to the other at a given instant of time is simply the rate of change of velocity with respect to x multiplied by the thickness of the element in the x direction:

$$\delta u = \left.\frac{\partial u}{\partial x}\right|_t \delta x \qquad (A.38)$$

This is the difference in velocity, not a difference in volume: effectively it is the velocity of one side of the fluid element relative to the opposite side. Hence the change in the length of the element during a small time period, δt, will be simply $\delta u \delta t$ and so the change in the volume of the element during the same small interval of time is just:

$$\delta V = \left.\frac{\partial u}{\partial x}\right|_t \delta x \delta y \delta z \delta t \qquad (A.39)$$

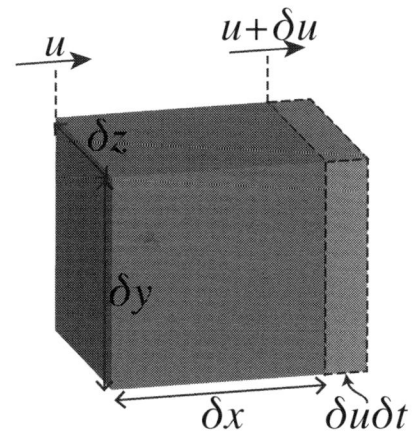

Fig. A.4: Small element of a fluid showing how the different fluid velocity at each end of the element leads to the element stretching over a small time period, δt.

The original volume of the element, V_0, is simply $\delta x \delta y \delta z$ and so we have:

$$\frac{\delta V}{V_0} = \left.\frac{\partial u}{\partial x}\right|_t \delta t \tag{A.40}$$

Now rearranging equation (A.37) we get:

$$\delta P = -B\frac{\delta V}{V_0} = -B\left.\frac{\partial u}{\partial x}\right|_t \delta t$$

$$\frac{\delta P}{\delta t} = -B\left.\frac{\partial u}{\partial x}\right|_t$$

and then we take the limit as $\delta t \to 0$ remembering that we are considering an element at a fixed position in x and that the pressure, P, is a function of both x and t so that instead of a full derivative we have a partial derivative with respect to time keeping x constant. This gives:

$$\left.\frac{\partial P}{\partial t}\right|_x = -B\left.\frac{\partial u}{\partial x}\right|_t \tag{A.41}$$

We now have two equations for the pressure in terms of the fluid velocity: (A.36) from Newton's second law and (A.41) from the bulk modulus. To combine these let's differentiate (A.36) with respect to x keeping t constant:

$$\frac{\partial^2 P}{\partial x^2} = -\rho\frac{\partial u}{\partial x \partial t} \tag{A.42}$$

Now differentiate (A.41) with respect to time while keeping x constant to get:

$$\frac{\partial^2 P}{\partial t^2} = -B\frac{\partial u}{\partial t \partial x} \tag{A.43}$$

Now the order of partial differentiation does not matter and so:

$$\frac{\partial u}{\partial x \partial t} = \frac{\partial u}{\partial t \partial x}$$

Hence we can combine equations (A.42) and (A.43) to get:

$$\frac{\partial^2 P}{\partial x^2} = \frac{\rho}{B}\frac{\partial^2 P}{\partial t^2} \tag{A.44}$$

which is the acoustic wave equation in terms of the absolute pressure, P, of the medium. Comparing this equation to the general one dimensional wave equation we can read off the phase velocity of the acoustic wave:

$$c = \sqrt{\frac{B}{\rho}} \tag{A.45}$$

For air the adiabatic bulk modulus is $1.42 \times 10^5 \, \mathrm{N\,m^{-2}}$ and the density is $1.2 \, \mathrm{kg\,m^{-3}}$ which gives a sound velocity of $344 \, \mathrm{m\,s^{-1}}$. If we compare this to the speed of sound in water then, with the far larger density of $1{,}000 \, \mathrm{kg\,m^{-3}}$ we might expect the speed of sound to be less in water. However water is far less compressible than air with a bulk modulus of $2.2 \times 10^9 \, \mathrm{N\,m^{-2}}$ which dominates giving a speed of sound in water of $1{,}483 \, \mathrm{m\,s^{-1}}$.

In this derivation we have used the absolute pressure of the medium, P. However when dealing with acoustic waves is it more typical to consider the deviation of the pressure from the undisturbed medium. This is easily achieved by defining:

$$p(x, t) = P(x, t) - P_0 \qquad (A.46)$$

where P_0 is the constant pressure of the undisturbed medium and p is the deviation of the pressure from the equilibrium pressure. Since P_0 is a constant any differential of P will be equal to the same differential of p. Hence we can easily rewrite the equation as:

$$\frac{\partial^2 p}{\partial x^2} = \frac{\rho}{B} \frac{\partial^2 p}{\partial t^2} \qquad (A.47)$$

where p is the pressure deviation.

A.3.2 3D Pressure Wave

In general acoustic waves are three dimensional and in this case the pressure is a function of x, y and z as well as time, t. If we apply Newton's second law in the y and z directions then we get:

$$\frac{\partial P}{\partial y} = -\rho \frac{\partial v}{\partial t} \quad \text{and} \quad \frac{\partial P}{\partial z} = -\rho \frac{\partial w}{\partial t} \qquad (A.48)$$

where v and w are the velocity of the fluid in the y and z directions respectively. If we now consider the change in volume of the element allowing for compression in the y and z directions as well then the total change in volume is a sum of the contributions from each of the three directions in the limit that δt is small: any corrections to this due to edge and corner effects will be of order $(\delta t)^2$ or higher and so negligible. Adding these terms into the bulk modulus equation gives:

$$\left. \frac{\partial P}{\partial t} \right|_x = -B \left[\frac{\partial u}{\partial x} + \frac{\partial v}{\partial y} + \frac{\partial w}{\partial z} \right] \qquad (A.49)$$

Using the same method that we used to combine the 1D acoustic wave equation now gives:

$$\frac{\partial^2 P}{\partial x^2} + \frac{\partial^2 P}{\partial y^2} + \frac{\partial^2 P}{\partial z^2} = \frac{\rho}{B} \frac{\partial^2 P}{\partial t^2} \qquad (A.50)$$

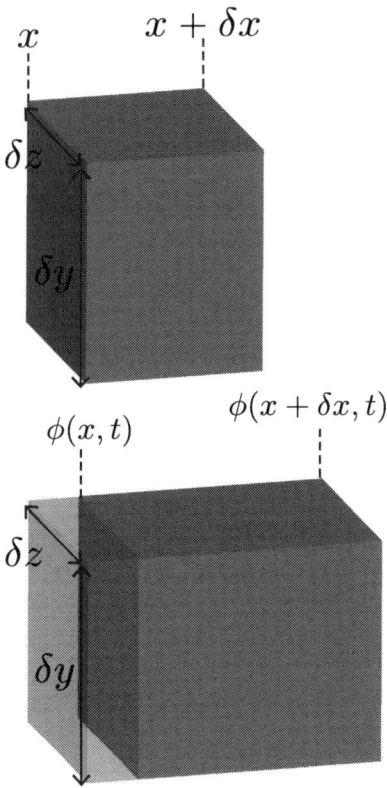

x $x + \delta x$

δz

δy

$\phi(x, t)$ $\phi(x + \delta x, t)$

δz

δy

Fig. A.5: Small element of a fluid showing how the variation in the displacement of the fluid, ϕ, as a function of the position, x, can lead to a stretching (or compression) of the fluid.

which is the full, three dimensional acoustic wave equation. The sum of all the second order derivatives with respect to the three directions occurs frequently in physics, particularly when dealing with waves, and so this is usually written as:

$$\nabla^2 P = \frac{\rho}{B} \frac{\partial^2 P}{\partial t^2} \quad \text{where} \quad \nabla^2 = \frac{\partial^2}{\partial x^2} + \frac{\partial^2}{\partial y^2} + \frac{\partial^2}{\partial z^2} \tag{A.51}$$

A.3.3 1D Displacement Wave

As before to keep things simple we will consider waves propagating in a single direction. Defining the displacement of the medium at any given point in space and time as $\phi(x, t)$ then we note that at any point in the medium this is related to the velocity of the medium, $u(x, t)$, by:

$$u(x, t) = \frac{\partial \phi(x, t)}{\partial t} \tag{A.52}$$

Starting with equation (A.36) which we derived using Newton's second law on an element of the fluid and using our expression for the velocity we just derived (A.52) we get:

$$\frac{\partial P}{\partial x} = -\rho \frac{\partial u}{\partial t} = -\rho \frac{\partial^2 \phi}{\partial t^2} = \frac{\partial p}{\partial x} \tag{A.53}$$

where we have chosen to rewrite it in terms of the pressure deviation, p.

Next we return to considering the compression of an element of the fluid. However instead of looking at the relative velocities of each end of an element of the fluid lets consider the displacement. Figure A.5 shows an element of the fluid and how it is affected by the wave. The change in the volume of the element compared to the undisturbed state is:

$$\delta V = \delta y \delta z \left[\phi(x + \delta x, t) - \phi(x, t) \right] \tag{A.54}$$

Since the original volume of the element was simply $\delta x \delta y \delta z$ then we have the fractional volume change of the element as:

$$\frac{\delta V}{V_0} = \frac{\left[\phi(x + \delta x, t) - \phi(x, t) \right]}{\delta x} \tag{A.55}$$

Now look again at our definition for the medium's bulk modulus given in equation (A.37). Here we are looking at the deformation of an element of the fluid compared to the undisturbed state of the fluid hence the change in pressure in this equation is just the deviation of the pressure from the undisturbed pressure, P_0 i.e. $\Delta P = p$. Hence we have:

$$p = -B\frac{\delta V}{V_0} = -B\frac{[\phi(x + \delta x, t) - \phi(x, t)]}{\delta x} \qquad \text{(A.56)}$$

All that remains is to take the limit as $\delta x \to 0$ which will convert the difference into a partial derivative since we are evaluating the difference at a fixed value of t:

$$p = -B\frac{\partial \phi}{\partial x} \qquad \text{(A.57)}$$

If we take the partial derivate with respect to x, keeping t constant, of both sides of this equation we get:

$$\frac{\partial p}{\partial x} = -B\frac{\partial^2 \phi}{\partial x^2} \qquad \text{(A.58)}$$

All that now remains is to combine this equation with our initial one from Newton's second law, (A.53), to eliminate the pressure term and we have:

$$\frac{\partial^2 \phi}{\partial x^2} = \frac{\rho}{B}\frac{\partial^2 \phi}{\partial t^2} \qquad \text{(A.59)}$$

Hence the displacement of the fluid also produces a wave equation which has the same phase velocity, $\sqrt{B/\rho}$, as the pressure wave. However it does not have the same phase as the pressure wave. It we write down the solution to equation (A.59) then we have:

$$\phi(x, t) = Ae^{i(kx - \omega t)} \qquad \text{(A.60)}$$

Now looking at equation (A.57) we see that the pressure deviation is:

$$p(x, t) = -B\frac{\partial \phi}{\partial x} = -ikBAe^{i(kx - \omega t)} = kBAe^{i(kx - \omega t - \pi/2)} \qquad \text{(A.61)}$$

where we used the relationship $-i = e^{-i\pi/2}$. This means that the pressure has a phase difference of $\pi/2$ compared to the displacement and that its phase lags behind that of the displacement. This is a somewhat surprising result because it means that the points of both maximum pressure (and density) and minimum pressure (and density) occur where there is zero displacement of the medium and not, as one might naively expect, when the displacement is at a maximum.

APPENDIX B

Blackbody Spectrum

To derive a mathematical expression for a blackbody emission spectrum we need to first understand the number of different electromagnetic standing waves in a three dimensional cavity and then apply some basic thermodynamics. Since we have only considered one dimensional standing waves so far we will first consider two dimensional standing waves before extrapolating to three dimensions.

A standing wave on a string is a one dimensional phenomenon. However waves can propagate in two and three dimensions and so it is possible to have two and three dimensional standing waves. To understand standing waves in higher dimensions we shall first consider two dimensions in detail and then expand this to three dimensions by extrapolation.

B.1 2D Standing Waves

In two dimensions the wave equation for any wave is:

$$\frac{\partial^2 \Psi}{\partial x^2} + \frac{\partial^2 \Psi}{\partial y^2} = \frac{1}{c^2} \frac{\partial^2 \Psi}{\partial t^2} \tag{B.1}$$

where Ψ is the displacement of the medium and so is a function of x, y and t: $\Psi(x, y, t)$. We can use the same method to solve this as we used for the one dimensional equation as shown in appendix A. To do this we start with assuming that we can separate the variables and write the solution in the form:

$$\Psi(x, y, t) = X(x)Y(y)T(t) \tag{B.2}$$

Substituting this into (B.1) we get:

$$YT\frac{d^2X}{dx^2} + XT\frac{d^2Y}{dy^2} = \frac{XY}{c^2}\frac{\partial^2 T}{\partial t^2} \qquad (B.3)$$

which we simply divide through by XYT to get:

$$\frac{1}{X}\frac{d^2X}{dx^2} + \frac{1}{Y}\frac{d^2Y}{dy^2} = \frac{1}{c^2T}\frac{d^2T}{dt^2} \qquad (B.4)$$

As before since we can chose x, y and t arbitrarily because they are free variables the only way that this equation can hold for all possible choices is if both sides are constant. If we call this constant $-k^2$ as before then we have two equations:

$$\frac{1}{X}\frac{d^2X}{dx^2} + \frac{1}{Y}\frac{d^2Y}{dy^2} = -k^2 \text{ and } \frac{d^2T}{dt^2} = -k^2c^2T \qquad (B.5)$$

The equation in t is the same as the one we found when solving the general 1D wave equation in appendix section A.2, equation (A.19) and which has a solution (A.20):

$$T(t) = A_t e^{i\omega t} + B_t e^{-i\omega t} \qquad (B.6)$$

where $\omega = ck$. For the other side of the equation we now need to separate the x and y terms and repeat:

$$\frac{1}{X}\frac{d^2X}{dx^2} = -k^2 - \frac{1}{Y}\frac{d^2Y}{dy^2} \qquad (B.7)$$

Again we can argue that since x and y are independent variables both sides must be equal to a constant which we will call $-k_x^2$. Looking at the x terms this gives:

$$\frac{1}{X}\frac{d^2X}{dx^2} = -k_x^2 \qquad (B.8)$$

which again we have already solved in appendix A using k instead of k_x. The solution is (A.20):

$$X(x) = A_x e^{ik_x x} + B_x e^{-ik_x x} \qquad (B.9)$$

This leaves the remaining y terms which give the equation:

$$\frac{1}{Y}\frac{d^2Y}{dy^2} = -k^2 + k_x^2 = -k_y^2 \qquad (B.10)$$

where we have defined the constant k_y such that:

$$k^2 = k_x^2 + k_y^2 \qquad (B.11)$$

By direct comparison to our equation for x given in (B.8) we can immediately solve this equation:

$$Y(y) = A_y e^{ik_y y} + B_y e^{-ik_y y} \tag{B.12}$$

Putting all these components together we have the full, two dimensional solution to the wave equation:

$$\Psi(x, y, t) = (A_x e^{ik_x x} + B_x e^{-ik_x x})(A_y e^{ik_y y} + B_y e^{-ik_y y})$$
$$\times (A_t e^{i\omega t} + B_t e^{-i\omega t}) \tag{B.13}$$

This contains all the possible solutions for a general two dimensional wave but we are interested in standing wave solutions. In two dimensions the equivalent of a string with a fixed length is a rectangular membrane which can vibrate up and down. Let this have the dimensions L_x and L_y in the x and y directions respectively and we will assume that it is clamped around the edges so that the displacement is required to be zero there. If one corner is placed at the origin, $(0, 0)$, and the diagonally opposite corner is at the coordinates (L_x, L_y) then this gives four boundary conditions, one for each edge as shown in figure B.1:

$$\Psi = 0 \quad \text{when } x = 0, x = L_x, y = 0 \text{ or } y = L_y \tag{B.14}$$

Lets look at the first of these conditions which can be written, using our separate functions for x, y and t as:

$$\Psi(0, y, t) = X(0)Y(y)T(t) = 0 \tag{B.15}$$

Since y and t are independent variables which we are free to choose the only way to ensure that this is true for all y and t is to require $X(0) = 0$. Putting this requirement into (B.16) gives the equation:

$$X(0) = A_x e^0 + B_x e^0 = 0 \implies A_x + B_x = 0 \implies B_x = -A_x \tag{B.16}$$

Putting this into equation (B.16) we get:

$$X(x) = A_x e^{ik_x x} - A_x e^{-ik_x x} = 2iA_x \sin(k_x x) \tag{B.17}$$

Looking at the second condition we again see that for this to be true for any value of y and t we require:

$$X(L_x) = 0 \implies \sin(k_x L_x) = 0 \tag{B.18}$$

For this to be true we require:

$$k_x L_x = n_x \pi \implies k_x = \frac{n_x \pi}{L_x} \quad \text{where } n_x = 1, 2, 3 \dots \tag{B.19}$$

and so, just like the one dimensional standing wave, we end up with only certain wavelengths, or in this case wavenumbers,

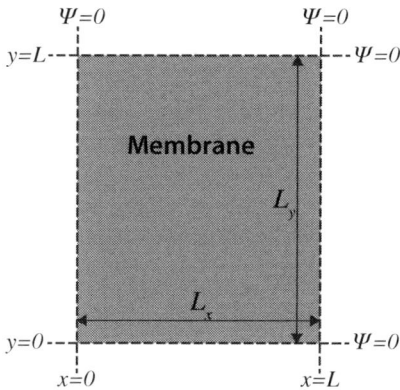

Fig. B.1: The boundary conditions on a 2D, rectangular membrane of dimensions L_x by L_y which is clamped at the edges.

being allowed. The difference is that this condition applies to just the wavenumber associated with the x direction. We can now repeat this exact procedure with the to boundary conditions that $\Psi = 0$ when $y = 0$ and $y = L_y$. Again we have to require that $Y(0) = 0$ which constrains us to a sine function and again the second condition at $y = L_y$ reduces us to certain wavenumbers in the y direction as given below:

$$Y(y) = 2iA_y \sin(k_y y) \text{ with } k_y = \frac{n_y \pi}{L_y} \text{ where } n_y = 1, 2, 3 \ldots$$

$$(\text{B.20})$$

Putting all this together we have a combined wavefunction that is:

$$\Psi(x, y, t) = -4A_x A_y \sin\left(\frac{n_x \pi}{L_x}x\right) \sin\left(\frac{n_y \pi}{L_y}y\right) (A_t e^{i\omega t} + B_t e^{-i\omega t})$$

$$(\text{B.21})$$

At this point we still have a complex solution: all the constants can be complex as are the complex exponentials of t. However the physical displacement of a membrane must be a real number and so $\Psi(x, y, t)$ must be real. As we have seen in the solutions for the simple harmonic oscillator when we require the sum of two complex exponentials to be real we end up with a sine or cosine function and two constants: an amplitude and initial phase. Hence with the requirement that the solution is real we get:

$$\Psi(x, y, t) = A \sin\left(\frac{n_x \pi}{L_x}x\right) \sin\left(\frac{n_y \pi}{L_y}y\right) \sin(\omega t + \phi) \qquad (\text{B.22})$$

where A is the amplitude and ϕ is the initial phase. This is the mathematical description of a standing wave in two dimensions. Instead of a single integer number n we now have two integer numbers: n_x and n_y which describe the x and y components of the wave.

If we consider just the x direction then $n_x - 1$ gives the number of nodes between the edges. However on a two dimensional sheet a node is now a line not a point. Similarly $n_y - 1$ gives the number of nodes in the y direction and so we can have a different number of nodes in x and y. Figure B.2 shows several different modes of the sheet for differing values of n_x and n_y.

B.2 Wave Vector

In deriving our two dimensional standing wave we discovered that instead of a single wave number we had two separate

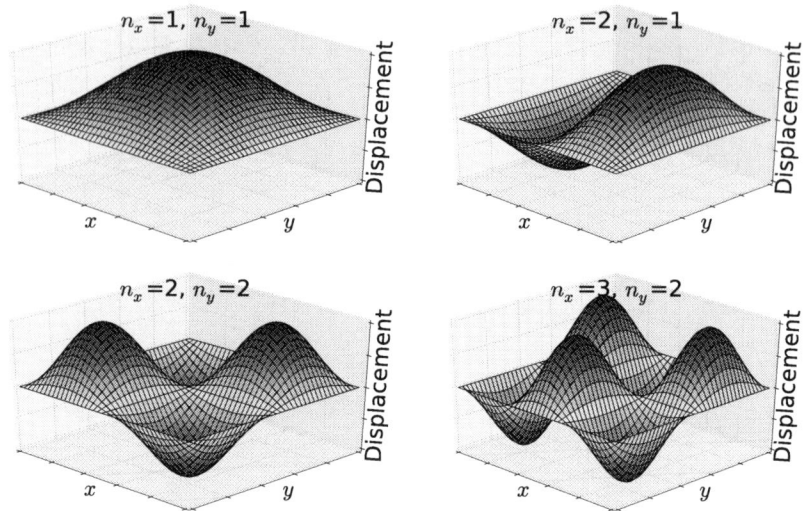

Fig. B.2: Several examples of two dimensional standing waves on a square sheet. The waves are each shown at maximum displacement. The frequency of the oscillations depends on the magnitude of the wave vector, $\omega = ck$, for the mode.

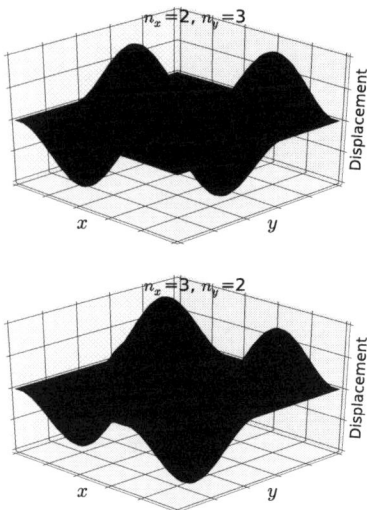

Fig. B.3: Two different, two dimensional standing wave modes which have the same wavenumber but different wave vectors.

wavenumbers: k_x and k_y which described the x and y components of the wave respectively. Furthermore we had a relationship between them and a third constant k given in equation (B.11). From this it is clear that we can treat k_x and k_y as the x and y components of a vector which has a magnitude k. This vector is known as the wave vector where:

$$\vec{k} = \begin{pmatrix} k_x \\ k_y \end{pmatrix} \tag{B.23}$$

The wavenumber of a wave is simply the magnitude of the wave vector and so the wavelength is now defined as:

$$\lambda = \frac{2\pi}{|\vec{k}|} \tag{B.24}$$

If the wave is propagating through a uniform, isotropic medium then the direction of the wave vector is the same as the direction of propagation of the wave. However in an anisotropic medium, such as light waves in an asymmetric crystal, this is not the case.

For standing waves there is no propagation and so the wave vector gives an indication of how the wave "fits" into the constrained medium. Typically there are multiple ways of doing this. For example if we consider the case of a two dimensional wave on a rectangular membrane discussed in the previous section then the states with $(n_x = 2, n_y = 3)$ and $(n_x = 3, n_y = 2)$, as seen in figure B.3, both have the same wavelength but different wave vectors. This shows that there are two ways to fit a wave with this wavelength into the medium.

B.3 3D Standing Waves

In three dimensions we have a wave equation which looks like:

$$\nabla^2 \Psi = \frac{\partial^2 \Psi}{\partial x^2} + \frac{\partial^2 \Psi}{\partial y^2} + \frac{\partial^2 \Psi}{\partial z^2} = \frac{1}{c^2}\frac{\partial^2 \Psi}{\partial t^2} \tag{B.25}$$

where again Ψ is the displacement of the wave and is now a function of x, y, z and t. Applying the same separation of variables technique which we used in two dimensions we can write:

$$\Psi(x, y, z, t) = X(x)Y(y)Z(z)T(t) \tag{B.26}$$

Again we put this into the wave equation and separate the variables. This will give us exactly what we had before only with the addition of a z component and so we end up with a third constant, k_z, which is the z component of the wave vector. Assuming that we now have a rectangular box with dimensions L_x, L_y and L_z this gives a solution which is:

$$\Psi(x, y, z, t) = A \sin\left(\frac{n_x \pi}{L_x}x\right) \sin\left(\frac{n_y \pi}{L_y}y\right) \sin\left(\frac{n_z \pi}{L_z}y\right) \sin(\omega t + \phi)$$

$$\tag{B.27}$$

where n_x, n_y and n_z are all positive integers. Each node is now a plane and the wave vector is:

$$\vec{k} = \begin{pmatrix} k_x \\ k_y \\ k_z \end{pmatrix} \text{ where } k_i = \frac{n_i \pi}{L_i} \tag{B.28}$$

and the wave number is:

$$k^2 = k_x^2 + k_y^2 + k_z^2 \tag{B.29}$$

B.4 Number of Modes

As we saw with two dimensional standing waves there are a variety of ways to fit a wave with a particular wavelength into the medium when we look at dimensions greater than one. This is even more noticeable when we expand to three dimensions since we have an extra degree of freedom.

The actual number of ways, or normal modes, that a given wavelength can exist in a cavity depends on both wavelength and the geometry of the cavity: some wavelengths will have many different modes available others will have none. This makes the precise number extremely difficult to calculate and highly dependent on even minor changes to the system. So instead we will calculate the average number of modes available as a function of wavenumber.

Let's start by considering the wave vector as written in equation (B.28) for a cube where $L = L_x = L_y = L_z$. This simplifies (B.28) to:

$$\vec{k} = \begin{pmatrix} n_x \\ n_y \\ n_z \end{pmatrix} \frac{\pi}{L} \qquad (B.30)$$

and so we can represent a particular mode as a point in a three dimensional space and since n_x, n_y and n_z are positive integers, these points will be evenly distributed in one octant of a 3D grid. The distance of a point representing a mode from the origin is simply the length of the wave vector which is just the wavenumber.

What we want to calculate is the number of modes as a function of the wavenumber which is the same as calculating the number of modes which lie on the surface of a sphere with a radius equal to that wavenumber. As we discussed this is a highly variable and hence not particularly useful number so instead we want to calculate the average number of modes on the surface of the sphere.

To do this we calculate the average density of modes, in other words the number of modes per unit volume in the wave vector space. Looking at (B.30) we see that each mode point is a distance π/L away in x, y and z from its nearest neighbour and so we have one mode in a volume of π^3/L^3. Hence the number of modes per units volume, the density of modes, is:

$$\rho_{\text{modes}} = \frac{L^3}{\pi^3} \qquad (B.31)$$

Now to calculate the number of modes with a particular wavenumber we consider the total number of states with a particular wavenumber or less. This is just the number of modes inside the octant of a sphere with a radius equal to the wavenumber, k, as shown in figure B.4. We calculate this number by multiplying the volume of one eighth of a sphere of radius k by the average number of modes per unit volume from (B.31):

$$N(k) = \frac{1}{8} \times \frac{4}{3}\pi k^3 \times \frac{L^3}{\pi^3} = \frac{L^3 k^3}{6\pi^2} \qquad (B.32)$$

This is the average total number of modes with a wavenumber of k or less which is very close to the true number when k is a lot greater than the grid spacing i.e. $k \gg \frac{\pi}{L}$. To find the average

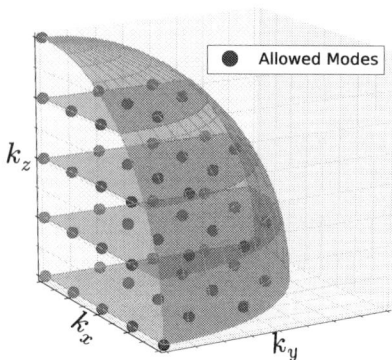

Fig. B.4: Plot of the allowed standing wave modes in a cubic cavity which have a wavenumber less than a particular value. The axes are the x, y and z components of the wave vector.

number of modes with a particular value of k we need to find the change in the total number of modes in this sphere when we increase the wavenumber from k to $k + \delta k$. This will be the number of modes which lies inside a thin shell a distance k from the origin.

To calculate this we consider the change in the total number of modes, δN, in our sphere when we change the radius of the sphere from k to $k + \delta k$, as shown for two dimensions in figure B.5, since this will be the number of modes inside the thin shell. In the limit $\delta k \to 0$ this will give the average number of modes with a particular wavenumber: $\frac{dN}{dk}$. So all we need to do to find this is differentiate (B.32) with respect to k:

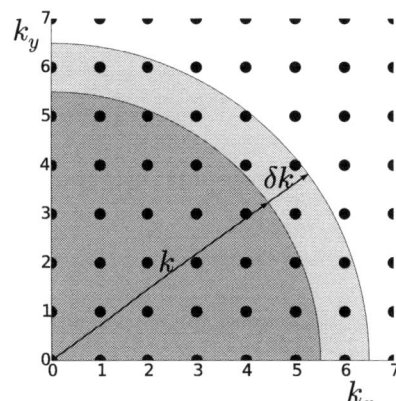

Fig. B.5: Plot of the allowed modes for a two dimensional sheet showing the number of modes which lie in the region k to $k + \delta k$. In the limit $\delta k \to 0$ this gives the average number of modes with a wavenumber of k.

$$\frac{dN}{dk} = \frac{L^3 k^2}{2\pi^2} \qquad (B.33)$$

The final step is to convert this into the average number as a function of wavelength. Starting with the definition of wavenumber we have:

$$k = \frac{2\pi}{\lambda} \implies \frac{dk}{d\lambda} = -\frac{2\pi}{\lambda^2} \qquad (B.34)$$

and so the rate of change of the total number of modes with respect to wavelength is:

$$\begin{aligned} \frac{dN}{d\lambda} &= \frac{dN}{dk} \times \frac{dk}{d\lambda} \\ &= \frac{L^3}{2\pi^2} \times \frac{4\pi^2}{\lambda^2} \times -\frac{2\pi}{\lambda^2} \\ &= -\frac{4\pi L^3}{\lambda^4} \qquad (B.35) \end{aligned}$$

where the negative sign indicates that the total number of modes, N, decreases with increasing wavelength as you would expect. This conversion calculation shows why it is important to remember that the number of modes for a particular wavelength, wavenumber or frequency is a differential quantity. Had we just written it down as a simple variable when we did the conversion from wavenumber to wavelength we would have omitted the additional $-2\pi/\lambda^2$ factor.

Now if we consider electromagnetic waves then we have two polarization states and so there are twice the number of modes. Also the volume of the cavity in which the waves are trapped is L^3 and so we can simplify things by considering the average number of EM modes per unit volume of the cavity, n_m, as a function of wavelength which is:

$$n_m(\lambda) = \left| \frac{dN}{d\lambda} \right| = 8\pi\lambda^{-4} \qquad (B.36)$$

where we ignore the negative sign since we are interested in the average number of modes with a particular wavelength rather than the change in the number of total modes.

To convert this into the number of modes as a function of frequency, v, we need to remember to convert the differential as well as just converting λ:

$$\lambda = \frac{c}{v} \implies \frac{d\lambda}{dv} = -\frac{c}{v^2} \tag{B.37}$$

and so the number of modes per unit volume as a function of frequency is:

$$\frac{dN}{dv} = \frac{dN}{d\lambda} \times \frac{d\lambda}{dv}$$

$$= -\frac{8\pi v^4}{c^4} \times -\frac{c}{v^2}$$

$$n_m(v) = \left| \frac{dN}{dv} \right| = \frac{8\pi v^2}{c^3} \tag{B.38}$$

This is the result that will be used to calculate the blackbody spectrum but there are a couple of points to note with it. First it is important to remember that this is an average number of modes: for any given wavelength or frequency there may be no available modes while for another there may be far more than this formula predicts. In the case where $\lambda \ll L$, i.e. the dimensions of the cavity are a lot larger than the wavelength being considered, the number of modes will be very large and so the statistical fluctuations for any given wavelength will be relatively small. Secondly although here we have only considered a cubic cavity, in the same limit where the wavelength is a lot smaller than the dimensions of the cavity, this becomes a general result for any volume of cavity i.e. the number of modes per unit volume does not depend on the precise configuration of the cavity walls.

B.5 Blackbody Spectrum

Our model for a perfect blackbody is an empty cavity with a very small opening. Any radiation entering the cavity will reflect multiple times off the walls inside until it is finally absorbed and any radiation emitted will be due to the radiation stored in the electromagnetic standing waves allowed inside the cavity. Hence the emission spectrum will depend on the energy stored in each frequency.

The equipartition theorem from thermodynamics tells us that for an object in thermal equilibrium the average energy stored in each standing wave mode, is kT where k is the Boltzmann constant and T is the temperature in kelvin. However it must be noted that this is the *average* energy and it is based on an assumption that the energy in each mode can have any value. To determine the average energy when each mode can only have certain allowed values for its energy we need to know the probability distribution of the energy in each mode. This is given by the Boltzmann distribution which states that the probability for any particular mode to have an energy E is:

$$P(E) \propto e^{-\frac{E}{kT}} \tag{B.39}$$

Now Planck stated that the energy in a particular mode is quantized in units of hv where h is the Planck constant and v is the frequency of the mode. To calculate the average energy in one particular mode we therefore need to sum the energy of every possible energy state weighted by the probability that the mode will have that much energy. For example if I have a system with two energy states which have 0 and 2 units of energy respectively then if the system has a 75% chance of being in the zero energy state the average energy of the system is:

$$\overline{E} = E_1 P(E_1) + E_2 P(E_2) = 0.75 \times 0 + 0.25 \times 2 = 0.5 \tag{B.40}$$

Applying this to calculating the average energy in a particular mode we get:

$$\overline{E} = \sum_{n=0}^{\infty} \left(nhv \frac{e^{-\frac{nhv}{kT}}}{N} \right) = \frac{1}{N} \sum_{n=0}^{\infty} \left(nhv e^{-\frac{nhv}{kT}} \right) \tag{B.41}$$

where $1/N$ is the constant of proportionality from equation (B.39). N is called the normalization factor and to determine it we need to use the fact there there is a total probability of 1.0 that the mode lies in some state i.e. the sum of all the probabilities for the individual states must add to give 1.0. Hence we have:

$$\sum_{n=0}^{\infty} \left(\frac{e^{-\frac{nhv}{kT}}}{N} \right) = 1 \implies N = \sum_{n=0}^{\infty} e^{-\frac{nhv}{kT}} \tag{B.42}$$

where we can take N out of the sum since it is a constant value. Putting this back into equation (B.41) we get:

$$\overline{E} = \frac{\sum_{n=0}^{\infty} \left(nhv e^{-\frac{nhv}{kT}} \right)}{\sum_{n=0}^{\infty} e^{-\frac{nhv}{kT}}} = hv \frac{\sum_{n=0}^{\infty} \left(ne^{-\frac{nhv}{kT}} \right)}{\sum_{n=0}^{\infty} e^{-\frac{nhv}{kT}}} \tag{B.43}$$

Now we make the substitution $u = e^{-\frac{hv}{kT}}$ which simplifies our expression to:

$$\overline{E} = hv \frac{\sum_{n=0}^{\infty} nu^n}{\sum_{n=0}^{\infty} u^n} = hv \frac{u + 2u^2 + 3u^3 + \cdots}{1 + u + u^2 + u^3 + \cdots} \tag{B.44}$$

We now have two series to sum. The series in the denominator is a simple geometric series and since $u = e^{-\frac{hv}{kT}}$ it is clear that $u < 1$ so we can sum this using the standard formula for a geometric series:

$$\sum_{k=0}^{\infty} ar^k = \frac{a}{1 - r} \implies 1 + u + u^2 + u^3 + \cdots = \frac{1}{1 - u} \qquad (B.45)$$

Summing the series in the numerator is a little bit trickier. To do this we need to recognize that this series can be obtained by differentiating another:

$$u + 2u^2 + 3u^3 + \cdots = u\frac{d}{du}\left(1 + u + u^2 + u^3 + \cdots\right) \qquad (B.46)$$

where the extra constant term will disappear in the differentiating and so have no effect other than to make but the sum of the series easier to differentiate. Now this series is again a geometric series which is easy to sum so we can sum this and then differentiate the result:

$$u\frac{d}{du}\left(1 + u + u^2 + u^3 + \cdots\right) = u\frac{d}{du}\left(\frac{1}{1-u}\right) = \frac{u}{(1-u)^2} \qquad (B.47)$$

Now putting these results for the sums of the series back into (B.43) we get:

$$\overline{E} = hv\frac{u}{1-u} = hv\frac{e^{-\frac{hv}{kT}}}{1 - e^{-\frac{hv}{kT}}} = \frac{hv}{e^{\frac{hv}{kT}} - 1} \qquad (B.48)$$

This is the average energy per mode with a frequency v and so to find the total energy per unit volume in the cavity as a function of frequency we need to multiply this by the total number of modes of frequency v per unit volume which we get from equation (B.38). Hence the total energy density stored in modes with a frequency v, which can be written as the rate of change of the total energy density in the cavity E with respect to frequency, is:

$$\frac{dE}{dv} = \frac{8h\pi v^3}{c^3}\frac{1}{e^{\frac{hv}{kT}} - 1} \qquad (B.49)$$

Since the frequencies emitted by the cavity depend on energy stored in each frequency this formula is proportional to the Planck spectrum for blackbody radiation. This spectrum is shown in figure B.6.

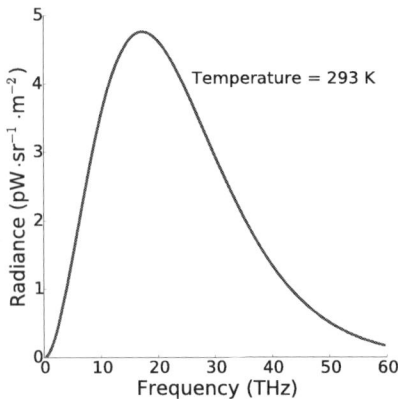

Fig. B.6: The Planck spectrum for a blackbody with a temperature of 293 K.

B.6 Wien's Displacement Law

Having now obtained something proportional to the emission spectrum our next step is to calculate the peak frequency of the spectrum. This will be at the stationary point in the spectrum where the rate of change of energy density per frequency with respect to frequency is zero which gives the following equation:

$$\frac{\mathrm{d}^2 E}{\mathrm{d}\nu^2} = 0 \implies \frac{\mathrm{d}}{\mathrm{d}\nu}\left(\frac{8h\pi\nu^3}{c^3}\frac{1}{e^{\frac{h\nu}{kT}}-1}\right) = 0 \qquad (B.50)$$

All that remains to to make the differentiation and solve the equation to find the value of ν where the emission is a maximum:

$$\frac{8h\pi}{c^3}\frac{\mathrm{d}}{\mathrm{d}\nu}\left(\frac{\nu^3}{e^{\frac{h\nu}{kT}}-1}\right) = 0$$

$$\left(\frac{3\nu^2}{e^{\frac{h\nu}{kT}}-1} - \frac{h\nu^3 e^{\frac{h\nu}{kT}}}{kT(e^{\frac{h\nu}{kT}}-1)^2}\right) = 0$$

$$3kT(e^{\frac{h\nu}{kT}}-1) = h\nu e^{\frac{h\nu}{kT}}$$

$$3kT(1 - e^{-\frac{h\nu}{kT}}) = h\nu \qquad (B.51)$$

Unfortunately this equation cannot be solved analytically. Applying numerical techniques we end up with a solution which is:

$$\nu_{\mathrm{max}} = 5.879 \times 10^{10}\,\mathrm{Hz\,K^{-1}} \cdot T \qquad (B.52)$$

or in terms of the peak wavelength we get:

$$\lambda_{\mathrm{max}} = \frac{2.89776829\,\mathrm{mm\,K}}{T} \qquad (B.53)$$

This is called Wien's displacement law and is named after Wilhelm Wien who derived it in 1893 by a purely thermodynamic

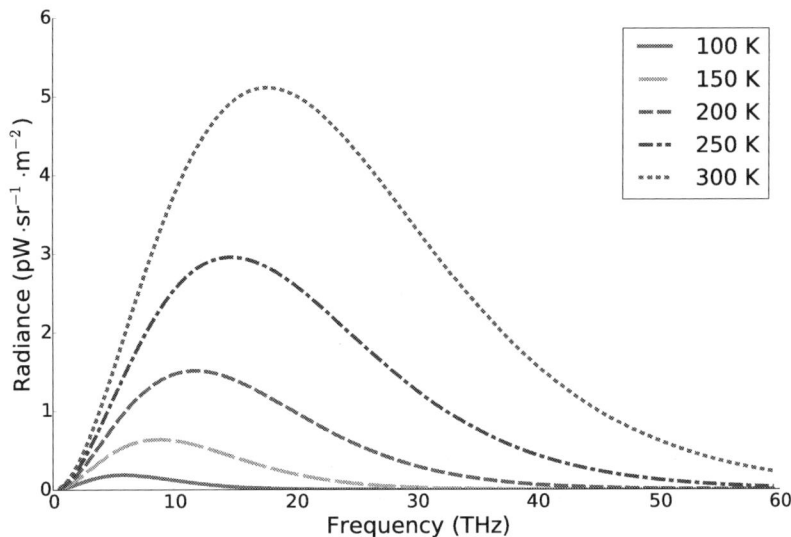

Fig. B.7: Planck spectra for blackbodies with differing temperatures showing how the peak emission frequency is proportional to temperature as described by Wien's displacement law.

approach which considered the adiabatic expansion of a cavity containing electromagnetic waves in thermodynamic equilibrium. The increase in the peak emission frequency with temperature can be clearly seen in figure B.7.

Glossary

absolute pressure An absolute pressure is one that is measured on a scale where zero pressure corresponds to the vacuum. 37, 38

achromatic doublet An achromatic doublet consists of two lenses placed together. For a net converging lens this consists of a weak, diverging lens made of a highly dispersive medium, such as flint glass, placed in front of a strong, converging lens make of a low dispersion material, such as crown glass. 178

aerodynamics Aerodynamics is the study of gas flow. Although the name suggests that it applies only to air the study of any gas flow can be referred to as aerodynamics. The non-constant density of gases can significantly affect the flow and is the reason for differentiating between studying gas and liquid flow. 45

amplitude The maximum displacement of an oscillator from its equilibrium position. It is often denoted by the upper case letter A. 61, 62, 66, 75, 78, 79, 83, 88, 91, 92, 95, 99–101

angle of incidence (optics) When a ray of light strikes a surface the angle between the normal to the surface and the direction of the ray of light is called the angle of incidence. 144–146, 184, 189

angle of reflection (optics) If a ray of light strikes a surface and is then reflected by that surface the angle between the reflected ray and the normal to the surface is called the angle of reflection. 145, 189

angle of refraction (optics) When a ray of light crosses the boundary between two media the angle of refraction is defined as the angle between the normal to the surface and the direction of the ray in the new medium. 145, 146, 184, 189

angular frequency The change in the phase of an oscillator per second. One complete cycle is a phase change of 2π. It is usually denoted by the lower case greek letter omega, ω. 62, 67, 76, 78–80, 83, 88, 107

angular magnification The angular magnification is defined as the ratio of the angular sizes of two images. This is generally more useful when describing the effective magnification seen by an observer because an image which is twice the size but twice as far away will appear to be the same size. 170–172, 174

anomalous dispersion Anomalous dispersion is the type of dispersion which occurs in a medium when the phase velocity of a wave is less than the group velocity. 187

anti-node An anti-node is the point on a standing wave where the amplitude of vibration is a maximum. They occur at a distance of one quarter wavelength from the nearest node and one half wavelength from the nearest anti-node. 132–134

aperture The aperture is the hole or opening through which light enters an optical instrument. Generally this is equal to the size of the primary optical element in the device. 176

apochromatic lens An achromatic lens is a lens which is corrected so that it has approximately the same focal length for all wavelengths of light. 178

Archimedes' principle Archimedes' principle gives the magnitude of the buoyancy force exerted by a fluid on an object immersed in a fluid. It can be simply stated as: "The upwards buoyancy force that is exerted on a body immersed in a liquid is equal to the weight of the fluid that the body displaces." 40

argand diagram An argand diagram is a means of representing complex numbers as points on a two dimensional plane, called the complex plane. The x coordinate of the point is the real component and the y coordinate gives the imaginary component. The x-axis is usually referred to as the real axis and the y-axis as the imaginary axis. 6, 7, 73, 75

argument (complex number) For a complex number $z = x + iy$ the argument is defined as $\arg(z) = \tan^{-1}(y/x)$. It corresponds to the angle between the positive real axis and line from the origin to the point representing the number on the complex plane. If a complex number is written in the form of an imaginary exponential, $z = re^{i\phi}$ then the argument is ϕ. 6

barometer A barometer is an instrument which measures absolute pressure using a vertical column of liquid underneath a vacuum. The external pressure on the fluid is determined by measuring the height of the fluid column. 37, 38

beats Beats is a phenomenon observed when two wave sources with slightly different frequencies interfere. The result is a wave which has a frequency equal to the average of the two sources and an amplitude which oscillates with a frequency equal to the frequency difference between the two sources. 129, 130

blackbody A black body in physics is an object which absorbs all wavelengths of radiation and emits radiation based solely on its temperature and not on the wavelengths of radiation it has absorbed. 216

Boltzmann distribution The Boltzmann distribution is the probability distribution that a system in thermal equilibrium will be in a state with a given energy. If the probability of the system having an energy, E, is $P(E)$ then the Boltzmann distribution says that $P(E) \propto e^{-E/kT}$ where k is the Boltzmann constant and T is the temperature in kelvin. 219, 255

boundary condition A condition placed on the solution to a differential equation at a particular set of coordinates, usual defining a boundary. If the variable is time this is equivalent to an initial condition. 72

breaking strength The breaking strength of a material is the maximum stress the material can withstand before it fails and cracks or separates. The breaking strength of a material will, in general, be different for each type of stress. 28, 29

brittle A material is said to be brittle if the elastic limit and the breaking strength have almost the same value. This will mean that once the elastic limit is exceeded the material will almost immediately fail and no plastic deformation is possible. Examples of brittle materials at room temperature are glass, china and some rigid plastics. 29

bulk modulus The bulk modulus of a material is defined as the negative of the rate of change of pressure (bulk stress) with respect to bulk strain: $K = -V\frac{\mathrm{d}p}{\mathrm{d}V}$ or, for regions where the bulk modulus is constant, $K = -\Delta p/(\Delta V/V_0)$. The addition of the negative sign is because a positive pressure results in a negative change in volume and the desire is to have a positive value to match that of the other moduli of elasticity. 26, 27, 114

bulk strain A bulk strain measures the deformation of a material due to a bulk stress. Since a bulk stress applies force from all directions at once this results in a change of volume for the material and so bulk strain is defined as the change in volume divided by the original volume. 26, 27

bulk stress A bulk stress occurs when an object is being crushed from all sides at once. It is different to a compressive stress since a bulk stress requires forces acting in all three dimensions at the same time whereas a compressive stress just requires crushing forces in one dimension. It is defined as the force per unit area acting on all surfaces of the material and is positive for a crushing force and hence is equal to the pressure which acts on the object. 26, 34

buoyancy When an object is immersed in a fluid the fluid will exert an upwards force on the object opposite to the direction of gravity. This force is the called the buoyancy force and its magnitude is determined by Archimedes' principle. 39, 40

chromatic aberration Chromatic aberration is observed in lenses which use refraction to focus light. The lensmaker's equation shows that the focal length is dependent on the refractive index of the material of the lens. However for real material this value is different for different frequencies of light, an effect called dispersion, with the result that each frequency will have a slightly different focal point. This corrupts an image by separating out the component colours of light. 178

coherent A coherent source of waves is one which emits light with a single phase. Two sources of waves are said to be coherent if they produce waves of the same frequency which have a constant, but not necessarily zero, phase relationship. 190

complex amplitude The complex amplitude of a simple harmonic oscillator is a complex constant which encodes both the physical amplitude and the initial phase into a single constant. The physical amplitude is the modulus of the complex amplitude and the initial phase is the argument. 75, 96, 98

complex number A complex number is one which has both a real and an imaginary part. 6, 64, 65, 73, 74

complex plane The complex plane is a two dimensional plane where each point on the plane represents a complex number. As such it is essentially synonymous with argand diagram. 6, 7

compound microscope A compound microscope is an optical instrument consisting of two converging lenses which is designed to provide a large magnification for objects which are close by. 172, 173

compound pendulum A rigid body which is free to rotate about a point other than its centre of gravity so that it can swing back and forth under the influence of gravity. 77, 79, 82

compressive strength The compressive strength of a material is the maximum compressive stress that the material can withstand before it fails. 29

compressive stress Compressive stress is a normal stress where the internal forces are perpendicular to the plane and act so as to compress the material. It is defined as the force perpendicular to the plane divided by the area of the plane and by convention is taken to be negative. 18, 22, 34

concave mirror A concave mirror is one where the centre of the reflecting surface is further from the incident light than the edges. Typically such mirrors are either parabolic or spherical in shape. A useful mnemonic to aid in remembering whether a mirror is concave or convex is to think that the concave mirror is shaped a little like a cave. 147, 150

conjugate (complex number) The conjugate of a complex number is a complex number with the same real part but with an inverted sign for the imaginary part. The conjugate is denoted by placing a line over the top of the complex number such that if $Z = x + iy$ then $\bar{z} = x - iy$. 5, 6, 74

continuity equation The continuity equation is an expression of conservation of mass for an ideal fluid flowing in a pipe. It states that the product of the fluid density, velocity and cross-sectional area of the pipe is a constant. 47, 51

critical angle (optics) When a ray of light is incident on a boundary with a medium that has a lower refractive index if the angle of incidence is greater than the critical angle the ray will be totally, internally reflected and will not refract. 146

critically damped A critically damped harmonic oscillator is one where the damping force is just sufficient to return the oscillator to the equilibrium position without subsequent oscillations. Any increase in damping will result in a slower return to the equilibrium position and any less will result in an overshoot and subsequent oscillation. A critically damped oscillator has a damping ratio of exactly one. 89, 91, 92

damping ratio A dimensionless number which measures the amount of damping for a damped harmonic oscillator. It is typically denoted by the lower case greek letter zeta, ζ. 88, 91, 96, 101

dark matter All that we currently know about dark matter comes from astrophysical observations. From these we know that: it is electrically neutral; it does not undergo strong, nuclear interactions; it exerts a gravitational field and it was produced moving at slow velocities by the Big Bang. This rules out all known fundamental particles and the true nature of dark matter is currently a major mystery of physics. 216

deformation Deformation is the change in the shape of an object in response to the application of stress. 21, 22

density Density is the mass per unit volume of a fluid or other material. Since the amount of fluid in a system is easily varied the mass of fluid may not be constant and so density, which is a property of the fluid regardless of the amount present, is a more useful quantity than mass. The SI units of density as kilograms per cubic metres ($kg\,m^{-3}$) and typically the symbol used for density is the greek letter rho, ρ. 33, 114

density Specific gravity is the ratio of a substance's density to the density of some reference substance. For liquids this is usually water at the temperature where its density is highest ($4\,°°C$). As a ratio of two densities specific gravity is dimensionless and has not units. 33

diffraction When a wave encounters an obstruction and is blocked the remaining wavefront will spread out into the region shadowed by the obstruction in a process which is called diffraction. At long distances from the obstruction the pattern of waves produced is determined by far-field, or Fraunhofer diffraction and near to the opening the near-field or Fresnel diffraction is used. 202, 203, 210

diffraction grating A diffraction grating is formed from many, closely spaced parallel slits. When waves are incident on such a grating each wavelength is diffracted through a diffraction angle unique to the wavelength and so the spectrum of frequencies in the incident waves can be analyzed. Diffraction gratings are rated by the number of slits, or lines, they have per metre. 208–211

diffraction pattern When light is diffracted after passing a obstruction the resulting arrangement of light and dark fringes on a screen is called a diffraction pattern. The diffraction pattern is entirely dependent on the shape of the obstruction causing it and the shape and dimensions of the obstruction and be determined by studying the diffraction pattern it produces. The term is not exclusive to diffraction of light waves: any type of wave which can be diffracted can produce a diffraction pattern. 205, 207, 208, 211

dispersion Dispersion occurs when the velocity of a wave in a medium varies for different frequencies of wave. This difference means that the phase velocity and group velocity of a wave will have different values. For light waves it also means that the refractive index varies with frequency and so white light being refracted at the boundary of such a medium will be split into separate colours since each colour will have a different angle of refraction. 178, 184

drag Drag is a force experienced by an object moving through a viscous fluid. It acts in the opposite direction to the velocity of the object relative to the fluid. 45

ductile A material is said to be ductile if there is a large difference between the elastic limit and the breaking strength. Such a gap means that the material can be plastically deformed into new shapes. Examples of ductile materials at room temperature are copper, aluminium and steel. 29

elastic deformation When a material's shape is changed by the application of a stress this is called a deformation. It is said to be an elastic deformation if the material then returns to its original shape after the stress is removed. An example of an elastic deformation would be stretching and then relaxing a rubber band. 23

elastic limit The elastic limit of a material is the largest stress the material can withstand while still undergoing an elastic deformation. Above this limit any further stress will cause either a plastic deformation or will cause the material to fail. 28, 29

electronvolt A unit of energy defined as the energy gained by an electron after moving through a potential difference of one volt. It uses the abbreviation 'eV' and is often used with SI prefixes e.g. MeV, GeV, TeV. It is typically used in the fields of atomic, nuclear and particle physics. 222

equation of motion Equations of motion are the equations which describe the motion of a physical system as a function of time. A simple example would be the constant acceleration equations which describe how the kinematic quantities of an option undergoing constant acceleration change with time. For more general cases they are differential equations whose solutions are the functions which describe how various dynamic quantities of a system vary with time. Most commonly this will be the displacement of a system but equations of motion for other dynamic variables such as momentum or angular momentum are also possible. 63, 65, 78, 82, 83, 88, 93, 95–99, 101, 102

equipartition theorem The equipartition theorem of thermodynamics states that in thermal equilibrium each degree of freedom of a system contains, on average, the same amount of energy. The energy is half of the product of Boltzmann's constant, k, with the temperature, T. However since oscillators have both kinetic and potential energy degrees of freedom they contain kT on average. 217, 254

evanescent wave A evanescent wave is a solution to the wave equation which results in a real exponential instead of the usual imaginary one. This represents a wave which is exponentially decaying. Classically such waves occur for light behind a surface where total internal reflection is occurring. In quantum mechanics evanescent quantum waves are how particles can tunnel through a potential barrier. 240

eyepiece The eyepiece is the lens in an optical instrument which is closest to the observer's eye. 172–175

focal length The distance between the centre of the mirror or lens and its focal points. 148, 156

focal point The point through which rays parallel to the optical axis of a mirror or lens will pass for a concave mirror or converging lens or the point from which they will appear to emanate from for a convex mirror or diverging lens. 148, 149, 155, 156

frequency The number of cycles which an oscillator completes per unit time. In the SI system this is measured in hertz where $1\,\text{Hz}=1\,\text{s}^{-1}$. It is normally denoted by the lower case letter f or by the lower case greek letter nu, ν. 62, 67, 72, 76, 78–80, 91, 96, 100

fundamental mode The fundamental mode of a wave medium with finite size is the longest wavelength, lowest frequency standing wave which the medium can support. 132

group velocity The group velocity of a wave is the velocity at which a change in amplitude of the wave propagates. For a non-dispersive medium this is equal to the phase velocity but for all real media it will either be greater or less than the phase velocity depending on the type of dispersion the medium exhibits. 108, 186, 187

harmonic An alternative name for normal mode which is commonly, but not exclusively, used when discussing standing waves in a musical context. 132, 133

hydrodynamics Hydrodynamics is the study of liquid flow. Although the name suggests that it is restricted to water the name applies equally to the study of any liquid. While liquids tend to have constant densities they also have a surface tension which is why the studies of liquid and gas flow tend to be differentiated. 45

hysteresis (elastic) Elastic hysteresis describes the difference between stress and strain relationships when a material is loaded and unloaded while undergoing elastic deformation. The effect can be shown by plotting stress vs. strain curves while loading and then unloading a material. For a material exhibiting hysteresis the result will be a closed loop with the area of the loop giving the energy absorbed by the material and usually converted into heat. 28

ideal fluid An ideal, or perfect, fluid is one which has no resistance to shear stresses and so flows without resistance. This means that at rest it can be completely characterized by its density and pressure. 46, 48, 50, 52

imaginary number An imaginary number is one that, when multiplied by itself, gives a negative number. 3, 64

lens equation The lens equation relates the position of the object and its associated image to the focal length of the lens. For converging lenses the focal length is positive and for diverging lenses it is negative. Similarly the distance to a real image or object is positive which the distance to a virtual image or object is negative. 168

lensmaker's equation The lensmaker's equation relates the radii of curvature of the two faces of a lens to the focal length of the lens. It is an approximate relationship which is only valid for thin lenses where the diameter of the lens is a lot less than the focal length. The radii of curvature are positive for convex faces and negative for concave faces which gives a focal length with the same sign convention as the lens equation: positive for converging lenses and negative for diverging ones. 168, 169

light ray A light ray is way to represent the propagation of plane waves of light using a line, called the ray, which is perpendicular to the wavefronts and which points in the direction of propagation of the waves. 144

Lissajous figure A closed curve defined by two oscillations one in the x direction and the other in the y direction. The frequencies do not have to be equal but one must be an integer multiple of the other. The shape can be altered by changing the constant phase offset between the two linked oscillations. The curves were first studied in 1857 by French physicist Jules Antoine Lissajous who, before the invention of oscilloscopes, used a light ray reflected off two mirrors attached to tuning forks to generate both the x and y oscillations. 70

longitudinal wave A longitudinal wave is one where the displacement of the medium or the direction of the excitation of the field is in the same direction of the propagation of the wave. 109, 133

luminiferous aether The luminiferous aether, of just aether, was the supposed medium for light waves which was believe to exist until 1887 when the Michelson-Morley experiment provided substantial evidence that it did not exist, a discovery which directly lead to the development of relativity. 198, 199

manometer A manometer is an instrument which measures pressure differences using a U-shaped tube which is filled with a liquid. The ends of the tube are exposed to the pressures whose difference should be measured and the relative pressure between them is determined by measuring the height difference between the surfaces of the liquid in the two arms of the tube. 38

mode (of vibration) A shortened form of normal mode and used to describe an allowed standing wave for a system. 132

modulus (complex number) For a complex number $z = x + iy$ the modulus is defined as $|z| = \sqrt{x^2 + y^2}$. It corresponds to the distance of the point representing the number on the complex plane from the origin. If a complex number is written in the form of an imaginary exponential, $z = re^{i\phi}$ then the modulus is r. 6

modulus of elasticity The modulus of elasticity (or elastic modulus) is defined as the rate of change of stress with respect to strain for an elastic deformation, or in other words the gradient of the stress vs. strain curve of the material. Each type of stress and strain pairing has its own modulus of elasticity and, in general, the modulus of elasticity is not constant. However for some types of material a modulus may be constant from zero stress up to some limit. In these regions the modulus of elasticity is equal to the ratio of the stress to the strain and for moduli which are only defined in these regions, such as Young's modulus, this is typically how they are defined. 23, 24

near point The near point is defined as the shortest distance at which the eye can generate a sharp, focussed image on the retina. 169–171

Newton's rings Newton's rings is a pattern of circular rings which appears when a lens is placed on a plane glass sheet and illuminated with monochromatic light. It is caused by the interference of light in the gap between the lens and the glass sheet. It is used when making lenses to check that the surface of the lens has been correctly ground to form the surface of a sphere since even tiny deviations will produce a very noticeable change in the pattern. 181, 195–197

node A node is the point on a standing wave where the amplitude of vibration is zero. They occur at a distance of one quarter wavelength from the nearest anti-node and one half wavelength from the nearest node. 132, 134

normal dispersion In a medium which exhibits normal dispersion the phase velocity of a wave is greater than the group velocity. 187

normal mode (of vibration) A wave medium of finite size can only support certain wavelengths of standing waves. Each of these allowed wavelengths is said to be a normal mode, or harmonic, of the system. 132, 133

normal strain Normal strain is a measure of the change in length of a material due to an applied normal stress. It is defined as the change in length divided by the original length and is positive for extensions in length and negative for reductions in length. 22, 23

normal stress A normal stress is one where the internal forces are perpendicular to the internal plane being considered. It is defined as the magnitude of the force perpendicular to the plane divided by the area of the plane. Normal stress can be further subdivided into tensile and compressive stress. 18–21, 25, 26

objective The objective is the lens in an optical instrument which is placed closest to the object being observed. 172–175

ocular lens The ocular lens is an alternative name for the eyepiece and is the lens in an optical instrument which is closest to the observer's eye. It derives from the latin *oculus* which means eye. 172

optical axis This line forming the symmetry axis of an optical system, typically one comprising of a mirror or lens. It passes through the centre of the mirror or lens and is perpendicular to the surface. 148–150, 152, 154, 155, 169

over damped An over damped harmonic oscillator is one which has a damping ratio greater than one. If such a system is disturbed from equilibrium it will slowly return to the equilibrium position and rest there. In these oscillators the damping force dominates and there will be no un-driven oscillations. 89, 90, 92

partial derivative A partial derivative is the derivative of a function of more than one variable with respect to one of the variables while the rest are held constant. 12, 111

Pascal's law Pascal's law regards applying an external pressure to an enclosed fluid and states that: "A change in pressure at any point in an enclosed fluid at rest is transmitted undiminished to all points in the fluid.". It is the principle used in hydraulics to magnify an applied force. 36

pendulum A pendulum is any object attached to a point in such a way that it is free to swing back and forth under the influence of gravity. 77

period The time it takes an oscillator to perform one complete cycle of its motion. It is normally denoted by the upper case letter T. 61, 62, 67, 78, 79

phase An angular quantity measured in radians pertaining to oscillators which describes where in the cycle the oscillator is. Two oscillators which are at the same point in their cycle, for example at their maximum positive displacement, at the same time are said to be "in phase" even if their amplitudes and even frequencies are different. However for the case of different frequencies such oscillators would find themselves in different phases immediately afterwards at which point they would be said to be "out of phase". One full cycle corresponds to a phase change of 2π. 68, 69, 75

phase velocity The phase velocity of a wave is the apparent velocity of a point of constant phase. As a velocity it has SI units of metres per second ($\mathrm{m\,s^{-1}}$) and is equal to the product of the frequency and wavelength i.e. $v_p = f\lambda$. 108, 112, 114, 115, 117, 135, 183, 186, 187, 243, 245

phaser A fictitious weapon invented for the universe of Star Trek and not to be confused with the physics concept called a phasor! 68

phasor A vector in the complex plane which represents a solution to the simple harmonic oscillator equation of motion. The length of the line gives the amplitude and the angle the line forms with the positive real axis is the phase. 68, 72

photoelectric effect Name for the effect that when light above a certain frequency shines on a metal it causes electrons to be emitted. 220, 221

photon The particle which transmits all electromagnetic radiation including light. 221

Planck constant The Planck constant is a fundamental, physical constant which gives the quantum of action for quantum physics. It was originally derived as the constant of proportionality between energy and frequency. It is often encountered as the constant $\hbar$, pronounced "h-bar", where $\hbar = h/2\pi$. The value of Planck's constant in SI units is 6.626×10^{-34} J s. 218, 223, 255

plastic deformation When a material's shape is changed by the application of a stress this is called a deformation. It is said to be an plastic deformation if, after the stress is removed, there is a permanent change in the material's shape. An example of plastic deformation is bending a paperclip. 28, 29

polarization The polarization of a wave is determined by the direction of the displacement of the medium. Transverse mechanical waves can have two polarization states one for each direction of vibration which is perpendicular to the motion of travel of the wave. Light waves also have two polarization states each one corresponding to the plane in which the electric field oscillates. 187–189

pressure Pressure is the force per unit area which a fluid exerts on the surface of any object placed in it. The pressure is the same regardless of the direction and orientation of the surface which makes pressure a scalar quantity. 26, 34, 35, 114

principal ray A principal ray is a light ray whose path is easily determined as is goes from the object, interacts with the optical device - usually a mirror or lens - and then forms the image. For both lenses and spherical mirrors there are three principal rays: one from the focal point, one parallel to the optical axis and one hitting the centre of the optical device being used. 149, 150

ray tracing Ray tracing is a technique used to determine the paths of light rays through a system. In physics this is used to determine where the image will be formed either by a lens or a reflecting surface. In computer graphics ray tracing is used to generate images by simulating the interaction of rays of light with various virtual objects before they strike the image plane. 147, 149, 150

real image An optical device, such as a mirror or lens, can cause the light rays emitted from an object to be deflected such that they are refocussed to a different location to produce an image of the object. This image is formed from actual light rays emanating from the object and so it is said to be a real image and it can be projected onto a screen. 150

refracting telescope A refracting telescope uses lenses to focus light from a distant object for viewing by the observer. The name derives from the fact the lenses work by the refraction of light. The first astronomical telescopes constructed were refracting telescopes. 173

refraction Refraction occurs when a ray of light moves from one medium to a other medium which has a difference refractive index . For rays which are not normal to the surface the refracted ray will be bent either towards or away from the normal to the surface depending on the difference in the refractive indices of the two materials. 144, 145

refractive index Refractive index is a property of a material which is used to determine how light is refracted when it enters the material from another medium. It is defined as the ratio of the speed of light in vacuo to the speed of light in the material. For normal materials this means that it has a value greater than or equal to one. 145, 146, 183–185, 189, 190, 192, 193

relative pressure A relative pressure is one which is measured on a scale where zero pressure is not the vacuum. Most typically the zero on such a scale will be atmospheric pressure. Relative pressure can sometimes be referred to as gauge pressure but this can be misleading since some pressure gauges will measure an absolute pressure. 38

resonance Resonance is a phenomena exhibited by some under damped harmonic oscillators when they are driven by an oscillating driving force. If such a system is driven at a certain frequency, called the resonant frequency, the steady-state amplitude of the system will be a maximum. Increasing or decreasing the driving frequency from this value will result in a smaller amplitude response. The condition for a system to show resonance is that the damping ratio, $\zeta < 1/\sqrt{2}$. 100, 101

resonant frequency The driving frequency at which the steady-state amplitude of a driven, damped harmonic oscillator has a maximum value. Note that not all damped harmonic oscillators have a resonant frequency. 100

separation of variables This is a technique to solve partial differential equations. The assumption is made that the solution can be written as a product of functions each of which only depends on a single variable and, if correct, this allows the partial differential equation to be separated into a series of ordinary differential equations which can be solved. 237, 251

shear modulus The shear modulus (or modulus of rigidity) of a material is the rate of change of shear stress with respect to shear strain. It is typically denoted with the symbol G. The higher the value of the shear modulus the more a material will resist shear stresss. 25

shear strain When a shear stress acts on a material the result is a shear strain. It is defined as the displacement of the material in the direction of the internal forces divided by the original perpendicular distance between the lines of action of the two opposing forces. While this gives a dimensionless quantity it is different from a normal strain in that the two distances used in the ratio are perpendicular to each other. plural 22, 23, 25, 26

shear strength The shear strength of a material is the maximum shear stress that the material can withstand before it fails. 29

shear stress Shear stress is a stress where the internal forces in a material are parallel to the plane being considered. This occurs when the lines of action of two equal and opposite external forces acting on a material are not the same and example of this would be the forces which scissors exert on paper when cutting it. plural 19–22, 25, 34

simple harmonic motion A type of repetitive motion where the period of the motion is independent of the amplitude. It is characterized by the acceleration of the system being proportional, and opposite to, its displacement from equilibrium. 62, 76, 82, 83

simple pendulum A point mass attached to the end of a light, inextensible string so that it is free to swing back and forth under the influence of gravity. 77, 79, 82

spherical aberration A spherical mirror only focusses parallel light to a single focal point in the limit of an infinitely large radius of curvature. For spherical mirrors with a finite radius of curvature parallel rays are not focussed to a point which results in a corruption of the image called spherical aberration. To avoid this a mirror needs to have a parabolic shape but such mirrors are harder to manufacture. The same effect exists for lenses where again the spherical surfaces and a finite thickness mean that parallel rays are not focussed to a single point. 177

standing wave A standing wave is one which is not moving, hence standing, but which still oscillates over time. They are created when a wave is trapped and confined to a region with finite boundaries. The boundary conditions typically result in only fixed, quantized values for wavelengths and frequencies being allowed. Standing waves are the physics behind musical instruments as well as electrons in an atom. 131, 133

steady-state solution (driven oscillator) The solution to the equation of motion for a driven, damped harmonic oscillator that remains after a long period of time has passed after the system is set in motion. 103

strain Strain is a measure of the deformation of a material in response to an applied stress. Strain is defined as a ratio between a change in a measured quantity of a material divided by an original quantity of the same dimension, typically a length of or a volume. Hence it is always a dimensionless quantity. 22–28

stress Stress is a measure of the deformation power of a force acting on a material. It is always defined in units of force per unit area and is typically subdivided into a variety of types. The SI units are newtons per square metre ($N\,m^{-2}$) or the pascal (Pa). If all types of stress are considered together the result is a tensor. 18, 21, 23–29, 34

surface tension Surface tension is a property of liquids which arises from the unbonded molecules on the surface of the liquid which are in a higher energy state that those in the bulk of the fluid. This results in an energy cost per unit area of surface which is why liquids typically have smooth surfaces to minimize the energy. The SI units for surface tension are newtons per metre ($N\,m^{-1}$) which is dimensionally the same as energy per unit area. 44, 45

tensile strength The tensile strength of a material is the maximum tensile stress that the material can withstand before it fails. 29

tensile stress Tensile stress is a normal stress where the internal forces are perpendicular to the plane and act so as to pull the material apart. It is defined as the force perpendicular to the plane divided by the area of the plane and by convention is taken to be positive. 18, 19, 22, 24, 25, 28, 34

terminal velocity The terminal velocity of an object is the constant velocity the object reaches when falling through a viscous fluid under the action of a force. As the force increases the object's velocity the viscous drag force opposing it increases too until at the terminal velocity the forces are perfectly balanced at the object moves at a constant velocity which is called the terminal velocity. 56, 57

transient solution (driven oscillator) The solution to the equation of motion for a driven, damped harmonic oscillator which decays exponentially with time after the system is first set in motion. 103

transverse wave A transverse wave is one where the displacement of the medium or the direction of the excitation of the field is perpendicular to the direction of the propagation of the wave. 109, 125, 133

ultraviolet catastrophe Refers to a problem in classical physics when calculating the spectrum of light emitted by a blackbody. In classical physics there is no minimum energy allowed in a single mode of oscillation and as the frequency increases the number of possible modes increases rapidly. This leads to a prediction that all objects should be emitting huge quantities of high energy gamma rays, a prediction not borne out by experiment! 218

under damped An under damped harmonic oscillator is one which has a damping ratio less than one. If such a system is disturbed from equilibrium it will oscillate with an exponentially decaying amplitude. 89, 91, 92, 100, 101

virtual image A virtual image is formed when the rays reflecting from a mirror or passing through a lens appear to emanate from an image which exists behind the mirror or lens. However because the rays do not actually originate from the image but only appear to the image is said to be virtual. Virtual images can only be viewed by looking at the mirror or lens which created them and they cannot be projected onto screens. 147, 150, 157

viscosity Viscosity is a measure of the resistance of a fluid to flow. It is caused by the friction between molecules of the liquid as they rub past one another when a fluid flows. 46, 52–54, 56

wave function A wave function is a mathematical function of both position and time which describes the displacement of the medium (or the value of the field) which makes up the wave. 111, 112, 125, 127, 131, 132, 236

wavelength The shortest distance in the direction of propagation between two points on the wave with the same phase. In all cases this is a physical distance and so has the SI units of metres. It is typically denoted using the lowercase greek letter lambda, λ. 107, 108, 110, 127, 129, 131–135

wavenumber The wavenumber is the rate of change of phase with respect to position for a wave. It is typically denoted with the symbol k and is related to the wavelength by the relationship: $k = 2\pi/\lambda$. 108, 128, 129, 132

work function The minimum amount of energy which is required to cause an electron to be emitted from a metal when light shines on it. The value of the work function depends on the metal. 221

Young's modulus Young's modulus, Y, is defined as the ratio of tensile stress to tensile strain of a material in the region around zero stress where this is has a constant value. In materials where the response to compressing stress is the same as the response to tensile stress the definition can be extended to include the ratio of compressive stress to compressive strain where the ratio remains constant. 24, 25

Manufactured by Amazon.ca
Bolton, ON